AF772586

Multivariable Mathematics

Multivariable Mathematics

Edited by:

Olga Moreira

ARCLER
PRESS

www.arclerpress.com

Multivariable Mathematics

Olga Moreira

Arcler Press
224 Shoreacres Road
Burlington, ON L7L 2H2
Canada
www.arclerpress.com
Email: orders@arclereducation.com

© 2021 Arcler Press

ISBN: 978-1-77407-700-9 (Hardcover)

This book contains information obtained from highly regarded resources. Reprinted material sources are indicated. Copyright for individual articles remains with the authors as indicated and published under Creative Commons License. A Wide variety of references are listed. Reasonable efforts have been made to publish reliable data and views articulated in the chapters are those of the individual contributors, and not necessarily those of the editors or publishers. Editors or publishers are not responsible for the accuracy of the information in the published chapters or consequences of their use. The publisher assumes no responsibility for any damage or grievance to the persons or property arising out of the use of any materials, instructions, methods or thoughts in the book. The editors and the publisher have attempted to trace the copyright holders of all material reproduced in this publication and apologize to copyright holders if permission has not been obtained. If any copyright holder has not been acknowledged, please write to us so we may rectify.

Notice: Registered trademark of products or corporate names are used only for explanation and identification without intent of infringement.

Arcler Press publishes wide variety of books and eBooks. For more information about Arcler Press and its products, visit our website at www.arclerpress.com

DECLARATION

Some content or chapters in this book are open access copyright free published research work, which is published under Creative Commons License and are indicated with the citation. We are thankful to the publishers and authors of the content and chapters as without them this book wouldn't have been possible.

ABOUT THE EDITOR

Olga Moreira obtained her Ph.D. in Astrophysics from the University of Liege (Belgium) in 2010, her BSc. in Physics and Applied Mathematics from the University of Porto (Portugal). Her post-graduate travels and international collaborations with the European Space Agency (ESA) and European Southern Observatory (ESO) led to great personal and professional growth as a scientist. Currently, she is working as an independent researcher, technical writer, and editor in the fields of Mathematics, Physics, Astronomy and Astrophysics.

TABLE OF CONTENTS

LIST OF CONTRIBUTORS

Yaping Hu
School of Science, East China University of Science and Technology, Shanghai 200237, China

Mahdi Teimouri
Department of Statistics, Faculty of Mathematics and Computer Science, Amirkabir University of Technology (Tehran Polytechnic), 424 Hafez Ave., Tehran 15914, Iran

Saeid Rezakhah
Department of Statistics, Faculty of Mathematics and Computer Science, Amirkabir University of Technology (Tehran Polytechnic), 424 Hafez Ave., Tehran 15914, Iran

Adel Mohammadpour
Department of Statistics, Faculty of Mathematics and Computer Science, Amirkabir University of Technology (Tehran Polytechnic), 424 Hafez Ave., Tehran 15914, Iran

Yanxia Tang
Department of Mathematics, Hebei North University, Zhangjiakou, 075000, China

JinyuGuan
Department of Mathematics, Hebei North University, Zhangjiakou, 075000, China

Yongchun Xu
Department of Mathematics, Hebei North University, Zhangjiakou, 075000, China

Yongfu Su
Department of Mathematics, Tianjin Polytechnic University, Tianjin, 300387, China

Milan Merkle
Department of Applied Mathematics, Faculty of Electrical Engineering, University of Belgrade, Bulevar Kralja Aleksandra 73, Belgrade, 11120, Serbia

Eric Houngla Adjakossa
Laboratoire de Probabilités et ModèlesAléatoires /Université Pierre et Marie Curie, Case courrier 188 - 4, Place Jussieu 75252 Paris cedex 05 France
University of Abomey-Calavi, 072 B.P. 50 Cotonou, Republic of Benin

Ibrahim Sadissou
Laboratoire de Biologieet de PhysiologieCellulaires /University of Abomey-Calavi, Cotonou, Republic of Benin
Centre d'Etudeet de Recherchesur le PaludismeAssocié à la Grossesse et à l'Enfance (CERPAGE), Cotonou, Republic of Benin

Mahouton Norbert Hounkonnou
University of Abomey-Calavi, 072 B.P. 50 Cotonou, Republic of Benin

Gregory Nuel
Laboratoire de Probabilités et ModèlesAléatoires /Université Pierre et Marie Curie, Case courrier 188 - 4, Place Jussieu 75252 Paris cedex 05 France

Pilar Ibarrola
Statistics Department, Universidad Complutense de Madrid, 28040 Madrid, Spain

Ricardo Vélez
Statistics Department, UNED, 28040 Madrid, Spain

ZhidongBai
KLASMOE and School of Mathematics and Statistics, Northeast Normal University, Changchun, China

Yongchang Hui
School of Mathematics and Statistics, Xi'an Jiaotong University, Xi'an, China

Dandan Jiang
School of Mathematics, Jilin University, Changchun, China

Zhihui Lv
KLASMOE and School of Mathematics and Statistics, Northeast Normal University, Changchun, China

Wing-Keung Wong
Department of Finance, Asia University, Taichung, Taiwan
Department of Economics, Lingnan University, Hong Kong, China

Shurong Zheng
KLASMOE and School of Mathematics and Statistics, Northeast Normal University, Changchun, China

Jimin Wang
College of Computer & Information, Hohai University, Nanjing 210098, China

Yuelong Zhu
College of Computer & Information, Hohai University, Nanjing 210098, China

Shijin Li
College of Computer & Information, Hohai University, Nanjing 210098, China

Dingsheng Wan
College of Computer & Information, Hohai University, Nanjing 210098, China

Pengcheng Zhang
College of Computer & Information, Hohai University, Nanjing 210098, China

Avraam Tapinos
School of Computer Science and Manchester Institute of Biotechnology, The University of Manchester, Manchester, United Kingdom,

Pedro Mendes
School of Computer Science and Manchester Institute of Biotechnology, The University of Manchester, Manchester, United Kingdom,
Virginia Bioinformatics Institute, Virginia Tech, Blacksburg, Virginia, United States of America

Lucas Lacasa
School of Mathematical Sciences, Queen Mary University of London, Mile End Road, London, E14NS, UK

Vincenzo Nicosia
School of Mathematical Sciences, Queen Mary University of London, Mile End Road, London, E14NS, UK

Vito Latora
School of Mathematical Sciences, Queen Mary University of London, Mile End Road, London, E14NS, UK

Wolfgang Mader
Institute of Physics, University of Freiburg, Germany
Freiburg Center of Data Analysis and Modeling, University of Freiburg, Germany

Malenka Mader
Institute of Physics, University of Freiburg, Germany
Freiburg Center of Data Analysis and Modeling, University of Freiburg, Germany
University Medical Center Freiburg, Germany

Jens Timmer
Institute of Physics, University of Freiburg, Germany
Freiburg Center of Data Analysis and Modeling, University of Freiburg, Germany
BIOSS, Center for Biological Signalling Studies, University of Freiburg, Germany

Marco Thiel
University of Aberdeen, UK

Björn Schelter
Institute of Physics, University of Freiburg, Germany
University of Aberdeen, UK

Lanlan Yan
School of Mathematics and Statistics, Central South University, Changsha 410083, China
School of Science, East China University of Technology, Nanchang 330013, China

Daniel Beale
University of Bath, Claverton Down, Bath, BA2 7AY, UK

Yong-Liang Yang
University of Bath, Claverton Down, Bath, BA2 7AY, UK

Neill Campbell
University of Bath, Claverton Down, Bath, BA2 7AY, UK

Darren Cosker
University of Bath, Claverton Down, Bath, BA2 7AY, UK

Peter Hall
University of Bath, Claverton Down, Bath, BA2 7AY, UK

Shan Chai
School of Transportation and Vehicle Engineering, Shandong University of Technology, Zibo, 255049, China

Xiang-Fei Ji
Key Laboratory of Electronic Equipment Structure Design, Xidian University, Xi'an, 710071, China

Li-Jun Li
School of Transportation and Vehicle Engineering, Shandong University of Technology, Zibo, 255049, China

Ming Guo
Shandong Linglong Tire Co., Ltd., Zhaoyuan, 265400, China

Osamu Saeki
Institute of Mathematics for Industry, Kyushu University, 744, Motooka, Nishi-ku, 819-0395, Fukuoka, Japan

Shigeo Takahashi
Graduate School of Information Science and Technology, The University of Tokyo, 7-3-1 Hongo, Bunkyo-ku, Tokyo, 133-8565, Japan

Hsuan-Ku Liu
Department of Mathematics and Information Education, National Taipei University of Education, Taipei, Taiwan

LIST OF ABBREVIATIONS

APCA	Adaptive Piecewise Constant Approximation
AUSLAN	Australian sign language
ARMA	Autoregressive Moving Average
BD	Brownian motion of Defects
CF	Characteristic function
CV	Coefficients of variation
CMLs	Coupled Map Lattices
CMP	Critical point model
DAX	Deutscher Aktien indeX
DFT	Discrete Fourier transform
DWT	Discrete wavelet transform
DTW	Dynamic time warping
DTE	Dynamic transmission error
EEG	Electroencephalogram
FAPs	Finitely additive probabilities
FDT	Fully Developed Turbulence
GLHT	General linear hypothesis testing
GEE	Generalized estimating equations
GIMME	Group Iterative Multiple Model Estimation
HMM	Hidden Markov model
HVG	Horizontal Visibility Graph
IMF	International Monetary Fund
LCSS	Longest Common Subsequence
MST	Maximal Spanning Trees
ML	Maximum likelihood
MSE	Mean Square Error
MRRSM	Multivariate Rational Function basis

MSG	Multivariate spectral gradient
MTS	Multivariate time series
NASDAQ	National Association of Securities Dealers Automated Quotations
NYSE	New York Stock Exchange
NURBS	Nonuniform rational B-spline
PIP	Perceptually important points
PAA	Piecewise Aggregate Approximation
PLA	Piecewise linear representation
PD	Point Distribution
PCA	Principal Component Analysis
REF	Robot execution failure
RMSE	Root mean-squared error
SQ	Sample quantile
SSE	Shanghai Stock Exchange
SDA	Shape description alphabet
STI	Spatio-temporal intermittency
SBML	Standard markup language
SAX	Symbolic aggregate approximation
VIM	Variational iteration method
VSA	Variational Shape Approximation
VAR	Vector Autoregressive
WSSVD	Weighted Sum Singular Value Decomposition

PREFACE

This book includes 16 open-access articles featuring a wide variety of subjects in multivariate mathematics. However, these can be divided into three major thematic groups. The first group is centered on multivariate extensions of univariate methods applied to optimization, statistical analysis, inference, and integration problems. The second group is centered on multivariate time series analysis and networks. The third group is centered on multivariate geometry, polynomial surfaces approximation techniques, differential equations and topology.

Chapter 1 to 7 fall under the first thematic group. The algorithm introduced in Chapter 1 is an extension of a multivariate spectral gradient (MSG) method for solving the non-smooth convex optimization problem. Chapter 2 proposes a multi-variate extended method for estimating tail index, masses, and location parameters of stable distributions based on a U-statistics. The estimation of spectral measure of multivariate stable distribution are important to many research fields including physics, economics, finance, insurance, and telecommunications. Chapter 3 describes a multivariate extension of the variational inequality problem with multiple applications in optimization theory, game theory, economic equilibrium, and mechanics. Chapter 4 proposes an axiomatic approach to solve multivariate mean value integrals. Chapter 5 includes a detailed multivariate analysis of linear mixed-effects models. Chapter 6 focuses on the problem of comparing the means of several multivariate Gaussian processes. Chapter 7 reviews the multivariate nonlinear Granger causality test as it is an important method for detecting causal relations between time series in economics and finance. This is the densest thematic group and requires an advanced experience in several areas of multivariate calculus and statistical analysis.

Chapter 8 to 11 fall under the second thematic group. Main phenomena in physics, biology, economic, multimedia and neural network research depends on the analysis of multivariate time series (MTS). Chapter 8 proposes a match-by-dimension approach to search similar sequences in MTS datasets. Chapter 9 introduces a new method called Semi Metric Ensemble Time Series (SMETS) for comparing groups of MTS of arbitrary dimensions. Chapter 10 proposes a non-parametric method for analyzing multivariate time series based on an MTS mapping into a multilayer network. Chapter 11 evaluates several multivariate measures for the analysis of networks.

Chapter 12 to 16 fall under the third group. Chapter 12 proposes the use of T-B-spline curves for the exact representation of ellipses and parabolas. Chapter 13 propose a method for quadratic fitting with a Bayesian prior. Chapter 14 proposes a polynomial-basis response surface method based on multivariate rational function basis and its application to the structural optimization of truss structures. Chapter 15 discusses the potential application of differential topology in the analysis of complicated data. Chapter 16 proposes a method for finding solutions on partial dynamic equations based on an extension of the variational iteration method to multivariable calculus on time scales.

1

Multivariate Spectral Gradient Algorithm for Nonsmooth Convex Optimization Problems

Yaping Hu

School of Science, East China University of Science and Technology, Shanghai 200237, China

ABSTRACT

We propose an extended multivariate spectral gradient algorithm to solve the nonsmooth convex optimization problem. First, by using Moreau-Yosida regularization, we convert the original objective function to a continuously differentiable function; then we use approximate function and gradient values of the Moreau-Yosida regularization to substitute the corresponding exact values in the algorithm. The global convergence is proved under suitable assumptions. Numerical experiments are presented to show the effectiveness of this algorithm.

INTRODUCTION

Consider the unconstrained minimization problem

Citation: Hu, Y. (2015). Multivariate spectral gradient algorithm for nonsmooth convex optimization problems. Mathematical Problems in Engineering, 2015. (8pages), DOI/URL: https://doi.org/10.1155/2015/145323.

Copyright: © 2015 Yaping Hu. This is an open access article distributed under the Creative Commons Attribution 3.0 Unported (CC BY 3.0) License.

$$\min_{x \in \Re^n} f(x), \tag{1}$$

where $f : \Re^n \to \Re$ is a nonsmooth convex function. The Moreau-Yosida regularization [1] of f at $x \in \Re^n$ associated with $z \in \Re^n$ is defined by

$$F(x) = \min_{z \in \Re^n} \left\{ f(z) + \frac{1}{2\lambda} \|z - x\|^2 \right\}, \tag{2}$$

where $\|\cdot\|$ is the Euclidean norm and λ is a positive parameter. The function minimized on the right-hand side is strongly convex and differentiable, so it has a unique minimizer for every $z \in \Re^n$. Under some reasonable conditions, the gradient function of (x) can be proved to be semismooth [2, 3], though generally $F(x)$ is not twice differentiable. It is widely known that the problem

$$\min_{x \in \Re^n} F(x) \tag{3}$$

and the original problem (1) are equivalent in the sense that the two corresponding solution sets coincidentally are the same.The following proposition shows some properties of the Moreau-Yosida regularization function $F(x)$.

Proposition 1 (see Chapter XV, Theorem , [1]). *The Moreau-Yosida regularization function F is convex, finitevalued, and differentiable everywhere with gradient*

$$g(x) \equiv \nabla F(x) = \frac{1}{\lambda}(x - p(x)), \tag{4}$$

Where

$$p(x) = \arg\min_{z \in \Re^n} \left\{ f(z) + \frac{1}{2\lambda} \|z - x\|^2 \right\} \tag{5}$$

is the unique minimizer in (2). Moreover, for all $x, y \in$ R , one has

$$\|g(x) - g(y)\| \leq \frac{1}{\lambda} \|x - y\|. \tag{6}$$

This proposition shows that the gradient function $: \Re^n \to \Re^n$ is Lipschitz continuous with modulus $1/\lambda$. In this case, the gradient function g is differentiable almost everywhere by the Rademacher theorem; then the Bsubdifferential [4] of g at $x \in \Re^n$ is defined by

$$\partial_B g(x) = \left\{ V \in \Re^{n \times n} : V = \lim_{x_k \to x} \nabla g(x_k),\ x_k \in D_g \right\}, \tag{7}$$

where$D_g = \{x \in \Re^n : g$ is differentiable at $x\}$, and the next property of BD-regularity holds [4–6].

Proposition 2. *If g is BD-regular at x, then*

(i) *all matrices $V \in \partial_B g(x)$ are nonsingular;*

(ii) *there exists a neighborhood N of $x \in \Re^n$, $\kappa_1 > 0$, and $\kappa_2 > 0$; for all $y \in N$, one has*

$$d^T V d \geq \kappa_1 \|d\|^2, \quad \|V^{-1}\| \leq \kappa_2, \quad \forall d \in \Re^n,\ V \in \partial_B g(y). \tag{8}$$

Instead of the corresponding exact values, we often use the approximate value of function (x) and gradient $g(x)$ in the practical computation, because $p(x)$ is difficult and sometimes impossible to be solved precisely. Suppose that, for any $\varepsilon > 0$ and for each $x \in \Re^n$, there exists an approximate vector $p^a(x, \varepsilon) \in \Re^n$ of the unique minimizer $p(x)$in (2)such that

$$f(p^a(x,\varepsilon)) + \frac{1}{2\lambda}\|p^a(x,\varepsilon) - x\|^2 \leq F(x) + \varepsilon. \tag{9}$$

The implementable algorithms to find such approximate vector $p^a(x, \varepsilon) \in \Re^n$ can be found, for example, in [7, 8]. The existence theorem of the approximate vector $p^a(x, \varepsilon)$ is presented as follows.

Proposition 3 (see Lemma in [7]). *Let $\{x_k\}$ be generated according to the formula*

$$x_{k+1} = x_k - \alpha_k v_k, \quad \textit{for } k = 1, 2, \ldots, \tag{10}$$

where$\alpha_k > 0$ is a stepsize and v_k is an approximate subgradient at x_k; that is,

$$v_k \in \partial_{\varepsilon_k} f(x_k) = \{v \mid f(v) - \langle v, x_k\rangle \leq f(x_k) + \varepsilon_k\}, \quad \textit{for } k = 1, 2, \ldots. \tag{11}$$

(i) If $v_k$satisfies

$$v_k \in \partial f(x_{k+1}), \quad \textit{for } k = 1, 2, \ldots, \tag{12}$$

then (11) holds with

$$\varepsilon_k = f(x_k) - f(x_{k+1}) - \alpha_k \|v_k\|^2 \geq 0. \tag{13}$$

(ii) Conversely, if(11) holds with ε_k given by (13), then (12) holds: $x_{k+1} = p^a(x_k, \varepsilon_k)$.

We use the approximate vector $p^a(x, \varepsilon)$ to define approximation function and gradient values of the Moreau-Yosida regularization, respectively, by

$$F^a(x,\varepsilon) = f(p^a(x,\varepsilon)) + \frac{1}{2\lambda}\|p^a(x,\varepsilon) - x\|^2, \tag{14}$$

$$g^a(x,\varepsilon) = \frac{(x - p^a(x,\varepsilon))}{\lambda}. \tag{15}$$

The following proposition is crucial in the convergence analysis. The proof of this proposition can be found in [2].

Proposition 4. *Let ε be arbitrary positive number and let* $p^a(x, \varepsilon)$ *be a vector satisfying (9). Then, one gets*

$$F(x) \le F^a(x,\varepsilon) \le F(x) + \varepsilon, \tag{16}$$

$$\|g^a(x,\varepsilon) - g(x)\| \le \sqrt{\frac{2\varepsilon}{\lambda}}, \tag{17}$$

$$\|p^a(x,\varepsilon) - p(x)\| \le \sqrt{2\lambda\varepsilon}. \tag{18}$$

Algorithms which combine the proximal techniques with Moreau-Yosida regularization for solving the nonsmooth problem (1) have been proved to be effective [7, 9, 10], and also some trust region algorithms for solving (1) have been proposed in [5, 11, 12], and so forth. Recently, Yuan et al. [13, 14] and Li [15] have extended the spectral gradient method and conjugate gradient-type method to solve (1), respectively.

Multivariate spectral gradient (MSG) method was first proposed by Han et al. [16] for optimization problems. This method has a nice property that it converges quadratically for objective function with positive definite diagonal Hessian matrix [16]. Further studies on such method for nonlinear equations and bound constrained optimization can be found, for instance, in [17, 18]. By using nonmonotone technique, some effective spectral gradient methods are presented in [13, 16, 17, 19]. In this paper, we extend the multivariate spectral gradient method by combining with a nonmonotone line search technique as well as the Moreau-Yosida regulation function to solve the nonsmooth problem (1) and do some numerical experiments to test its efficiency.

The rest of this paper is organized as follows. In Section 2, we propose multivariate spectral gradient algorithm to solve (1). In Section 3, we prove the global convergence of the proposed algorithm; then some numerical results are presented in Section 4. Finally, we have a conclusion section.

ALGORITHM

In this section, we present the multivariate spectral gradient algorithm to solve the nonsmooth convex unconstrained optimization problem (1). Our approach is using the tool of the Moreau-Yosida regularization to smoothen the nonsmooth function and then make use of the approximate values of function F and gradient g in multivariate spectral gradient algorithm.

We first recall the multivariate spectral gradient algorithm [16] for smooth optimization problem:

$$\min\{f(x) \mid x \in \Re^n\}, \qquad (19)$$

where $f: \Re^n \to \Re$ is continuously differentiable and its gradient is denoted by g. Let x_k be the current iteration; multivariate spectral gradient algorithm is defined by

$$x_{k+1} = x_k - \operatorname{diag}\left\{\frac{1}{\lambda_k^1}, \frac{1}{\lambda_k^2}, \ldots, \frac{1}{\lambda_k^n}\right\} g_k, \qquad (20)$$

where g_k is the gradient vector of f at xk and $\operatorname{diag}\{\lambda_k^1, \lambda_k^2, \ldots, \lambda_k^n\}$ is solved by minimizing

$$\left\|\operatorname{diag}\{\lambda^1, \lambda^2, \ldots, \lambda^n\} s_{k-1} - u_{k-1}\right\|_2 \qquad (21)$$

with respect to $\{\lambda^i\}_{i=1}^n$, where $s_{k-1} = x_k - x_{k-1}$, $u_{k-1} = g_k - g_{k-1}$.

Denote the ith element of s_k and y_k by s_k^i and y_k^i, respectively. We present the following multivariate spectral gradient (MSG) algorithm.

Algorithm 5. Set $x_0 \in \Re^n$, $\sigma \in (0,1)$, $\beta > 0$, $\lambda > 0$, $\gamma \geq 0$, $\delta > 0$, $\rho \in [0,1]$, $\epsilon \in (0,1)$, $E_0 = 1$, and $\tau_0 \in (0,1]$; $\{\tau_k\}$ is a strictly decreasing sequence with $\lim_{k\to 0}\tau_k = 0$, $k := 0$.

Step 1. Set $\varepsilon_0 = \tau_0$. Calculate $F^a(x_0, \varepsilon_0)$ by (14) as well as $g^a(x_0, \varepsilon_0)$ by (15). Let $J_0 = F^a(x_0, \varepsilon_0)$, $d_0 = -g^a(x_0, \varepsilon_0)$.

Step 2. Stop if $\|g^a(x_k, \varepsilon_k)\| = 0$. Otherwise, go to Step 3.

Step 3. Choose ε_{k+1} satisfying $0 < \varepsilon_{k+1} \leq \min\{\tau_k, \tau_k\|g^a(x_k, \varepsilon_k)\|^2\}$; find α_k which satisfies

$$F^a(x_k + \alpha_k d_k, \varepsilon_{k+1}) - J_k \leq \sigma\alpha_k g^a(x_k, \varepsilon_k)^T d_k, \qquad (22)$$

where $\alpha_k = \beta 2^{-i_k}$ and i_k is the smallest nonnegative integer such that (22) holds.

Step 4. Let $x_{k+1} = x_k + \alpha_k d_k$. Stop if $\|g^a(x_{k+1}, \varepsilon_{k+1})\| = 0$.

Step 5. Update J_{k+1} by the following formula:

$$E_{k+1} = \rho E_k + 1,$$

$$J_{k+1} = \frac{\rho E_k J_k + F^a(x_k + \alpha_k d_k, \varepsilon_{k+1})}{E_{k+1}}. \tag{23}$$

Step 6. Compute the search direction d_{k+1} by the following:

(a) If $y_k^i / s_k^i > 0$, then set $\lambda_{k+1}^i = y_k^i / s_k^i$; otherwise set $\lambda_{k+1}^i = s_k^T y_k / s_k^T s_k$ for $i = 1, 2, \ldots, n$, where $y_k = g^a(x_{k+1}, \varepsilon_{k+1}) - g^a(x_k, \varepsilon_k) + \gamma s_k$, $s_k = x_{k+1} - x_k$.

(b) If $\lambda_{k+1}^i \le \epsilon$ or $\lambda_{k+1}^i \ge 1/\epsilon$, then set $\lambda_{k+1}^i = \delta$ for $i = 1, 2, \ldots, n$.

Let $d_{k+1} = -\text{diag}\{1/\lambda_{k+1}^1, 1/\lambda_{k+1}^2, \ldots, 1/\lambda_{k+1}^n\} g^a(x_{k+1}, \varepsilon_{k+1})$.

Step 7. Set $:= k + 1$; go back to Step 2.

Remarks. (i) The definition of $\varepsilon_{k+1} = o(\|g^a(x_k, \varepsilon_k)\|^2)$ in Algorithm 5, together with (15) and Proposition 3, deduces that

$$\varepsilon_{k+1} = o\left(\left\|x_k - p^a(x_k, \varepsilon_k)\right\|^2\right) = o\left(\left\|x_k - x_{k+1}\right\|^2\right)$$
$$= o\left(\alpha_k^2 \left\|d_k\right\|^2\right); \tag{24}$$

then, with the decreasing property of ε_{k+1}, the assumed condition $\varepsilon_k = o(\alpha_k^2 \|d_k\|^2)$ in Lemma 7 holds.

(ii) From the nonmonotone line search technique (22), we can see that J_{k+1} is a convex combination of the function value $F^a(x_{k+1}, \varepsilon_{k+1})$ and J_k. Also J_k is a convex combination of the function values $F^a(x_k, \varepsilon_k), \ldots, F^a(x_1, \varepsilon_1), F^a(x_0, \varepsilon_0)$ as $J_0 = F^a(x_0, \varepsilon_0)$. ρ is a positive value that plays an important role in manipulating the degree of nonmonotonicity in the nonmonotone line search technique, with $\rho = 0$ yielding a strictly monotone scheme and with $\rho = 1$ yielding $J_k = C_k$, where

$$C_k = \frac{1}{k+1}\sum_{i=0}^{k} F^a(x_i, \varepsilon_i) \tag{25}$$

is the average function value.

(iii) From Step 6, we can obtain that

$$\min\left\{\epsilon, \frac{1}{\delta}\right\} \|g^a(x_k, \varepsilon_k)\| \le \|d_k\|$$

$$\le \max\left\{\frac{1}{\epsilon}, \frac{1}{\delta}\right\} \|g^a(x_k, \varepsilon_k)\|; \tag{26}$$

then there is a positive constant μ such that, for all k,

$$g^a(x_k, \varepsilon_k)^T d_k \le -\mu \|g^a(x_k, \varepsilon_k)\|^2, \tag{27}$$

which shows that the proposed multivariate spectral gradient algorithm possesses the sufficient descent property.

GLOBAL CONVERGENCE

In this section, we provide a global convergence analysis for the multivariate spectral gradient algorithm. To begin with, we make the following assumptions which have been given in [5, 12–14].

Assumption A. (i) F is bounded from below.

(ii) The sequence $\{V_k\}$, $V_k \in \partial_B(x_k)$, is bounded; that is, there exists a constant $M > 0$ such that, for all k,

$$\|V_k\| \le M. \tag{28}$$

The following two lemmas play crucial roles in establishing the convergence theorem for the proposed algorithm. By using (26) and (27) and Assumption A, similar to Lemma in [20], we can get the next lemma which shows that Algorithm 5 is well defined. The proof ideas of this lemma and Lemma in [20] are similar, hence omitted.

Lemma 6. *Let $\{F^a(x_k, \varepsilon_k)\}$ be the sequence generated by Algorithm 5. Suppose that Assumption A holds and C_k is defined by (25). Then one has $F^a(x_k, \varepsilon_k) \le J_k \le C_k$ for all k. Also, there exists a stepsize α_k satisfying the nonmonotone line search condition.*

Lemma 7. *Let $\{(x_k, \varepsilon_k)\}$ be the sequence generated by Algorithm 5. Suppose that Assumption A and $\varepsilon_k = o(\alpha_k^2 \|d_k\|^2)$ hold. Then, for all k, one has*

$$\alpha_k \geq m_0, \tag{29}$$

where $m_0 > 0$ is a constant.

Proof (Proof by Contradiction). Let αk satisfy the nonmonotone Armijo-type line search (22). Assume on the contrary that $\liminf_{k\to\infty}\alpha_k = 0$ does hold; then there exists a subsequence $\{\alpha_k\}'$ such that $\alpha_k \to 0$ as $k \to \infty$. From the nonmonotone line search rule (22), $\alpha'_k = \alpha_k/2$ satisfies

$$F^a\left(x_k + \alpha'_k d_k, \varepsilon_{k+1}\right) - J_k > \sigma\alpha'_k g^a\left(x_k, \varepsilon_k\right)^T d_k; \tag{30}$$

together with $F^a(x_k, \varepsilon_k) \leq J_k \leq C_k$ in Lemma 6, we have

$$\begin{aligned} & F^a\left(x_k + \alpha'_k d_k, \varepsilon_{k+1}\right) - F^a\left(x_k, \varepsilon_k\right) \\ & \geq F^a\left(x_k + \alpha'_k d_k, \varepsilon_{k+1}\right) - J_k > \sigma\alpha'_k g^a\left(x_k, \varepsilon_k\right)^T d_k. \end{aligned} \tag{31}$$

By (28) and (31) and Proposition 4 and using Taylor's formula, there is

$$\begin{aligned} \sigma\alpha'_k g^a\left(x_k, \varepsilon_k\right)^T d_k &< F^a\left(x_k + \alpha'_k d_k, \varepsilon_{k+1}\right) \\ &\quad - F^a\left(x_k, \varepsilon_k\right) \\ &\leq F\left(x_k + \alpha'_k d_k\right) - F\left(x_k\right) + \varepsilon_{k+1} \\ &= \alpha'_k d_k^T g\left(x_k\right) \\ &\quad + \frac{1}{2}\left(\alpha'_k\right)^2 d_k^T V\left(u_k\right) d_k + \varepsilon_{k+1} \\ &\leq \alpha'_k d_k^T g\left(x_k\right) + \frac{M}{2}\left(\alpha'_k\right)^2 \left\|d_k\right\|^2 \\ &\quad + \varepsilon_{k+1}, \end{aligned} \tag{32}$$

where $u_k \in (x_k, x_{k+1})$. From (32) and Proposition 4, we have

$$\begin{aligned} \frac{\alpha_k}{2} = \alpha'_k &> \left[\frac{\left(g^a\left(x_k, \varepsilon_k\right) - g\left(x_k\right)\right)^T d_k - (1-\sigma) g^a\left(x_k, \varepsilon_k\right)^T d_k - \varepsilon_{k+1}/\alpha'_k}{\left\|d_k\right\|^2}\right]\frac{2}{M} \\ &\geq \left[\frac{\mu(1-\sigma)\left\|g^a\left(x_k, \varepsilon_k\right)\right\|^2 - \sqrt{2\varepsilon_k/\lambda}\left\|d_k\right\| - \varepsilon_k/\alpha'_k}{\left\|d_k\right\|^2}\right]\frac{2}{M} = \left[\frac{\mu(1-\sigma)\left\|g^a\left(x_k, \varepsilon_k\right)\right\|^2}{\left\|d_k\right\|^2} - \frac{o\left(\alpha_k\right)}{\sqrt{\lambda}} - o\left(\alpha_k\right)\right]\frac{2}{M} \\ &\geq \left[\frac{\mu(1-\sigma)}{\left(\max\{1/\epsilon, 1/\delta\}\right)^2} - \frac{o\left(\alpha_k\right)}{\sqrt{\lambda}} - o\left(\alpha_k\right)\right]\frac{2}{M}, \end{aligned} \tag{33}$$

where the second inequality follows from (26), Part 3 in Proposition 4, and $\varepsilon_{k+1} \leq \varepsilon_k$, the equality follows from $\varepsilon k = o(\alpha_k^2\|d_k\|^2)$, and the last inequality follows from (27). Dividing each side by α_k and letting $k \to \infty$ in the above inequality, we can deduce that

$$\frac{1}{2} \geq \lim_{k\to\infty}\left(\frac{2\mu(1-\sigma)}{(\max\{1/\epsilon, 1/\delta\})^2 M}\right)\frac{1}{\alpha_k} = +\infty, \quad (34)$$

which is impossible, so the conclusion is obtained.

By using the above lemmas, we are now ready to prove the global convergence of Algorithm 5.

Theorem 8. *Let $\{x_k\}$ be generated by Algorithm 5 and suppose that the conditions of Lemma 7 hold. Then one has*

$$\lim_{k\to\infty} \|g(x_k)\| = 0; \quad (35)$$

sequence $\{x_k\}_{k=0}^{\infty}$ has accumulation point, and every accumulation point of $\{x_k\}_{k=0}^{\infty}$ is optimal solution of problem (1).

Proof. Suppose that there exist $\epsilon_0 > 0$ and $k_0 > 0$ such that

$$\|g^a(x_k, \varepsilon_k)\| \geq \epsilon_0, \quad \forall k > k_0. \quad (36)$$

From (22), (26), and (29), we get

$$\begin{aligned} F^a(x_k + \alpha_k d_k, \varepsilon_{k+1}) - J_k &\leq \sigma\alpha_k g^a(x_k, \varepsilon_k)^T d_k \\ &\leq -\sigma\alpha_k \min\left\{\epsilon, \frac{1}{\delta}\right\}\|g^a(x_k, \varepsilon_k)\|^2 \\ &\leq -\sigma m_0\epsilon_0 \min\left\{\epsilon, \frac{1}{\delta}\right\}, \quad \forall k > k_0. \end{aligned} \quad (37)$$

Therefore, it follows from the definition of J_{k+1} and (23) that

$$\begin{aligned} J_{k+1} &= \frac{\rho E_k J_k + F^a(x_k + \alpha_k d_k, \varepsilon_{k+1})}{E_{k+1}} \\ &\leq \frac{\rho E_k J_k + J_k - \sigma m_0\epsilon_0 \min\{\epsilon, 1/\delta\}}{E_{k+1}} \\ &\leq J_k - \frac{\sigma m_0\epsilon_0 \min\{\epsilon, 1/\delta\}}{E_{k+1}}. \end{aligned} \quad (38)$$

By Assumption A, F is bounded from below. Further by Proposition 4, $(x_k) \leq F^a(x_k, \varepsilon_k)$ for all k, we see that $F^a(x_k, \varepsilon_k)$ is bounded from below. Together with $F^a(x_k, \varepsilon_k) \leq J_k$ for all k from Lemma 6, it shows that J_k is also bounded from below. By (38), we obtain

$$\sum_{k=k_0}^{\infty} \frac{\sigma m_0 \epsilon_0 \min\{\epsilon, 1/\delta\}}{E_{k+1}} < \infty. \tag{39}$$

On the other hand, the definition of E_{k+1} implies that $E_{k+1} \leq k + 2$, and it follows that

$$\sum_{k=k_0}^{\infty} \frac{\sigma m_0 \epsilon_0 \min\{\epsilon, 1/\delta\}}{E_{k+1}} \geq \sum_{k=k_0}^{\infty} \frac{\sigma m_0 \epsilon_0 \min\{\epsilon, 1/\delta\}}{k+2}$$
$$= +\infty. \tag{40}$$

This is a contradiction. Therefore, we should have

$$\lim_{k \to \infty} \left\| g^a\left(x_k, \varepsilon_k\right) \right\| = 0. \tag{41}$$

From (17) in Proposition 4 together with ε_k as $k \to \infty$, which comes from the definition of ε_k and $\lim_{k \to 0} \tau_k = 0$ in Algorithm 5, we obtain

$$\lim_{k \to \infty} \left\| g\left(x_k\right) \right\| = 0. \tag{42}$$

Set S* as an accumulation point of sequence $\{x_k\}_{k=0}^{\infty}$; there is a convergent subsequence $\{x_{k_l}\}_{l=0}^{\infty}$ such that

$$\lim_{l \to \infty} x_{k_l} = x^*. \tag{43}$$

From (4) we know that $g(x_k) = (x_k - p(x_k))/\lambda$. Consequently, (42) and (43) show that $x^* = (x^*)$. Hence, x^* is an optimal solution of problem (1).

NUMERICAL RESULTS

This section presents some numerical results from experiments using our multivariate spectral gradient algorithm for the given test nonsmooth problems which come from [21]. We also list the results of [14] (modified Polak-Ribiere- ` Polyak gradient method, MPRP) and [22] (proximal bundle method, PBL) to make a comparison with the result of Algorithm 5. All codes were written in MATLAB R2010a and were implemented on a PC with 2.8 GHz CPU, 2 GB of memory, and Windows 8. We set $\beta = \lambda = 1$, $\sigma = 0.9$, $\epsilon = 10^{-10}$, and $\gamma = 0.01$, and the parameter δ is chosen as

$$\delta = \begin{cases} 1 & \text{if } \|g^a(x_k, \varepsilon_k)\| > 1, \\ \|g^a(x_k, \varepsilon_k)\|^{-1} & \text{if } 10^{-5} \le \|g^a(x_k, \varepsilon_k)\| \le 1, \\ 10^{-5} & \text{if } \|g^a(x_k, \varepsilon_k)\| < 10^{-5}; \end{cases} \quad (44)$$

then we adopt the termination condition $\|g^a(x_k, \varepsilon_k)\| \le 10^{-10}$. For subproblem (5), the classical PRP CG method (called subalgorithm) is used to solve it; the algorithm stops if $\|\partial f(x_k)\| \le 10^{-4}$ or $f(x_{k+1}) - f(x_k) + \|\partial f(x_{k+1})\|^2 - \|\partial f(x_k)\|^2 \le 10^{-3}$ holds, where $\partial f(x_k)$ is the subgradient of (x) at the point x_k. The subalgorithm will also stop if the iteration number is larger than fifteen. In its line search, the Armijo line search technique is used and the step length is accepted if the search number is larger than five. Table 1 contains problem names, problem dimensions, and the optimal values.

Table 1: Test problems

Nr.	Problems	Dim.	$f_{ops}(x)$
1	Rosenbrock	2	0
2	Crescent	2	0
3	CB2	2	1.9522245
4	CB3	2	2.0
5	DEM	2	−3
6	QL	2	7.20
7	LQ	2	−1.4142136
8	Mifflin 1	2	−1.0
9	Mifflin 2	2	−1.0
10	Wolfe	2	−8.0
11	Rosen-Suzuki	4	−44
12	Shor	5	22.600162

The summary of the test results is presented in Tables 2-3, where "Nr." denotes the name of the tested problem, "NF" denotes the number of function evaluations, "NI" denotes the number of iterations, and "f(x)" denotes the function value at the final iteration.

Table 2: Results on Rosenbrock with different ρ and ε

τ_k	$\rho = 0$ NI/NF/$f(x)$	Time	$\rho = 0.75$ NI/NF/$f(x)$	Time
$1/2k^2$	$30/46/1.581752e-9$	1.794	$29/30/7.778992e-9$	1.076
$1/3(k+2)^3$	$28/38/5.207744e-9$	1.420	$26/27/6.541087e-9$	1.023
$1/4(k+2)^4$	$29/37/1.502034e-9$	1.388	$27/28/5.112699e-9$	1.030
$1/5(k+2)^5$	$27/37/1.903969e-9$	1.451	$27/28/6.329141e-9$	1.092
$1/6(k+2)^6$	$27/36/4.859901e-9$	1.376	$27/28/6.073222e-9$	1.025

Table 3: Numerical results for MSG/MPRP/PBL on problems 1–12

Nr.	MSG NI/NF/$f(x)$	MPRP NI/NF/$f(x)$	PBL NI/NF/$f(x)$	$f_{ops}(x)$
1	$29/30/7.778992e-9$	$46/48/7.091824e-7$	$42/45/0.381e-6$	0
2	$9/10/1.450669e-5$	$11/13/6.735123e-5$	$18/20/0.679e-6$	0
3	9/10/1.9522245	12/14/1.952225	32/34/1.9522245	1.9522245
4	4/9/2.000009	2/6/2.000098	14/16/2.0000000	2.0
5	3/4/−2.999949	4/6/−2.999866	17/19/−3.0000000	−3
6	11/12/7.200000	10/12/7.200011	13/15/7.2000015	7.20
7	3/4/−1.4142136	2/3/−1.414214	11/12/−1.4142136	−1.4142136
8	9/10/−0.9999638	4/6/−0.9919815	66/68/−0.9999994	−1.0
9	12/13/−0.9999978	20/23/−0.9999925	13/15/−1.0000000	−1.0
10	5/6/−7.999999	—	43/46/−8.0000000	−8.0
11	6/7/−43.99797	28/58/−43.99986	43/45/−43.999999	−44
12	12/13/2.260017	33/91/22.60023	27/29/22.600162	22.600162

The value of ρ controls the nonmonotonicity of line search which may affect the performance of the MSG algorithm. Table 2 shows the results for different parameter ρ, as well as different values of the parameter τ_k ranging from $1/6(k+2)^6$ to $1/2k^2$ on problem Rosenbrock, respectively. We can conclude from the table that the proposed algorithm works reasonably well for all the test cases. This table also illustrates that the value of ρ can influence the performance of the algorithm significantly if the value of ε is within a certain range, and the choice $\rho = 0.75$ is better than $\rho = 0$.

Then, we compare the performance of MSG to that of the algorithms MPRP and PBL. In this test, we fix $\tau_k = 1/2k^2$ and $\rho = 0.75$. To illustrate the performance of each algorithm more specifically, we present three comparison results in terms of number of iterations, number of function evaluations, and the final objective function value in Table 3.

The numerical results indicate that Algorithm 5 can successfully solve the test problems. From the number of iterations in Table 3, we see that Algorithm 5 performs best among these three methods, and the final function value obtained by Algorithm 5 is closer to the optimal function value than those obtained by MPRP and PBL. In a word, the numerical experiments show that the proposed algorithm provides an efficient approach to solve nonsmooth problems.

CONCLUSIONS

We extend the multivariate spectral gradient algorithm to solve nonsmooth convex optimization problems. The proposed algorithm combines a nonmonotone line search technique and the idea of Moreau-Yosida regularization. The algorithm satisfies the sufficient descent property and its global convergence can be established. Numerical results show the efficiency of the proposed algorithm.

ACKNOWLEDGMENTS

The author would like to thank the anonymous referees for their valuable comments and suggestions which help a lot to improve the paper greatly. The author also thanks Professor Gong-lin Yuan for his kind offer of the source BB codes on nonsmooth problems. This work is supported by the National Natural Science Foundation of China (Grant no. 11161003).

REFERENCES

1. J. B. Hiriart-Urruty and C. Lemaréchal, *Convex Analysis and Minimization Algorithms*, Springer, Berlin, Germany, 1993.
2. M. Fukushima and L. Qi, "A globally and superlinearly convergent algorithm for nonsmooth convex minimization," *SIAM Journal on Optimization*, vol. 6, no. 4, pp. 1106–1120, 1996.
3. L. Q. Qi and J. Sun, "A nonsmooth version of Newton's method," *Mathematical Programming*, vol. 58, no. 3, pp. 353–367, 1993.
4. F. H. Clarke, *Optimization and Nonsmooth Analysis*, Wiley, New York, NY, USA, 1983.
5. S. Lu, Z. Wei, and L. Li, "A trust region algorithm with adaptive cubic regularization methods for nonsmooth convex minimization," *Computational Optimization and Applications*, vol. 51, no. 2, pp. 551–573, 2012.
6. L. Q. Qi, "Convergence analysis of some algorithms for solving nonsmooth equations," *Mathematics of Operations Research*, vol. 18, no. 1, pp. 227–244, 1993.
7. R. Correa and C. Lemaréchal, "Convergence of some algorithms for convex minimization," *Mathematical Programming*, vol. 62, no. 1–3, pp. 261–275, 1993.
8. M. Fukushima, "A descent algorithm for nonsmooth convex optimization," *Mathematical Programming*, vol. 30, no. 2, pp. 163–175, 1984.
9. J. R. Birge, L. Qi, and Z. Wei, "Convergence analysis of some methods for minimizing a nonsmooth convex function," *Journal of Optimization Theory and Applications*, vol. 97, no. 2, pp. 357–383, 1998.
10. Z. Wei, L. Qi, and J. R. Birge, "A new method for nonsmooth convex optimization," *Journal of Inequalities and Applications*, vol. 2, no. 2, pp. 157–179, 1998.
11. N. Sagara and M. Fukushima, "A trust region method for nonsmooth convex optimization," *Journal of Industrial and Management Optimization*, vol. 1, no. 2, pp. 171–180, 2005.
12. G. Yuan, Z. Wei, and Z. Wang, "Gradient trust region algorithm with limited memory BFGS update for nonsmooth convex minimization," *Computational Optimization and Applications*, vol. 54, no. 1, pp. 45–64, 2013.

13. G. Yuan and Z. Wei, "The Barzilai and Borwein gradient method with nonmonotone line search for nonsmooth convex optimization problems," *Mathematical Modelling and Analysis*, vol. 17, no. 2, pp. 203–216, 2012.
14. G. Yuan, Z. Wei, and G. Li, "A modified Polak-Ribière-Polyak conjugate gradient algorithm for nonsmooth convex programs," *Journal of Computational and Applied Mathematics*, vol. 255, pp. 86–96, 2014.
15. Q. Li, "Conjugate gradient type methods for the nondifferentiable convex minimization," *Optimization Letters*, vol. 7, no. 3, pp. 533–545, 2013.
16. L. Han, G. Yu, and L. Guan, "Multivariate spectral gradient method for unconstrained optimization," *Applied Mathematics and Computation*, vol. 201, no. 1-2, pp. 621–630, 2008.
17. G. Yu, S. Niu, and J. Ma, "Multivariate spectral gradient projection method for nonlinear monotone equations with convex constraints," *Journal of Industrial and Management Optimization*, vol. 9, no. 1, pp. 117–129, 2013.
18. Z. Yu, J. Sun, and Y. Qin, "A multivariate spectral projected gradient method for bound constrained optimization," *Journal of Computational and Applied Mathematics*, vol. 235, no. 8, pp. 2263–2269, 2011.
19. Y. Xiao and Q. Hu, "Subspace Barzilai-Borwein gradient method for large-scale bound constrained optimization," *Applied Mathematics and Optimization*, vol. 58, no. 2, pp. 275–290, 2008.
20. H. Zhang and W. W. Hager, "A nonmonotone line search technique and its application to unconstrained optimization," *SIAM Journal on Optimization*, vol. 14, no. 4, pp. 1043–1056, 2004.
21. L. Lukšan and J. Vlček, "Test problems for nonsmooth unconstrained and linearly constrained optimization," Tech. Rep. 798, Institute of Computer Science, Academy of Sciences of the Czech Republic, Praha, Czech Republic, 2000.
22. L. Lukšan and J. Vlček, "A bundle-Newton method for nonsmooth unconstrained minimization," *Mathematical Programming*, vol. 83, no. 3, pp. 373–391, 1998.

2

U-Statistic for Multivariate Stable Distributions

Mahdi Teimouri[1] ,Saeid Rezakhah[1] , and Adel Mohammadpour[1]

[1]Department of Statistics, Faculty of Mathematics and Computer Science, Amirkabir University of Technology (Tehran Polytechnic), 424 Hafez Ave., Tehran 15914, Iran

ABSTRACT

A U-statistic for the tail index of a multivariate stable random vector is given as an extension of the univariate case introduced by Fan (2006). Asymptotic normality and consistency of the proposed U-statistic for the tail index are proved theoretically. The proposed estimator is used to estimate the spectral measure. The performance of both introduced tail index and spectral measure estimators is compared with the known estimators by comprehensive simulations and real datasets.

INTRODUCTION

In recent years, stable distributions have received extensive use in a vast number of fields including physics, economics, finance, insurance, and

Citation: Teimouri, M., Rezakhah, S., &Mohammadpour, A. (2017). -Statistic for Multivariate Stable Distributions. Journal of Probability and Statistics, 2017. (13 pages), DOI/URL: https://doi.org/10.1155/2017/3483827.

Copyright: © 2017 Mahdi Teimouri et al. This is an open access article distributed under the Creative Commons Attribution 4.0 International (CC BY 4.0) License.

telecommunications. Different sorts of data found in applications arise from heavy tailed or asymmetric distribution, where normal models are clearly inappropriate. In fact, stable distributions have theoretical underpinnings to accurately model a wide variety of processes. Stable distribution has originated with the work of Lévy [1]. There are a variety of ways to introduce a stable random vector. In the following, two definitions are proposed for a stable random vector; see Samorodnitsky and Taqqu [2].

Definition 1. A random vector $X = (X_1,...,X_d)^T$ is said to be stable in $\mathbb{R}^d$ if for any positive numbers A and B there are a positive number C and a vector $D \in \mathbb{R}^d$ such that

$$A\mathbf{X}_1 + B\mathbf{X}_2 \stackrel{d}{=} C\mathbf{X} + \mathbf{D}, \qquad (1)$$

where X_1 and X_2 are independent and identical copies of X and $C = (A^\alpha + B^\alpha)^{1/\alpha}$.

Definition 2. Let $0<\alpha<2$. Then X is a non-Gaussian α- stable random vector in $\mathbb{R}^d$ if there exist a finite measure Γ on the unit sphere $\mathbb{S}^d = \{x = (x_1,...,_d)^T \in \mathbb{R}^d \mid \langle x, x \rangle = 1\}$ and a vector $\mu = (\mu_1,...,\mu_d)^T \in \mathbb{R}^d$ such that

$$\varphi_{\mathbf{X}}(\mathbf{t}) = \log E(\exp(i\langle \mathbf{t}, \mathbf{X}\rangle)) = \begin{cases} -\int_{\mathbb{S}^d} |\langle \mathbf{t}, \mathbf{s}\rangle|^\alpha \left[1 - i \operatorname{sgn}\langle \mathbf{t}, \mathbf{s}\rangle \tan\left(\frac{\pi\alpha}{2}\right)\right] \Gamma(d\mathbf{s}) + i\langle \mathbf{t}, \boldsymbol{\mu}\rangle, & \alpha \neq 1, \\ -\int_{\mathbb{S}^d} |\langle \mathbf{t}, \mathbf{s}\rangle| \left[1 + i \operatorname{sgn}\langle \mathbf{t}, \mathbf{s}\rangle \frac{2}{\pi} \log|\langle \mathbf{t}, \mathbf{s}\rangle|\right] \Gamma(d\mathbf{s}) + i\langle \mathbf{t}, \boldsymbol{\mu}\rangle, & \alpha = 1, \end{cases} \qquad (2)$$

where $\langle \mathbf{t}, \mathbf{s}\rangle = \sum_{i=1}^d t_i s_i$ for $\mathbf{t} = (t_1, \ldots, t_d)^T$, $\mathbf{s} = (s_1, \ldots, s_d)^T$, $i^2 = -1$, and $\operatorname{sgn}(\cdot)$ denotes the sign function. The pair (Γ, μ) is unique.

The parameter α, in Definitions 1 and 2, is called tail index. A random vector X is said to be a strictly α-stable randomvector in $\mathbb{R}^d$ if $\mu = 0$ for $\alpha/\neq 1$; see Samorodnitsky and Taqqu [2]. We note that X is strictly α-stable, in the sense of Definition 1, if $D = 0$. Throughout we assume that X is strictly α-stable and $\alpha \neq 1$. The probability density function of a stable distribution has no closed-form expression and moments with orders greater than or equal to α are not finite finite for the members of this class. The two aforementioned difficulties make statistical inference about the parameters of a stable distribution hard. However, a series of contributions has permitted inference about the parameters of univariate and multivariate stable distributions. For example, in the univariate case, maximum likelihood (ML) estimation was studied first by DuMouchel (1971) and then by Nolan [3]. Although the ML approach leads to an efficient estimate for samples of large size, it involves numerical complexities. A program, called STABLE uses a cubic spline interpolation of stable densities for this purpose; see Nolan [4]. STABLE estimates all four parameters of a stable distribution for $\alpha \geq 0.4$. Sample quantile (SQ) technique is another approach proposed by McCulloch [5]. The results are simple and consistent estimators of all four parameters based on five sample quantiles. The empirical characteristic function (CF) is suggested by Kogon and Williams [6]. The CF and SQ methods work well but are not as efficient as the ML method. As the last approach considered here, U-statistics for the tail index and scale parameters of a univariate strictly stable distribution are introduced by Fan [7]. In multivariate case, the focus of interest is the spectral measure estimation. Among them, we refer to Nolan et al. [8], Pivato and Seco [9], Ogata [10], and Mohammadi et al. [11].

The structure of the paper is as follows. In Section 2, new estimators for the tail index and spectral measure of a strictly stable distribution are presented which is an extension of the U-statistic proposed by Fan [7] for the univariate case. A comprehensive simulation study is performed in Section 3 to compare the performance of the introduced estimators and the known estimators. Two real data sets are analyzed in this section to illustrate the performance of the proposed method.

NEW ESTIMATORS

This section consists of two subsections. Firstly, we propose an estimator for the tail index. Secondly, an estimator for the spectral measure is given.

Estimation of Tail Index

The main result of this section is given in Theorem 4, which gives U-statistic for the inverse of tail index of a strictly stable distribution. We present the main result in the light of Lemma 3 given as follows.

Lemma 3. *Let* $X = (X_1,\ldots,_d)^T$ *be a d-dimensional strictly stable random vector. Then,* $Var \log \|X\|$ *is finite, where* $\|\cdot\|$ *denotes the Euclidean norm.*

Theorem 4. *Let* $x^1,\ldots, x^n$ *be a sequence of n observations from a d-dimensional strictly stable random vector. Then*

$$U_n = \binom{n}{2}^{-1} \sum_{1 \leq i < j \leq n} H\left(\mathbf{x}_i, \mathbf{x}_j\right), \quad (3)$$

where

$$H(\mathbf{x}_i, \mathbf{x}_j) = \frac{\log\|\mathbf{x}_i + \mathbf{x}_j\|}{\log 2} - \frac{\log\|\mathbf{x}_i\| + \log\|\mathbf{x}_j\|}{2\log 2} \quad (4)$$

is the U-statistic for $1/\alpha$.

As it is seen, from Theorem 4, the introduced U-statistic is an unbiased estimator for $1/\alpha$. Hereafter, we write $\hat{\alpha}_{MU} = 1/U_n$ as introduced estimator for α. Here, subscript MU indicates that $\hat{\alpha}_{MU}$ is constructed based on multivariate U- statistic defined in Theorem 4. It should be noted that when the true value of α is near two, the kernel given in (4) could be less than 0.5. So, $\hat{\alpha}_{MU}$ is greater than two. In this case, we set $\hat{\alpha}_{MU} = 2$.

Spectral Measure Estimation

We use $\hat{\alpha}_{MU}$ to estimate an$m$-point discrete approximation to the exact spectral measure of the form

$$\Gamma(\cdot) = \sum_{j=1}^{m} \gamma_j I_{\mathbf{s}_j}(\cdot), \quad (5)$$

where γ_j is a mass at point s_j in the unit sphere $\mathbb{S}^d$ and $I_{\mathbf{s}_j}(\cdot)$ is an indicator function at point s_j; for $j = 1, \ldots, m$, see Byczkowski et al. [12]. To estimate $\Gamma(\cdot)$, we replace Definition 2 for a strictly d-dimensional stable random vector with

$$\begin{aligned} \varphi_X(\mathbf{t}) &= -\sum_{j=1}^{m} |\langle \mathbf{t}, \mathbf{s}_j \rangle|^{\alpha} \left[1 - i \operatorname{sgn}\langle \mathbf{t}, \mathbf{s}_j \rangle G(\alpha, \mathbf{t}, \mathbf{s}_j)\right] \gamma_j, \\ &= -\sum_{j=1}^{m} \psi(\mathbf{t}, \mathbf{s}_j, \alpha) \gamma_j, \end{aligned} \quad (6)$$

where$G(\alpha,\mathrm{t},\mathrm{s}_j) = \tan(\pi\alpha/2)$ for $\alpha /= 1$ and $-2/\pi \log|\langle \mathrm{t},\mathrm{s}_j\rangle|$ for $\alpha=1$. Define

$$\Lambda = \begin{pmatrix} \psi(\mathbf{t}_1, \mathbf{s}_1, \alpha) & \psi(\mathbf{t}_1, \mathbf{s}_2, \alpha) & \cdots & \psi(\mathbf{t}_1, \mathbf{s}_m, \alpha) \\ \psi(\mathbf{t}_2, \mathbf{s}_1, \alpha) & \psi(\mathbf{t}_2, \mathbf{s}_2, \alpha) & \cdots & \psi(\mathbf{t}_2, \mathbf{s}_m, \alpha) \\ \vdots & \vdots & \ddots & \vdots \\ \psi(\mathbf{t}_m, \mathbf{s}_1, \alpha) & \psi(\mathbf{t}_m, \mathbf{s}_2, \alpha) & \cdots & \psi(\mathbf{t}_m, \mathbf{s}_m, \alpha) \end{pmatrix}, \quad (7)$$

$$\mathbf{V} = (-\log\varphi_X(\mathbf{t}_1), \ldots, -\log\varphi_X(\mathbf{t}_m))^T, \quad (8)$$

where $\mathbf{t}_j = (t_{j1}, \ldots, t_{jd})^T \in \mathbb{S}^d$, for $j = 1, \ldots, m$. Using (7) and (8), both sides of (6) are connected together through the following linear system:

$$\mathbf{V} = \Lambda\gamma, \quad (9)$$

where$\gamma = (\gamma_1, \ldots, \gamma_m)^T$. Assuming that Λ in (9) is nonsingular, then $\gamma = \Lambda^{-1}\mathrm{V}$. Hence, we estimate the vector of the masses as

$$\widehat{\gamma} = \widehat{\Lambda}^{-1}\widehat{\mathbf{V}}, \quad (10)$$

where

$$\widehat{\Lambda} = \begin{pmatrix} \psi(\mathbf{t}_1, \mathbf{s}_1, \widehat{\alpha}_{\mathrm{MU}}) & \psi(\mathbf{t}_1, \mathbf{s}_2, \widehat{\alpha}_{\mathrm{MU}}) & \cdots & \psi(\mathbf{t}_1, \mathbf{s}_m, \widehat{\alpha}_{\mathrm{MU}}) \\ \psi(\mathbf{t}_2, \mathbf{s}_1, \widehat{\alpha}_{\mathrm{MU}}) & \psi(\mathbf{t}_2, \mathbf{s}_2, \widehat{\alpha}_{\mathrm{MU}}) & \cdots & \psi(\mathbf{t}_2, \mathbf{s}_m, \widehat{\alpha}_{\mathrm{MU}}) \\ \vdots & \vdots & \ddots & \vdots \\ \psi(\mathbf{t}_m, \mathbf{s}_1, \widehat{\alpha}_{\mathrm{MU}}) & \psi(\mathbf{t}_m, \mathbf{s}_2, \widehat{\alpha}_{\mathrm{MU}}) & \cdots & \psi(\mathbf{t}_m, \mathbf{s}_m, \widehat{\alpha}_{\mathrm{MU}}) \end{pmatrix},$$

$$\widehat{\mathbf{V}} = \left(-\log \frac{1}{n}\sum_{i=1}^{n} \exp\{i\langle \mathbf{t}_1, \mathbf{x}_i\rangle\}, \ldots, -\log \frac{1}{n}\sum_{i=1}^{n} \exp\{i\langle \mathbf{t}_m, \mathbf{x}_i\rangle\} \right)^T, \quad (11)$$

in which x_i is i-th vector observation in random sample of size n.

Due to the standard error of $\widehat{\mathrm{V}}$, we have two problems with direct use of (10). Firstly, $\hat{\gamma}$ may be complex, and secondly, its real part may be quite negative. Since $\widehat{\Lambda}$ and $\widehat{\mathrm{V}}$ are complex while gamma is constrained to be real (and nonnegative), the Euclidean norm used by McCulloch [13] and Nolan et al. [8] must be replaced with the complex modulus to solve both problems in a novel way. For this, we use the nnls(·) library in the R package. In the next section, the estimated spectral measure $\hat{\gamma}$, based on $\hat{\alpha}_{\mathrm{MU}}$, is shown by $\hat{\gamma}_{\mathrm{MU}}$. We note that another estimator of γ can be constructed by separating both of the real and imaginary parts in the structure of $\widehat{\mathrm{V}}$. But simulation results show that constructed estimator gives the same performance.

SIMULATION STUDY

This section is in three parts. Firstly, we study the performance of the proposed estimator with the known ones for estimating the tail index. Secondly, we compare the performance of the spectral measure estimator developed through the introduced tail index estimator with the known approaches. In the last subsection, we give a real data example to illustrate the efficiency of the proposed estimators.

Performance Analysis of the Tail Index Estimators

Here, we perform a simulation study to compare the performance of $\widehat{\alpha}_{\mathrm{MU}}$ and four other estimators for α, including (1) $\widehat{\alpha}_{\mathrm{ML}}$, (2) $\widehat{\alpha}_{\mathrm{SQ}}$, (3) $\widehat{\alpha}_{\mathrm{CF}}$, and (4) $\widehat{\alpha}_{\mathrm{MM}}$. The first three competitors are ML, SQ, and CF estimations for the tail index, respectively. Each of three competitors is obtained as $\widehat{\alpha}_{\mathrm{PROJ}} = 1/m \sum_{j=1}^{m} \widehat{\alpha}(\mathbf{u}_j)$ after projecting the d-dimensional stable random vector using $\langle_{\mathrm{u}} j,\mathrm{X}\rangle$. Here, m is the number of masses, $\mathrm{u}_j = (u_{j1},\ldots,_{jd})^T$ is an arbitrary unit vector, and X is the d-dimensional stable random vector. It is worth noting that the first three competitors are computed by the help of STABLE software after projecting. The fourth estimator, that is, $\widehat{\alpha}_{\mathrm{MM}}$, is the second estimator for tail index proposed by Mohammadi et al. [11]. We compare both the bias and root mean-squared error (RMSE) of estimators for 500 replications of samples of size n = 500 and 5000 of a bivariate stable random vector generated by the method given in Modarres and Nolan [14]. We use two settings for discrete spectral measure with m=8 masses, including γ_1 = (0.1, 0.2, 0.3, 0.4, 0.1, 0.2, 0.3, 0.4) and γ_2 = (0, 0.1, 0.7, 0.3, 0.7, 0.3, 0.7, 0.1)^T. In both cases, masses are concentrated on points s_j = $(\cos(2\pi(j-1)/m),\sin(2\pi(j-1)/m))^T$for $j = 1, \ldots, 8$. In the first case that data are coming from a stable distribution with $\gamma1$, we generate tj from a uniform distribution on the unit sphere Sd. For the second case, we set t_j = s_j. Biases and RMSEsfor α = (0.1 : 0.1 : 0.9, 0.95, 1.05, 1.1 : 0.1 : 1.9, 1.95, 2) are shown in Figures 1 and 2. As Figure 1 shows, when γ_1 = (0.1, 0.2, 0.3, 0.4, 0.1, 0.2, 0.3, 0.4), we observe that $\widehat{\alpha}_{\mathrm{MU}}$ is more efficient than $\widehat{\alpha}_{\mathrm{SQ}}$ for n = 5000. Also, it works better than $\widehat{\alpha}_{\mathrm{MM}}$ in terms of RMSE (for $\alpha \leq 1.8$). Based on Figure 2, when γ_2 = (0, 0.1, 0.7, 0.3, 0.7, 0.3, 0.7, 0.1), we observe that $\widehat{\alpha}_{\mathrm{MU}}$ is more efficient than other methods when α< 1.4 and n = 5000 in the sense of RMSE. Also, when α< 1.7 and n = 500, $\widehat{\alpha}_{\mathrm{MU}}$ is more efficient than $\widehat{\alpha}_{\mathrm{SQ}}$, $\widehat{\alpha}_{\mathrm{CF}}$, and $\widehat{\alpha}_{\mathrm{MM}}$ with respect to RMSE.

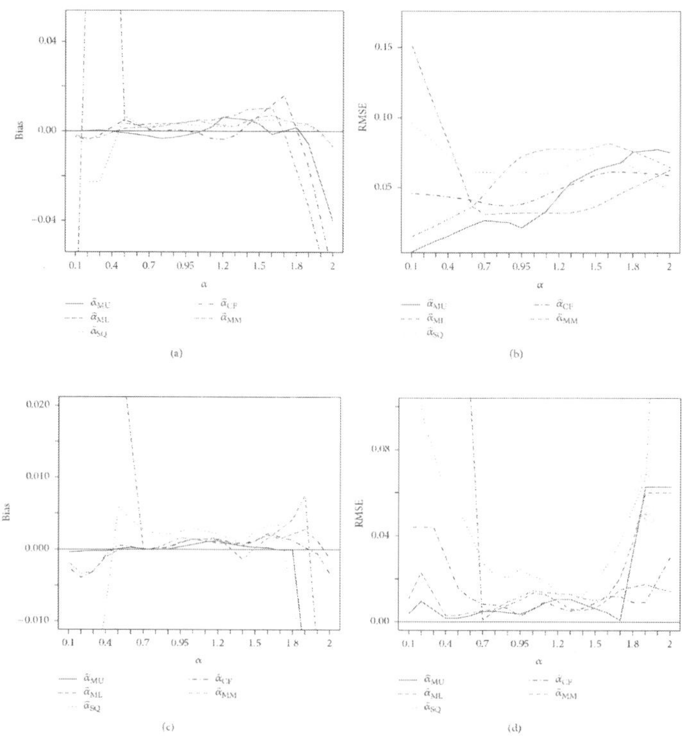

Figure 1: Biases and RMSEs of estimators when data are generated from a strictly stable distribution with discrete spectral measure $\gamma_1 = (0.1, 0.2, 0.3, 0.4, 0.1, 0.2, 0.3, 0.4)$. (a) Bias when $n = 500$, (b) RMSE when $n = 500$, (c) bias when $n = 5000$, and (d) RMSE when $n = 5000$.

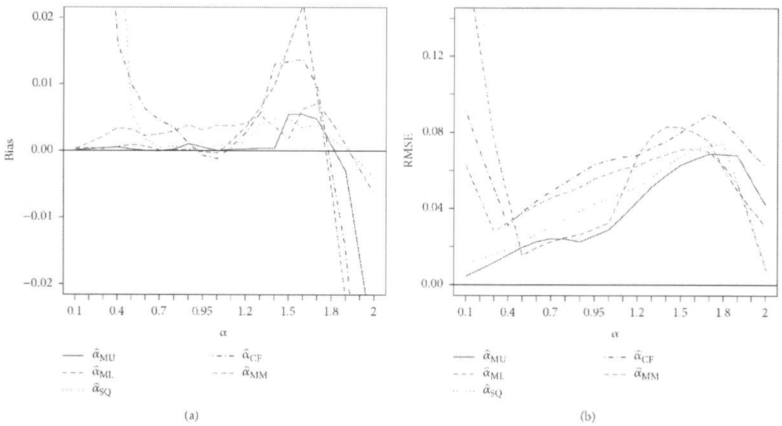

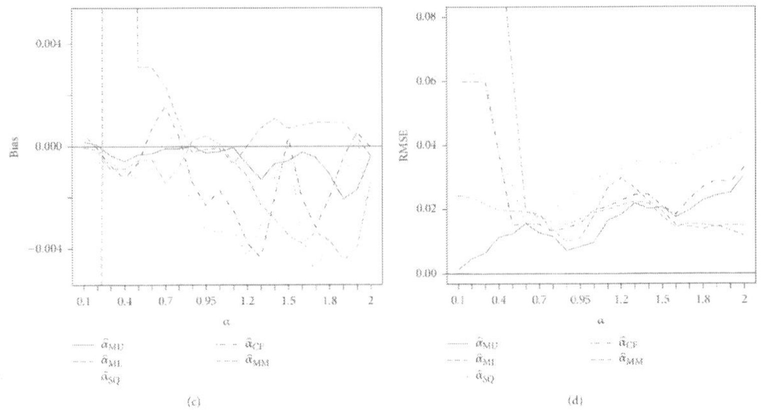

Figure 2: Biases and RMSEs of estimators when data are generated from a strictly stable distribution with discrete spectral measure $\gamma_2 = (0, 0.1, 0.7, 0.3, 0.7, 0.3, 0.7, 0.1)$. (a) Bias when $n = 500$, (b) RMSE when $n = 500$, (c) bias when $n = 5000$, and (d) RMSE when $n = 5000$.

Performance Analysis of the Spectral Measure Estimators

Here, we compare the performance of the estimator for masses of spectral measure $\gamma = (\gamma_1,...,_m)^T$ constructed based on U-statistic, $\widehat{\gamma}_{\text{MU}}$ with the other four known estimators for the spectral measure. The competitors are three types of estimators for γ based on empirical characteristic function method: (1) $\widehat{\gamma}_{\text{MLE-cf}}$; (2) $\widehat{\gamma}_{\text{SQ-cf}}$; (3) $\widehat{\gamma}_{\text{CF-cf}}$; and (4) Mohammadi et al. [11] estimator for γ, $\widehat{\gamma}_{\text{MM}}$. For computing $\widehat{\gamma}_{\text{MLE-cf}}, \widehat{\gamma}_{\text{SQ-cf}}, \text{and } \widehat{\gamma}_{\text{CF-cf}}$, we use command mvstable.fit(x, nspectral, method1d, method2d, param) in the STABLE program, where x is data vector, nspectral is number of spectral measure masses, method1d is the method to use for estimating parameters of univariate stable distribution, that is, MLE, SQ, and CF (corresponding codes in STABLE are 1, 2, and 3, respectively), method2d is the method to use for estimating parameters of bivariate stable distribution (we set method2d = 2 which corresponds to empirical characteristic function approach, cf), and param refers to kind of parameterization. Here, we set param = 1 since we are using the characteristic function in (2). More information about the first three competitors is given in Robust Analysis Inc. [15]. The estimators $\widehat{\gamma}_{\text{MLE-cf}}, \widehat{\gamma}_{\text{SQ-cf}}, \text{and } \widehat{\gamma}_{\text{CF-cf}}$, and $\widehat{\gamma}_{\text{MM}}$ are obtained by substituting $\hat{\alpha}_{\text{ML}}, \hat{\alpha}_{\text{SQ}} \hat{\alpha}_{\text{CF}}, \text{and } \hat{\alpha}_{\text{MM}}$ into (7) and then solving linear system (10), respectively. Comparisons are based on the RMSE of $\hat{\gamma}_j$, for $j = 1, \ldots, m$, which is defined as $\sqrt{1/N \sum_{i=1}^{N} (\hat{\gamma}_{ij} - \gamma_{ij})^2}$, where N is the number of iterations and $\hat{\gamma}ij$ is the estimation of jth component of $\hat{\gamma}$ at ith iteration. We consider five scenarios for the structure of discrete spectral measure as follows.

(1) Independent case: $\gamma = (1/4, 0, 1/4, 0, 1/4, 0, 1/4, 0)$.

(2) Symmetric case: $\gamma = (0.1, 0.2, 0.3, 0.4, 0.1, 0.2, 0.3, 0.4)$.

(3) Uniform case: $\gamma = (1/8, 1/8, 1/8, 1/8, 1/8, 1/8, 1/8, 1/8)$.

(4) Triangle case: $\gamma = (0, 0.1, 0.7, 0.3, 0.7, 0.3, 0.7, 0.1)$.

(5) Exchangeable case: $\gamma = (0.1, 0.2, 0.1, 0.4, 0.3, 0.2, 0.3, 0.4)$.

We note that the first and the third scenarios above are similar to Examples 2 and 1 of Nolan et al. [8], respectively. The fourth scenario is called Triangle since corresponding density contour plot is similar to a triangle. For each of the above five scenarios, we arrange the settings of simulation as m=8, α = 1.25; 1.75; n = 2000; 5000 (n is sample size),and N = 500. It should be noted that masses are located at $s_j = (\cos(2\pi(j-1)/m), \sin(2\pi(j-1)/m))^T$, for $j = 1, \ldots, m$, and components of $t_j = (t_{j1}, \ldots, t_{jd})^T$ are generated from a uniform distribution on the unit sphere $\mathbb{S}^d$. The results of simulations are given in Figures 3–6. As it is seen, $\hat{\gamma}_{MU}$ shows better performance than $\hat{\gamma}_{MM}$.

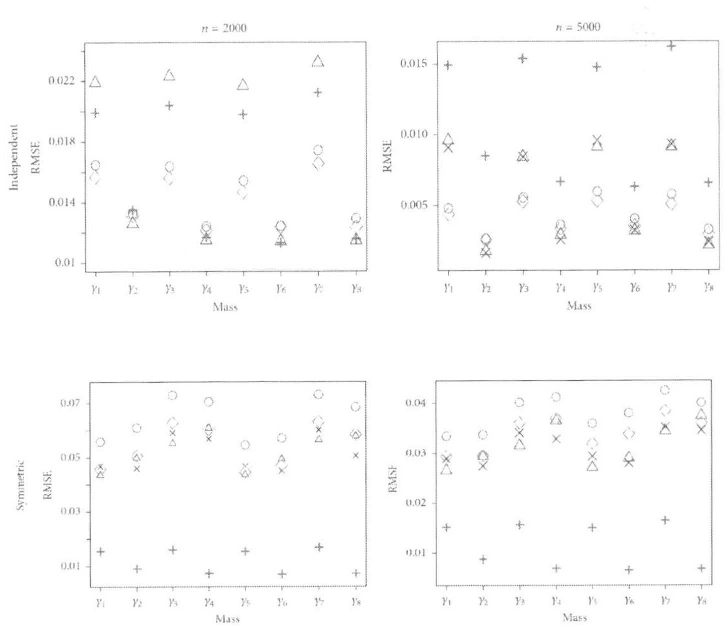

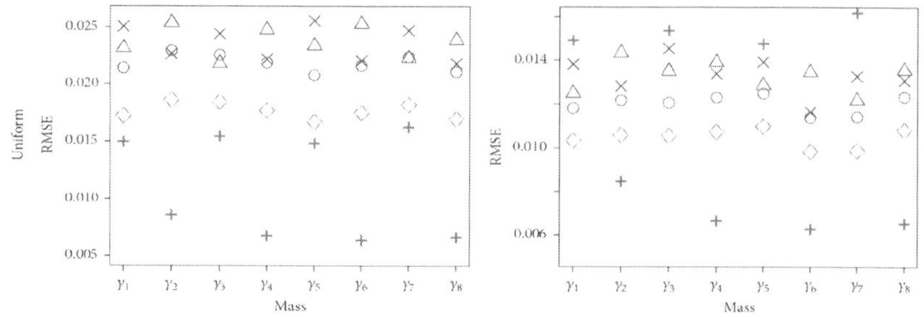

Figure 3: RMSEs of $\hat{\gamma}$ under different scenarios when $\alpha = 1.25$. We use the following symbol scheme: ◇ for $\widehat{\gamma}_{\mathrm{UM}}$, ○ for $\widehat{\gamma}_{\mathrm{MM}}$, + for $\widehat{\gamma}_{\mathrm{ML\text{-}cf}}$, × for $\widehat{\gamma}_{\mathrm{SQ\text{-}cf}}$, and △ for $\widehat{\gamma}_{\mathrm{CF\text{-}cf}}$.

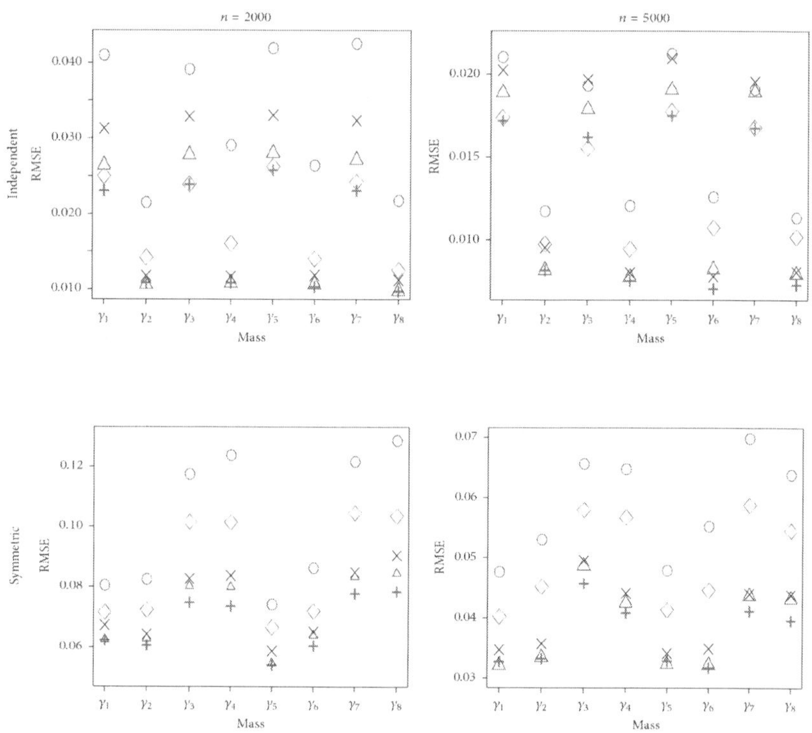

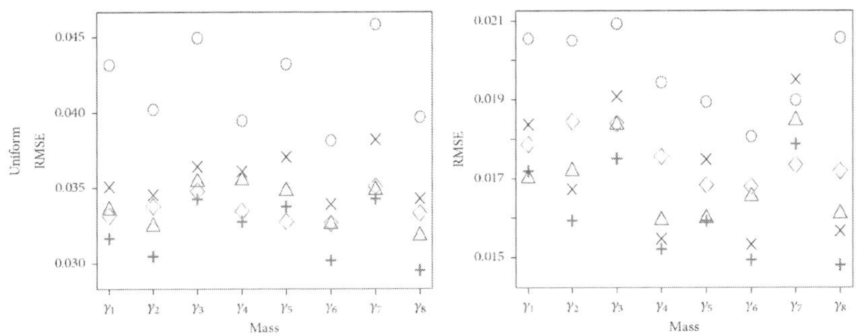

Figure 4: RMSEs of $\hat{\gamma}$ under different scenarios when α = 1.75. We use the following symbol scheme: ◇ for $\widehat{\gamma}_{\text{UM}}$, ○ for $\widehat{\gamma}_{\text{MM}}$, + for $\widehat{\gamma}_{\text{ML-cf}}$, × for $\widehat{\gamma}_{\text{SQ-cf}}$, and △ for $\widehat{\gamma}_{\text{CF-cf}}$.

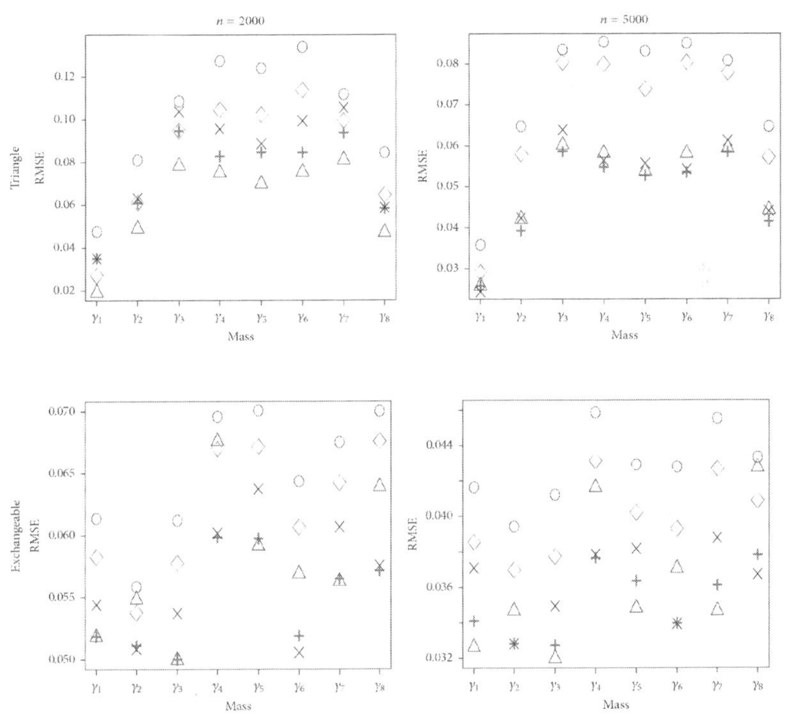

Figure 5: RMSEs of $\hat{\gamma}$ under different scenarios when α = 1.25. We use the following symbol scheme: ◇ for $\widehat{\gamma}_{\text{UM}}$, ○ for $\widehat{\gamma}_{\text{MM}}$, + for $\widehat{\gamma}_{\text{ML-cf}}$, × for $\widehat{\gamma}_{\text{SQ-cf}}$, and △ for $\widehat{\gamma}_{\text{CF-cf}}$.

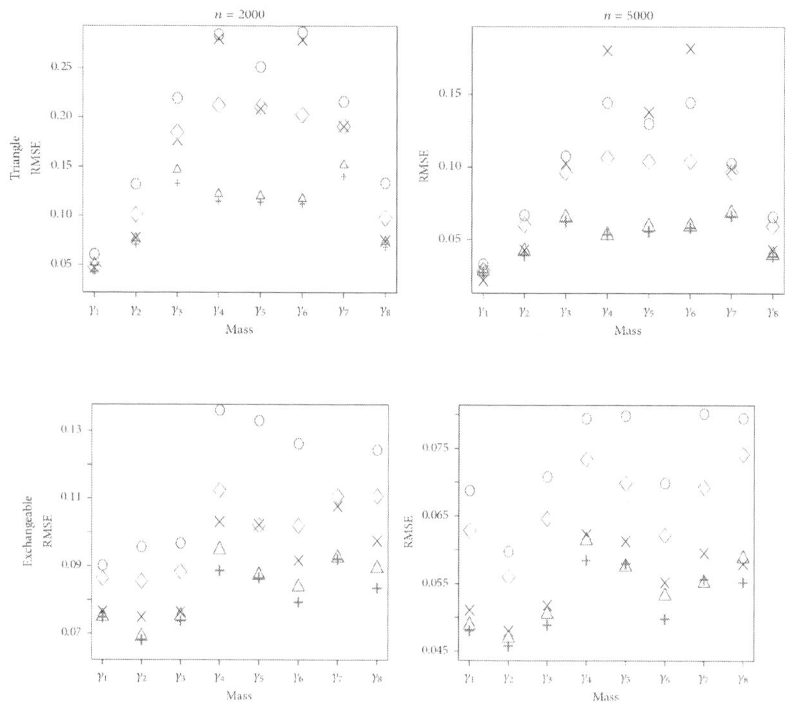

Figure 6: RMSEs of $\hat{\gamma}$ under different scenarios when α = 1.75. We use the following symbol scheme: ◇ for $\hat{\gamma}_{\text{UM}}$, O for $\hat{\gamma}_{\text{MM}}$, + for $\hat{\gamma}_{\text{ML-cf}}$, × for $\hat{\gamma}_{\text{SQ-cf}}$, and △ for $\hat{\gamma}_{\text{CF-cf}}$.

Real Data Analysis

Here, we give two examples. In the first example, adjusted daily log-return (in percent) for the 30 stocks at the Dow Jones index is collected between January 3, 2000, and December 31, 2004. The log-return percent of 1247 closing prices has been computed for AXP (American Express Company) and MRK (Merck & Co. Inc.) stocks after multiplying the daily log-return by 100; see Nolan [16]. The scatter plot of AXP and MRK stocks log-return percent values, X = (AXP, MRK) ^{T}, is shown in Figure 7. We use a bivariate α-stable distribution with m = 12 points of masses for spectral measure addressed by $s_j = (\cos(2\pi(j-1)/m), \sin(2\pi(j-1)/m))^T$, for $j = 1, \ldots, 12$. We estimate thelocation parameter as $\hat{\mu}_{\text{ML-cf}} = (-3.438E-07, -2.402E-07)$. So, a strictly α-stable distribution is fitted to the Y = (X − $\hat{\mu}_{\text{ML-cf}}$) T. For this, we set $t_j = s_j$, for $j = 1, \ldots, 12$. Table 1 shows the results for modelling data through five methods. We note that estimated tail indices are $\widehat{\alpha}_{\text{MU}} = 1.581$, $\widehat{\alpha}_{\text{MM}} = 1.734$, $\widehat{\alpha}_{\text{ML-cf}} = 1.618$, $\widehat{\alpha}_{\text{SQ-cf}} = 1.493$, and $\widehat{\alpha}_{\text{CF-cf}} = 1.723$. As it is seen, estimated tail indices through estimators $\widehat{\alpha}_{\text{MU}}$ and $\widehat{\alpha}_{\text{ML-cf}}$ are closer together than the other estimators. In the second example, we focus on the cubic-root of the monthly average of river discharge. We choose discharge of the Odra and Wisla rivers in Poland during 1901 to 1986 (raw data are in m^3 /s. They are available at https://nelson.wisc.edu/sage/data-and-models/riverdata/). The scatter plot for cubic-root of Odra river discharge versus cubic-root of Wisla river discharge is shown in Figure 8. Setting $m=8, s_j = (\cos(2\pi(j-1)/m), \sin(2\pi(j-1)/m))^T$, and $t_j = s_j$, for $j = 1, \ldots, 8$, we obtain $\widehat{\mu}_{\text{ML-cf}} = (7.9837, 9.9947)^T$. After fitting a strictly α-stable distribution to the shifted data, results for estimating spectral measure are given in Table 2. Estimated tail indices are $\widehat{\alpha}_{\text{MU}} = 1.860$, $\widehat{\alpha}_{\text{MM}} = 1.312$, $\widehat{\alpha}_{\text{ML-cf}} = 1.813$, $\widehat{\alpha}_{\text{SQ-cf}} = 1.936$, and $\widehat{\alpha}_{\text{CF-cf}} = 1.962$. Based on results given in Table 2, estimated masses through estimators $\widehat{\gamma}_{\text{MU}}$, $\widehat{\gamma}_{\text{ML-cf}}$, and $\widehat{\gamma}_{\text{SQ-cf}}$ are closer together than the other estimators. We compare here $\widehat{\gamma}_{\text{MU}}$ with $\widehat{\gamma}_{\text{ML-cf}}$ and $\widehat{\gamma}_{\text{SQ-cf}}$ since the latter estimators are among the best estimators for the masses as shown in the previous subsection.

Table 1: Estimation results after fitting a strictly bivariate α-stable distribution to AXP and MRK stocks data

Estimator	$\widehat{\gamma}_1$	$\widehat{\gamma}_2$	$\widehat{\gamma}_3$	$\widehat{\gamma}_4$	$\widehat{\gamma}_5$	$\widehat{\gamma}_6$	$\widehat{\gamma}_7$	$\widehat{\gamma}_8$	$\widehat{\gamma}_9$	$\widehat{\gamma}_{10}$	$\widehat{\gamma}_{11}$	$\widehat{\gamma}_{12}$
$\widehat{\gamma}_{\text{MU}}$	0.396	0.070	0.077	0.433	0.087	0	0.231	0.378	0	0.439	0.078	0
$\widehat{\gamma}_{\text{MM}}$	0.373	0	0	0.463	0.021	0	0.333	0.479	0	0.563	0.100	0
$\widehat{\gamma}_{\text{ML-cf}}$	0.338	0.156	0.162	0.412	0	0	0.490	0.206	0.055	0.425	0.122	0
$\widehat{\gamma}_{\text{SQ-cf}}$	0.232	0.216	0.089	0.404	0.012	0	0.480	0.115	0.177	0.254	0.156	0
$\widehat{\gamma}_{\text{CF-cf}}$	0.421	0.129	0.237	0.434	0	0	0.537	0.209	0	0.527	0.089	0

Table 2: Estimation results after fitting a strictly bivariate α-stable distribution to Odra and Wisla discharge data

Estimator	$\widehat{\gamma}_1$	$\widehat{\gamma}_2$	$\widehat{\gamma}_3$	$\widehat{\gamma}_4$	$\widehat{\gamma}_5$	$\widehat{\gamma}_6$	$\widehat{\gamma}_7$	$\widehat{\gamma}_8$
$\widehat{\gamma}_{\text{MU}}$	0.006	1.459	0.498	0	0	0	0	0
$\widehat{\gamma}_{\text{MM}}$	0	0.800	0.257	0	0.026	0.483	0.165	0
$\widehat{\gamma}_{\text{ML-cf}}$	0	1.380	0.508	0	0	0	0	0
$\widehat{\gamma}_{\text{SQ-cf}}$	0	1.490	0.532	0	0	0	0	0
$\widehat{\gamma}_{\text{CF-cf}}$	0	0.196	0.199	0	0.009	1.283	0.339	0

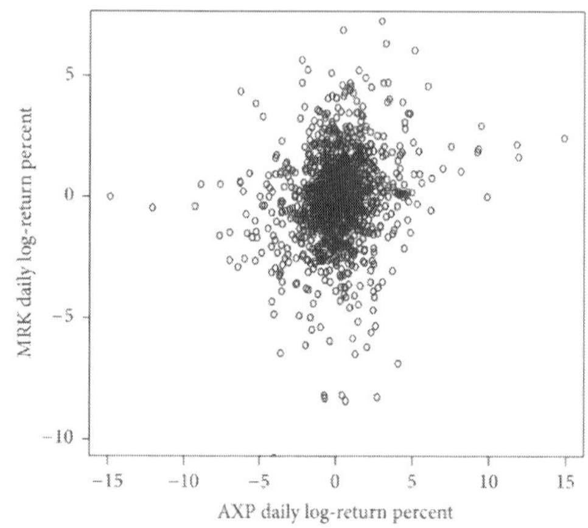

Figure 7: Scatter plot for AXP versus MRK daily log-return percent, X = (AXP, MRK) ^{T}.

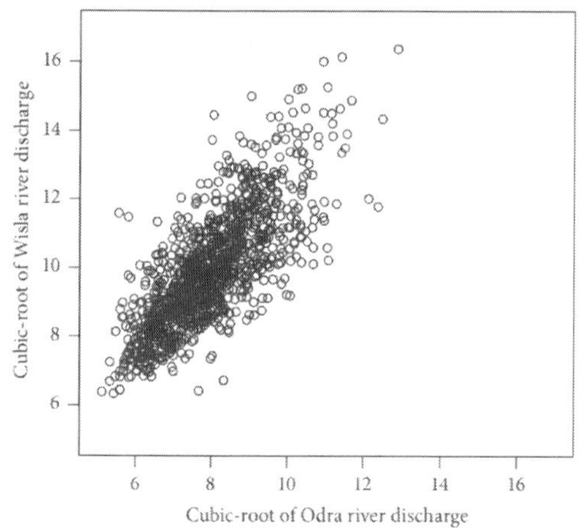

Figure 8: Scatter plot for cubic-root of Odra and Wisla rivers discharge.

CONCLUSION

We compare the performance of the introduced U-statistic for the tail index with the well-known methods, including maximum likelihood, empirical characteristic function, sample quantile, and that introduced in Mohammadi et al. [11] through a simulation study. In the sense of root meansquared error, it is proved that proposed tail index estimator always outperforms Mohammadi et al. [11] and SQ methods when $\alpha \leq 1.4$. This is while ML and

CF methods show better performance than the proposed estimator for large α, say $\alpha > 1.4$ in terms of root mean-squared error. Simulation studies for estimating the discrete spectral measure γ under five scenarios prove that estimator of γ based on introduced U-statistic shows, in terms of root mean-squared error, better performance than Mohammadi et al. [11] estimator. Analysis of two sets of real data reveals that estimator of the tail index and γ based on U-statistic shows expedient performance. As some possible future works, firstly, we aim to introduce a U-statistic for the case of a nonzero location parameter. Secondly, we look for methodology possibly based on a U- statistic, to estimate tail, masses, and location parameters simultaneously. Finally, recalling that the approach employed in this work is based on characteristic function, the discrete spectral measure using $\widehat{\alpha}_{\mathrm{MU}}$ can be estimated through projection approach.

REFERENCES

1. P. Lévy, "Théorie des erreurs. La loi de Gauss et les loisexceptionnelles," *Bulletin De La SociétéMathématique De France*, vol. 52, pp. 49–85, 1924.
2. G. Samorodnitsky and M. S. Taqqu, *Stable Non-Gaussian Random Processes: Stochastic Models and Infinite Variance*, Stochastic Modeling, Chapman & Hall, New York, NY, USA, 1994.
3. J. P. Nolan, "Maximum likelihood estimation of stable parameters," in *Lévy Processes: Theory and Applications*, O. E. Barndorff-Nielsen, T. Mikosch, and I. Resnick, Eds., pp. 379–400, Birkhöuser, Boston, Mass, USA, 2001.
4. J. P. Nolan, "Numerical calculation of stable densities and distribution functions," *Communications in Statistics. Stochastic Models*, vol. 13, no. 4, pp. 759–774, 1997.
5. J. H. McCulloch, "Simple consistent estimators of stable distribution parameters," *Communications in Statistics—Simulation and Computation*, vol. 15, no. 4, pp. 1109–1136, 1986.
6. S. M. Kogon and D. B. Williams, "Characteristic function based estimation of stable parameters," in *A Practical Guide to Heavy Tailed Data*, R. Adler, R. Feldman, and M. Taqqu, Eds., pp. 311–338, Birkhäuser, Boston, Mass, USA, 1998.
7. Z. Fan, "Parameter estimation of stable distributions," *Communications in Statistics. Theory and Methods*, vol. 35, no. 1-3, pp. 245–255, 2006.
8. J. P. Nolan, A. K. Panorska, and J. H. McCulloch, "Estimation of stable spectral measures," *Mathematical and Computer Modelling*, vol. 34, no. 9–11, pp. 1113–1122, 2001.
9. M. Pivato and L. Seco, "Estimating the spectral measure of a multivariate stable distribution via spherical harmonic analysis," *Journal of Multivariate Analysis*, vol. 87, no. 2, pp. 219–240, 2003.
10. H. Ogata, "Estimation for multivariate stable distributions with generalized empirical likelihood," *Journal of Econometrics*, vol. 172, no. 2, pp. 248–254, 2013.
11. M. Mohammadi, A. Mohammadpour, and H. Ogata, "On estimating the tail index and the spectral measure of multivariate $\alpha\alpha$-stable distributions," *Metrika*, vol. 78, no. 5, pp. 549–561, 2015.

12. T. Byczkowski, J. P. Nolan, and B. Rajput, "Approximation of multidimensional stable densities," *Journal of Multivariate Analysis*, vol. 46, no. 1, pp. 13–31, 1993.
13. J. H. McCulloch, "Estimation of the bivariate stable spectral representation by the projection method," *Computational Economics*, vol. 16, no. 1-2, pp. 47–62, 2000.
14. R. Modarres and J. P. Nolan, "A method for simulating stable random vectors," *Computational Statistics*, vol. 9, no. 1, pp. 11–19, 1994.
15. Robust Analysis Inc, *User Manual for STABLE 5.0*, Software and User Manual, 2010, http://www.robustanalysis.com.
16. J. P. Nolan, "Multivariate elliptically contoured stable distributions: theory and estimation," *Computational Statistics*, vol. 28, no. 5, pp. 2067–2089, 2013.
17. E. E. Kuruoglu, "Density parameter estimation of skewed /spl alpha/-stable distributions," *IEEE Transactions on Signal Processing*, vol. 49, no. 10, pp. 2192–2201, 2001.

3

A Kind of System of Multivariate Variational Inequalities and the Existence Theorem of Solutions

Yanxia Tang[1], JinyuGuan[1] ,Yongchun Xu[1] and Yongfu Su[2]

[1]Department of Mathematics, Hebei North University, Zhangjiakou, 075000, China
[2]Department of Mathematics, Tianjin Polytechnic University, Tianjin, 300387, China

ABSTRACT

Let K be a nonempty closed convex and bounded subset of a reflexive Banach space X. Let $A_1, A_2, \ldots, A_N$ be N-variables monotone demi-continuous mappings from K^N into X. Then: (1) the system of multivariate variational inequalities

$$\begin{cases} \langle A_1(x_1,x_2,\ldots,x_N), y_1 - x_1\rangle \geq 0, & \forall y_1 \in K, \\ \langle A_2(x_1,x_2,\ldots,x_N), y_2 - x_2\rangle \geq 0, & \forall y_2 \in K, \\ \ldots \\ \langle A_N(x_1,x_2,\ldots,x_N), y_N - x_N\rangle \geq 0, & \forall y_N \in K, \end{cases}$$

has a solution $(x_1^*, x_2^*, \ldots, x_N^*) \in K^N$; (2) the set of solutions of this system of multivariate variational inequalities is closed convex in K^N; (3) if $A_1, A_2, \ldots, A_N$ are also strictly monotone, this system of multivariate variational inequalities has a unique solution.

Citation: Tang, Y., Guan, J., Xu, Y., & Su, Y. (2017). A kind of system of multivariate variational inequalities and the existence theorem of solutions. Journal of inequalities and applications, 2017(1), 1-10. (10 pages), DOI: https://doi.org/10.1186/s13660-017-1486-9.

Copyright: © Open Access. This article is distributed under the terms of the Creative Commons Attribution 4.0 International (CC BY 4.0) License.

Keywords: variational inequality, system of variational inequalities, reflexive Banach space, dual space, solution

INTRODUCTION

Let X be a Banach space with the dual space X^* and let $\langle \cdot , \cdot \rangle$ denote the duality pairing of X and X^*. Let K be a nonempty closed convex subset of X, $A{:}K \to X^*$ a mapping. The classical variational inequality problem is to find $x \in K$ such that

$$\langle Ax, y - x \rangle \geq 0, \quad \forall y \in K. \tag{1.1}$$

The variational inequality problem has been recognized as one of the suitable mathematical models for dealing with many problems arising in different fields, such as optimization theory, game theory, economic equilibrium, mechanics. In the last four decades, since the time of the celebrated Hartman-Stampacchia theorem (see [1, 2]), the existence of a solution of a variational inequality and other related problems has become a basic research topic, which continues to attract attention of researchers in applied mathematics (see *e.g.*, [3–14] and the references therein).

In 1966, Hartman and Stampacchia [2] proved the following result.

Theorem 1.1

[2]

Let K be a nonempty closed convex and bounded subset of R^n. Let $A{:}K \to R^n$ be a continuous mapping. Then the variational inequality (1.1) *has a solution $x^* \in K$.*

In 1967, Browder proved the following more general result (see [15]).

Theorem 1.2

[15]

Let K be a nonempty compact convex subset of a locally convex topological vector space X. Let $A{:}K \to X^*$ be a continuous mapping. Then the variational inequality (1.1) *has a solution $x^* \in K$.*

The variational inequality (1.1) is called the Hartman-Stampacchia variational inequality. It is an important classical variational inequality which is also a classical and powerful tool in nonlinear analysis and other mathematical fields.

Definition 1.3

[15]

Let X be a normed space, $A:X \to X^*$ a mapping, $x_0 \in X$. A is said to be demi-continuous at x_0, if for any given $y \in X$, $A(x_0 + t_n y)$ weak* converges to $A(x_0)$ wherever $t_n \to 0$, $t_n \geq 0$.

In 1991, Chang [15] proved the following result in reflexive Banach spaces.

Theorem 1.4

[15]

Let K be a nonempty closed convex and bounded subset of a reflexive Banach space X. Let $A:K \to X^*$ be a monotone demi-continuous mapping. Then

- *the variational inequality* (1.1) *has a solution* $x^* \in K$;
- *the set of solutions of* (1.1) *is closed convex*;
- *if A is strictly monotone, then* (1.1) *has a unique solution.*

In 2010, Plubtieng and Sombut [16] proved the following result.

Theorem 1.5

Let X be a reflexive Banach space, let K be a compact convex subset of X, and let $A, B:K \to X^*$ be two continuous mappings. Then the system of variational inequalities

$$\begin{cases} \langle A(x), z-y \rangle \geq 0, & \forall z \in K, \\ \langle B(y), z-x \rangle \geq 0, & \forall z \in K, \end{cases} \tag{1.2}$$

has a solution $(x, y) \in K \times K$ *and the set of solutions of* (1.2) *is closed.*

Multivariate calculus is a more general mathematical branch which paly a more important role in mathematical and applied fields. In recently, multivariate fixed point theorems and the system of N-variables nonlinear operators have been studied by some authors. Many interesting results and the applications have also been given. In 2016, Su *et al.* [17] presented the concept of multivariate fixed point and proved a multivariate fixed point theorem for N-variables contraction mappings which further generalizes Banach contraction mapping principle. In 2016, Luo *et al.* [18] presented the concept of multivariate best proximity point and proved the multivariate best proximity point theorems in metric spaces for N-variables contraction mappings. In 2017, Xu *et al.* [19] presented the concept of multivariate contraction mapping in a locally convex topological vector spaces and proved the multivariate contraction mapping principle in such spaces. In 2017, Guan *et al.* [20] studied a kind of system of N-variables pseudo contractive operator equations and proved the existence theorem of solutions.

The purpose of this paper is to study a kind of system of multivariate variational inequalities and to prove the existence theorem of solutions. The results of this paper improve and extend the results of [15, 16] in reflexive Banach spaces. In order to get the expected results, an ingenious mathematical method is used in this paper.

PRELIMINARIES

Let us introduce some conclusions which will be useful for our main results.

Lemma 2.1

[20]

Let X be a Banach space with the norm $\|\cdot\|$. We consider on the Cartesian product space $X^N = X \times X \times \cdots \times X$ the following functional:

$$\|x\|_* = \sqrt{\sum_{i=1}^{N} \|x_i\|^2}, \quad \forall x = (x_1, x_2, \ldots, x_N) \in X^N.$$

Then $(X^N, \|\cdot\|_*)$ is a Banach space.

Lemma 2.2

[20]

$(X^N, \|\cdot\|_*)^* = ((X, \|\cdot\|)^*)^N$.

Lemma 2.3

[20]

Let X be a reflexive Banach space with the norm $\|\cdot\|$, $X^N = X \times X \times \cdots \times X$ be Cartesian product space of X. Let

$$\|x\|_* = \sqrt{\sum_{i=1}^{N} \|x_i\|^2}, \quad \forall x = (x_1, x_2, \ldots, x_N) \in X^N.$$

Then $(X^N, \|\cdot\|_*)$ is a reflexive Banach space.

MAIN RESULTS

Let K be a nonempty subset of a normed space X, $A_i{:}K^N \to X^*$ a N-variables mapping for all $i = 1, 2, \ldots, N$. We consider the following system of multivariate variational inequalities:

$$\begin{cases} \langle A_1(x_1, x_2, \ldots, x_N), y_1 - x_1 \rangle \geq 0, & \forall y_1 \in K, \\ \langle A_2(x_1, x_2, \ldots, x_N), y_2 - x_2 \rangle \geq 0, & \forall y_2 \in K, \\ \ldots \\ \langle A_N(x_1, x_2, \ldots, x_N), y_N - x_N \rangle \geq 0, & \forall y_N \in K. \end{cases} \quad (3.1)$$

Definition 3.1

Let K be a nonempty subset of a normed space X. A N-variables mapping $A{:}K^N \to X^*$ is said to be monotone, if

$$\langle A(x_1, x_2, \ldots, x_N) - A(y_1, y_2, \ldots, y_N), x_i - y_i \rangle \geq 0, \ \forall i = 1, 2, \ldots, N,$$

for all $(x_1, x_2, \ldots, x_N), (y_1, y_2, \ldots, y_N) \in K^N$. A N-variables monotone mapping is said to be strictly monotone, if

$$\langle A(x_1, x_2, \ldots, x_N) - A(y_1, y_2, \ldots, y_N), x_i - y_i \rangle = 0, \ \forall i = 1, 2, \ldots, N,$$

implies $(x_1, x_2, \ldots, x_N) = (y_1, y_2, \ldots, y_N)$.

The following is the main result of this paper.

Theorem 3.2

Let K be a nonempty closed convex and bounded subset of a reflexive Banach space X. Let $A_i{:}K^N \to X^*$ be a N-variables monotone demi-continuous mapping for all $i = 1, 2, \ldots, N$. Then:

- *the system of variational inequalities* (3.1) *has a solution* (x∗1,x∗2,…,x∗N)∈KN;
- *the set of solutions of* (3.1) *is closed convex in* K^N;
- *if A_i is strictly monotone for all $i = 1, 2, \ldots, N$, then* (3.1) *has a unique solution.*

Proof

Let $A^*{:}K^N \to (X^*)^N$ be a mapping defined by

$A^*(x_1, x_2, \ldots, x_N) = (A_1(x), A_2(x), \ldots, A_N(x)),$

for any $x = (x_1, x_2, \ldots, x_N) \in K^N$, where

$$A_1(x) = A_1(x_1, x_2, \ldots, x_N),$$

$$A_2(x) = A_2(x_1, x_2, \ldots, x_N),$$

$$\ldots$$

$$A_N(x) = A_N(x_1, x_2, \ldots, x_N).$$

From Lemma 2.1 and Lemma 2.2, we know $(X^*)^N = (X^N)^*$ and hence A^* is a mapping from K^N into $(X^N, \|\cdot\|_*)^*$.

Next, we prove that A^* is a monotone mapping from K^N into $(X^N, \|\cdot\|_*)^*$. Since A_i is a monotone mapping from K into $(X, \|\cdot\|)^*$ for all $i = 1, 2, \ldots, N$,

$$\langle A^*x - A^*y, x - y\rangle$$
$$= \langle (A_1(x) - A_1(y), A_2(x) - A_2(y), \ldots, A_N(x) - A_N(y)), x - y\rangle$$
$$= \sum_{i=1}^{N} \langle A_i(x) - A_i(y), x_i - y_i\rangle \geq 0,$$

for any $x = (x_1, x_2, \ldots, x_N)$, $y = (y_1, y_2, \ldots, y_N) \in K^N$.

We also need to prove A^* is demi-continuous on K^N. For any given $x_0 \in K^N$ and any given $y = (y_1, y_2, \ldots, y_N) \in K^N$ such that $x_0 + t_n y \in K^N$, we have

$$A^*(x_0 + t_n y)(x) = \langle (A_1(x_0 + t_n y), A_2(x_0 + t_n y), \ldots, A_N(x_0 + t_n y)), x\rangle$$
$$= \sum_{i=1}^{N} \langle A_i(x_0 + t_n y), x_i\rangle$$
$$\to \sum_{i=1}^{N} \langle A_i(x_0), x_i\rangle \quad (t_n \to 0)$$
$$= A^*(x_0)(x), \quad \forall x = (x_1, x_2, \ldots, x_N) \in K^N.$$

Then A^* is demi-continuous on K^N.

It is easy to see that K^N is a nonempty closed convex and bounded subset of Banach space $(X^N, \|\cdot\|_*)$. By using Theorem 1.4, we know that the following variational inequality:

$$\langle A^*x, y - x\rangle \geq 0, \quad \forall y = (y_1, y_2, \ldots, y_N) \in K^N, \tag{3.2}$$

has a solution $x^* = (x_1^*, x_2^*, \ldots, x_N^*) \in K^N$. That is,

$$\langle A^* x^*, y - x^* \rangle \geq 0, \quad \forall y = (y_1, y_2, \ldots, y_N) \in K^N. \tag{3.3}$$

We rewrite (3.3) as follows:

$$\sum_{i=1}^{N} \langle A_i(x_1^*, x_2^*, \ldots, x_N^*), y_i - x_i^* \rangle \geq 0, \quad \forall y = (y_1, y_2, \ldots, y_N) \in K^N. \tag{3.4}$$

For any $y \in K$, let $y_i = (y, x_2^*, \ldots, x_N^*) \in K^N$ in (3.4), we get

$$\langle A_1(x_1^*, x_2^*, \ldots, x_N^*), y - x_1^* \rangle \geq 0, \quad \forall y \in K. \tag{3.5}$$

For any $y \in K$, let $y_i = (x_1^*, \ldots, x_{j-1}^*, y, x_{j+1}^*, \ldots, x_N^*) \in K^N$ in (3.4), we get

$$\langle A_j(x_1^*, x_2^*, \ldots, x_N^*), y - x_j^* \rangle \geq 0, \quad \forall y \in K, \tag{3.6}$$

for all $j = 2, 3, \ldots, N-1$. For any $y \in K$, let $y_i = (x_1^*, x_2^*, \ldots, x_{N-1}^*, y) \in K^N$ in (3.4), we get

$$\langle A_N(x_1^*, x_2^*, \ldots, x_N^*), y - x_N^* \rangle \geq 0, \quad \forall y \in K. \tag{3.7}$$

From (3.5)-(3.7), we know that $x^* = (x_1^*, x_2^*, \ldots, x_N^*)$ is a solution of (3.1). This completes the proof of conclusion (1).

On the other hand, let $x = (x_1, x_2, \ldots, x_N)$ be an arbitrary solution of (3.1). We have

$$\begin{cases} \langle A_1(x_1, x_2, \ldots, x_N), y_1 - x_1 \rangle \geq 0, & \forall y_1 \in K, \\ \langle A_2(x_1, x_2, \ldots, x_N), y_2 - x_2 \rangle \geq 0, & \forall y_2 \in K, \\ \cdots & \\ \langle A_N(x_1, x_2, \ldots, x_N), y_N - x_N \rangle \geq 0, & \forall y_N \in K, \end{cases}$$

which implies

$$\begin{aligned} \langle A^* x, y - x \rangle &= \langle (A_1 x, A_2 x, \ldots, A_N x), y - x \rangle \\ &= \sum_{i=1}^{N} \langle A_i(x_1, x_2, \ldots, x_N), y_i - x_i \rangle \geq 0, \quad \forall y \in K^N. \end{aligned}$$

Then $x = (x_1, x_2, \ldots, x_N)$ is a solution of the variational inequality (3.2) in reflexive Banach space $(X^N, \|\cdot\|_*)$. From the above, we claim that the system of multivariate variational inequalities (3.1) is equivalent to the variational inequality (3.2). By using Theorem 1.4, we know that the set of solutions of the variational inequality (3.2) is closed convex. This completes the proof of conclusion (2).

Finally, if A is strictly monotone, then

$$\langle A^* x - A^* y, x - y\rangle = 0,$$

implies $x = y$. Hence A^* is also strictly monotone. By using Theorem 1.4, the variational inequality (3.2) has a unique solution and hence the multivariate variational inequalities (3.1) has a unique solution. This completes the proof.

Corollary 3.3

Let K be a nonempty closed convex and bounded subset of a reflexive Banach space X. Let $A{:}K \to X^*$ be a N-variables monotone semi-continuous mapping. Then:

- the multivariate variational inequalities

$$\langle A(x_1, x_2, \ldots, x_N), y - x_i\rangle \geq 0, \quad \forall y \in K, \forall i = 1, 2, \ldots, N \tag{3.8}$$

has a solution $(x_1^*, x_2^*, \ldots, x_N^*) \in K^N$;

- *the set of solutions of* (3.8) *is closed convex in* K^N;
- *if A is strictly monotone, then* (3.8) *has a unique solution.*

Proof

Let $A_i = A$ for all $i = 1, 2, \ldots, N$ in Theorem 3.2, we can get the conclusion.

Corollary 3.4

Let K be a nonempty closed convex and bounded subset of a reflexive Banach space X. Let $A{:}K \to X^*$ be a N-variables monotone semi-continuous mapping. Then

- the multivariate variational inequalities

$$\left\langle A(x_1, x_2, \ldots, x_N), y - \frac{1}{N}\sum_{i=1}^{N} x_i \right\rangle \geq 0, \quad \forall y \in K, \tag{3.9}$$

has a solution $(x_1^*, x_2^*, \ldots, x_N^*) \in K^N$;

- *the set of solutions of* (3.9) *is closed convex in* K^N;

- *if A is strictly monotone, then* (3.9) *has a unique solution.*

Proof

From (3.8), we have

$$\frac{1}{N}\sum_{i=1}^{N}\langle A(x_1,x_2,\ldots,x_N),y-x_i\rangle \geq 0, \quad \forall y \in K.$$

That is,

$$\left\langle A(x_1,x_2,\ldots,x_N),y-\frac{1}{N}\sum_{i=1}^{N}x_i\right\rangle \geq 0, \quad \forall y \in K.$$

This completes the proof.

Next, we prove an existence theorem of solutions for the system of variational inequalities (1.2) in normed spaces.

Theorem 3.5

Let X be a normed space, let K be a compact convex subset of X, and let A, $B{:}K \to X^*$ be two continuous mappings. Then the system of variational inequalities (1.2) *has a solution* $(x, y) \in K \times K$ *and the set of solutions of* (1.2) *is closed.*

Proof

Let $A(x, y) = A(y)$, $B(x, y) = B(x)$ for all $(x, y) \in K \times K$, then the system of variational inequalities (1.2) is equivalent to

$$\begin{cases} \langle A(y,x),z-y\rangle \geq 0, & \forall z \in K, \\ \langle B(y,x),z-x\rangle \geq 0, & \forall z \in K. \end{cases} \tag{3.10}$$

Let $C^*{:}K \times K \to X^* \times X^* = (X{\times}X)^*$ be defined by

$C^*(x, y) = (A(x, y), B(x, y))$

for all $(x, y) \in K \times K$. It is easy to see that C^* is a continuous mapping from the nonempty compact convex subset $K \times K$ into the dual space $(X{\times}X)^*$ of normed space $X \times X$. By using Theorem 1.2, there exists an element (x^*, y^*) $\in K \times K$ such that

$\langle C^*(x^*,y^*),(z_1,z_2)-(x^*,y^*)\rangle_* \geq 0, \ \forall(z_1, z_2) \in K \times K,$

where $\langle\cdot,\cdot\rangle_*$ denotes the duality pairing of $X \times X$ and $X^* \times X^* = (X \times X)^*$. This implies

$\langle (A(x^*,y^*),B(x^*,y^*)),(z_1-x^*,z_2-y^*)\rangle_* \geq 0,\ \forall (z_1, z_2) \in K \times K.$

Hence

$\langle (A(y^*),B(x^*)),(z_1-x^*,z_2-y^*)\rangle_* \geq 0,\ \forall (z_1, z_2) \in K \times K.$

From the definition of $\langle\cdot,\cdot\rangle_*$, we have

$$\langle A(y^*), z_1 - x^*\rangle + \langle B(x^*), z_2 - y^*\rangle \geq 0, \quad \forall (z_1, z_2) \in K \times K. \tag{3.11}$$

Let $z_2 = y^*$ and $z_1 = x^*$ in (3.11), respectively, we get

$$\begin{cases} \langle A(y^*), z_1 - x^*\rangle \geq 0, & \forall z_1 \in K, \\ \langle B(x^*), z_2 - y^*\rangle \geq 0, & \forall z_2 \in K. \end{cases}$$

Then $(x^*, y^*) \in K \times K$ is a solution of the system of variational inequalities (1.2). Since A, B are continuous, so the set of solutions of (1.2) is closed. This completes the proof.

It is obvious that Theorem 1.5 is a special form of Theorem 3.5 in reflexive Banach spaces.

Corollary 3.6

Theorem 1.5

Let X be a reflexive Banach space, let K be a compact convex subset of X, and let A, B:$K \to X^*$ be two continuous mappings. Then the system of variational inequalities

$$\begin{cases} \langle A(x), z - y\rangle \geq 0, & \forall z \in K, \\ \langle B(y), z - x\rangle \geq 0, & \forall z \in K, \end{cases}$$

has a solution $(x, y) \in K \times K$ and the set of solutions of (1.2) is closed.

We give an example to show the mathematical and physical significance of the main results of this paper.

Example 3.7

Let $R = (-\infty, +\infty)$, $K = [a, b]$. Let $f(x_1, x_2, \ldots, x_N)$ be a continuous real N-variables function with $f \in C^{(1)}(K^N, R)$. Then there exists an element $x_0 =$

$(x_{0,1}, x_{0,2}, \ldots, x_{0,N}) \in K^N$ such that

$$f(x_{0,1}, x_{0,2}, \ldots, x_{0,N}) = \min_{(x_1, x_2, \ldots, x_N) \in K^N} f(x_1, x_2, \ldots, x_N).$$

This element x_0 must be a solution of the following system of multivariate variational inequalities:

$$\begin{cases} \langle \frac{\partial f}{\partial x_1}(x_1, x_2, \ldots, x_N), y_1 - x_1 \rangle \geq 0, & \forall y_1 \in K, \\ \langle \frac{\partial f}{\partial x_2}(x_1, x_2, \ldots, x_N), y_2 - x_2 \rangle \geq 0, & \forall y_2 \in K, \\ \ldots \\ \langle \frac{\partial f}{\partial x_N}(x_1, x_2, \ldots, x_N), y_N - x_N \rangle \geq 0, & \forall y_N \in K. \end{cases} \quad (3.12)$$

In fact, we have

$$\frac{\partial f}{\partial x_i}(x_{0,1}, x_{0,2}, \ldots, x_{0,N}) \begin{cases} = 0, & x_{0,i} \in (a, b), \\ \geq 0, & x_{0,i} = a, \\ \leq 0, & x_{0,i} = b, \end{cases}$$

for all $i = 1, 2, \ldots, N$. Hence x_0 must satisfy (3.12). In addition, the system of multivariate variational inequalities (3.12) is equivalent to

$$\langle \mathrm{grad} f(x), y - x \rangle \geq 0, \quad \forall y \in K^N, \quad (3.13)$$

where

$$\mathrm{grad} f(x) = \left(\frac{\partial f}{\partial x_1}, \frac{\partial f}{\partial x_2}, \ldots, \frac{\partial f}{\partial x_N} \right).$$

This example is actually a practical background of Theorem 3.2, where $A_i = \frac{\partial f}{\partial x_i}$ for all $i = 1, 2, \ldots, N$ and $\mathrm{grad} f(x) = A^*$.

CONCLUSION

In this article, we use an ingenious mathematical method to prove the existence theorem of solutions for a kind of system of multivariate variational inequalities:

$$\begin{cases} \langle A_1(x_1, x_2, \ldots, x_N), y_1 - x_1 \rangle \geq 0, & \forall y_1 \in K, \\ \langle A_2(x_1, x_2, \ldots, x_N), y_2 - x_2 \rangle \geq 0, & \forall y_2 \in K, \\ \ldots \\ \langle A_N(x_1, x_2, \ldots, x_N), y_N - x_N \rangle \geq 0, & \forall y_N \in K. \end{cases}$$

Here K is a nonempty closed convex and bounded subset of a reflexive Banach space X and A_1, A_2, ..., A_N are N-variables monotone demi-continuous mappings from K^N into X^*. This system of multivariate variational inequalities has a solution. The set of solutions of this system of multivariate variational inequalities is closed convex in K^N. If A_1, A_2, ..., A_N are also strictly monotone, this system of multivariate variational inequalities has a unique solution.

ACKNOWLEDGEMENTS

This project is supported by the major project of Hebei North University under grant No. ZD201304.

AUTHORS' CONTRIBUTIONS

All authors contributed equally and significantly in writing this article. All authors read and approved the final manuscript.

REFERENCES

1. Kinderlehrer D, Stampacchia G. An Introduction to Variational Inequalities and Their Applications. New York: Academic Press; 1980.
2. Hartman P, Stampacchia G. On some non-linear elliptic differential-functional equations. Acta Math. 1966;115:153–188. doi: 10.1007/BF02392210.
3. Chang S, Joseph Lee H, Chan C. A new method for solving equilibrium problem fixed point problem and variational inequality problem with application to optimization. Nonlinear Anal., Theory Methods Appl. 2009;70:3307–3319. doi: 10.1016/j.na.2008.04.035.
4. Chang S, Lee B, Chen Y. Variational inequalities for monotone operators in nonreflexiveBanach spaces. Appl. Math. Lett. 1995;8:29–34. doi: 10.1016/0893-9659(95)00081-Z.
5. Fang F, Huang N. Variational-like inequalities with generalized monotone mappings in Banach spaces. J. Optim. Theory Appl. 2003;118:327–338. doi: 10.1023/A:1025499305742.
6. Huang N, Fang Y. Fixed point theorems and a new system of multivalued generalized order complementarity problems. Positivity. 2003;7:257–265. doi: 10.1023/A:1026222030596.
7. Isac G. Complementarity Problems. Berlin: Springer; 1992.
8. Isac G, Sehgal V, Singh S. An alternate version of a variational inequality. Indian J. Math. 1999;41:25–31.
9. Junlouchai P, Plubtieng S. Existence of solutions for generalized variational inequality problems in Banach spaces. Nonlinear Anal. 2011;74:999–1004. doi: 10.1016/j.na.2010.09.058.
10. Nadezhkina N, Takahashi W. Weak convergence theorem by an extragradient method for nonexpansive mappings and monotone mappings. J. Optim. Theory Appl. 2006;128:191–201. doi: 10.1007/s10957-005-7564-z.
11. Takahashi W, Toyoda M. Weak convergence theorems for nonexpansive mappings and monotone mappings. J. Optim. Theory Appl. 2003;118:417–428. doi: 10.1023/A:1025407607560.
12. Verma R. On a new system of nonlinear variational inequalities and associated iterative algorithms. Math. Sci. Res. Hot-Line. 1999;3:65–68.
13. Verma R. Iterative algorithms and a new system of nonlinear

quasivariational inequalities. Adv. Nonlinear Var. Inequal. 2001;4:117–124.

14. Yao J, Chadli O. Pseudomonotone complementarity problems and variational inequalities. In: Crouzeix JP, Haddjissas N, Schaible S, editors. Handbook of Generalized Convexity and Generalized Monotonicity. New York: Springer; 2005. pp. 501–558.
15. Chang S. Variational Inequality and Complementarity Problem Theory with Applications. Shanghai: Shanghai Science and Technology Press; 1991.
16. Plubtieng S, Sombut K. Existence results for system of variational inequality problems with semimonotone operators. J. Inequal. Appl. 2010;2010:14.
17. Su Y, Petruşel A, Yao J. Multivariate fixed point theorems for contractions and nonexpansive mappings with applications. Fixed Point Theory Appl. 2016;2016:9. doi: 10.1186/s13663-015-0493-0.
18. Luo Y, Su Y, Gao W. Multivariate best proximity point theorems in metric spaces. J. Nonlinear Sci. Appl. 2016;9:5756–5765.
19. Xu Y, Guan J, Tang Y, Su Y. Multivariate contraction mapping principle with the error estimate formulas in locally convex topological vector spaces and application. J. Nonlinear Sci. Appl. 2017;10:1064–1074. doi: 10.22436/jnsa.010.03.17.
20. Guan J, Tang Y, Xu Y, Su Y. System of *N* fixed point operator equations with *N*-pseudo-contractive mapping in reflexive Banach spaces. J. Nonlinear Sci. Appl. 2017;10:2457–2470. doi: 10.22436/jnsa.010.05.16.

4

An Axiomatic Integral and a Multivariate Mean Value Theorem

Milan Merkle

Department of Applied Mathematics, Faculty of Electrical Engineering, University of Belgrade, Bulevar Kralja Aleksandra 73, Belgrade, 11120, Serbia

ABSTRACT

In order to investigate minimal sufficient conditions for an abstract integral to belong to the convex hull of the integrand, we propose a system of axioms under which it happens. If the integrand is a continuous R^n-valued function over a path-connected topological space, we prove that any such integral can be represented as a convex combination of values of the integrand in at most n points, which yields an ultimate multivariate mean value theorem.

INTRODUCTION AND MOTIVATION

The basic integral mean value theorem states that for a function X which is continuous on the interval [a, b], there exists a point $t \in (a, b)$ such that

Citation: Merkle, M. (2015). An axiomatic integral and a multivariate mean value theorem. Journal of Inequalities and Applications, 2015(1), 346. (11 pages)

Copyright: © Open Access. This article is distributed under the terms of the Creative Commons Attribution 4.0 International (CC BY 4.0) License.

$$\frac{1}{b-a}\int_a^b X(s)\,ds = X(t). \qquad 1$$

To show that the assumption of continuity is crucial for validity of this theorem, we can take the interval [−1,1] and define X(s)=−1 for s∈[−1,0) and X(s)=1 for s∈[0,1]. Hence here we do not have a single point t∈(−1,1) for (1) to be satisfied. However, we can achieve a similar result with a convex combination of values of *f* in two points:

$$\frac{1}{b-a}\int_a^b X(s)\,ds = \frac{1}{2}X(t_1) + \frac{1}{2}X(t_2), \quad t_1 \in (-1,0), t_2 \in (0,1). \qquad 2$$

It turns out that the difference of one extra point for non-continuous functions remains in a much more general case and for a very broad class of integrals in higher dimensions. This is the topic of this article.

In multivariate case, with X∈R^n, n≥1, there is an old and seemingly forgotten mean value theorem by Kowalewski [1, 2] as follows.

Theorem A [1] Let $x_1,\ldots,x_n$ be continuous functions in a variable t∈[a,b]. There exist real numbers $t_1,\ldots,t_n$ in [a,b] and non-negative numbers $\lambda_1,\ldots,\lambda_n$, with $\sum_{i=1}^n \lambda_i = b-a$, such that

$$\int_a^b X_k(t)\,dt = \lambda_1 X_k(t_1) + \cdots + \lambda_n X_k(t_n) \quad \textit{for each } k = 1,2,\ldots,n,$$

or in more compact notation,

$$\int_a^b X(s)\,ds = \lambda_1 X(t_1) + \cdots + \lambda_n X(t_n), \qquad 3$$

where $X=(X_1,\ldots,X_n)$

A recent generalization of Theorem A is proved in [3] using the following modification of classical Carathéodory's convex hull theorem.

Theorem B [3] Let $C : s \mapsto X(s),\ s \in I$, be a continuous curve in R^n, where I⊂R is an interval, and let K be the convex hull of the curve C. Then each v ∈ K can be represented as a convex combination of n or fewer points of the curve C.

In this way, a more general mean value theorem for n-dimensional functions is obtained directly from the fact that the normalized integral should belong to the convex hull of the set of values of the integrand. This is proved in the following theorem for Lebesgue integral with a probability measure.

Theorem C ([3], Lemma 1)

Let (S,F,μ) be a probability space, and let X_i: S→R, i=1,… ,n, be μ-integrable functions. Let $X(t)=(X_1(t),\ldots,X_n(t))$ for every t∈S. Then $\int_S X(t)\,d\mu(t) \in \mathbf{R}^n$ is in the convex hull of the set $X(S)=\{X(t)|t\in S\}\subset R^n$.

Finally, the main result of [3] reads as follows.

Theorem D ([3], Theorem 2)

For an interval I⊆R, let μ be a finite positive measure on the Borel sigma-field of I. Let X_k, k=1,…,n, n≥1, be continuous functions on I integrable on I with respect to the measure μ. Then there exist points $t_1,\ldots,t_n$ in I and non-negative numbers $\lambda_1,\ldots,\lambda_n$, with $\sum_{i=1}^{n} \lambda_i = \mu(I)$, such that

$$\int_I X_k(s)\,d\mu(s) = \sum_{i=1}^{n} \lambda_i X_k(t_i), \quad k = 1,\ldots,n. \tag{4}$$

Let us note that without the continuity assumption we still may use Carathéodory's convex hull theorem which would yield (4) with n+1 points titi and the same number of λi's, which shows that the example at the beginning of the text well describes the general situation in R^n.

In this paper we give a more general theorem of the type (4), tracing the steps of the proof in [3] in a much more general context. In Section 3 we show that a result like (4) holds if the integral over I is replaced with a general linear functional on some function space, under a system of axioms, whereas the interval I can be replaced by a topological space which is path connected.

To reach this goal, we need to extend Theorem C. In Section 2 we show that Theorem C holds for any linear functional which satisfies a condition slightly stronger than positivity, and where functions Xk are defined over an arbitrary non-empty set.

Applications of such a very general mean value theorem are numerous, and we are not discussing particular applications in this paper. Let us just mention that as shown in [3], the theorems of this type can be considered as an aid to construct quadrature rules or their approximative versions. See also a recent paper [4] for another application related to integrals. Another advantage of the approach presented in this paper is that the results are widely applicable to different kinds of integrals treated as linear functionals over some space of functions.

AXIOMS AND THEIR CONSEQUENCES

We start with an arbitrary non-empty set S with an algebra F (may be a sigma algebra as well) of its subsets. Therefore, F contains S and if a set A is a result of finitely many set operations over sets in F, then A∈F.

Let SS be a family of functions X:S→R which satisfies the following conditions:

(C1): If X1,X2∈S, then aX_1+bX_2∈S for all a,b∈R.

(C2): For B∈F, the indicator function IB(·) belongs to S.

(C3): For X∈S and any interval J, the set {s∈S:X(s)∈J} is in F.

(C4): For X∈S such that X(s)≥0 for all s∈S, it holds that X·IX∈J∈S for any interval J.

Note that from (C1) and (C2) it follows that all constants are in SS. Let us also note that functions in S are not assumed to be bounded.

Let E be a functional defined on SS and taking values in **R** such that the following axioms hold.

(A1): Ec=c for any constant c;

(A2): $\mathrm{E}(\sum_{i=1}^{m}\alpha_i X_i) = \sum_{i=1}^{m}\alpha_i \mathrm{E}(X_i)$ for a_j∈R and for Xi∈S, i=1,…,m.

Now we may define a set function P as

$$P(B) = \mathrm{E}\big(I_B(\cdot)\big), \quad B \in \mathcal{F}, \qquad (5)$$

and consider yet another condition related to P:

(C5): If X∈S and P (N)=0, then X·IN∈ S, and E(X·IN)= 0.

The last axiom that we propose is as follows.

(A3): For X∈S, if EX=0, then either P(X=0)= 1 or there exist s_1,s_2∈ S such that $X(s_1)X(s_2)< 0$,

or equivalently (see Lemma 2.3 below)

(A3′): For X∈S, if X(s)≥0 for all s∈S and EX=0, then P(X=0)= 1.

Finally we extend the functional E to act on functions with values in R^n. Let X=(X1,…Xn) be a function from S to R^n, where we assume that Xi, i =1,…,n, satisfy axioms and conditions above, then we define

$E(X):=(EX_1,\dots,EX_n)$.

The central result of this section is Theorem 2.10, where we show that under (A1)-(A2)-(A3), E(X) belongs to the convex hull of X(S).

A similar axiomatic approach is applied in Daniell's integral, and there are other axiomatic systems in the literature for different purposes like in [5] for Riemann integrals in connection to evaluating the length of a curve, general means in [6], finitely additive probabilities (FAPs) in [7] and applications in a recent article [8]. The system of axioms applied in this article differs from others in the literature in the conditions that allow non-absolute integrals, as well as in axiom (A3), which is a slightly stronger condition than the usual positivity. The reason for introducing this system of axioms is that it provides conditions under which EX belongs to the convex hull of X(S) (Theorem 2.10) independently of the kind of integrals that are considered.

Now we are going to derive some additional properties as consequences from the axioms.

Lemma 2.1. Under the system of axioms (A1)-(A2)-(A3) or (A1)-(A2)-(A3′), assuming also conditions(C1)-(C2), the set function P defined on F with (5) is a finitely additive probability on(S,F).

Proof

Since IS(s)=1 for all s∈S, we have that P(S)=1. For disjoint sets B1,..., Bm, using (A2) we get additivity

$$P\left(\bigcup_{i=1}^{m} B_i\right) = \mathrm{E}\left(\sum_{i=1}^{m} I_{B_i}\right) = \sum_{i=1}^{m} P(B_i).$$

Let us now show that P(B)≥0 for all B∈F. Indeed, suppose that P(B)=−ε for some ε>0. This implies (by (A1) and (A2)) that E(IB+ε)= 0; on the other hand, IB(s)+ε>0for all s∈S, which contradicts both (A3) and (A3′), and this ends the proof.

From now on, the quintuplet (S,F,S,E,P) will be assumed to be as defined above in the framework of axioms (A1)-(A2)-(A3) and conditions (C1)-(C5) (if not specified differently). The letter X will be reserved for elements of S, and P is a set function derived from E as in (5).

Lemma 2.2. (Positivity) If for all s∈S, X(s)≥ 0, then EX≥0.

Proof

Suppose X(s)≥ 0 for all s∈ S, and EX=− ε for some ε>0. Then, by (A1) and (A2), E(X+ε)=0, whereas X+ε>0. This contradicts (A3). Therefore, if X≥0, then EX≥0.

Lemma 2.3. Assuming that (A1) and (A2) hold, axioms (A3) and (A3′) are equivalent.

Proof

In Lemma 2.1 we already proved the property that P is a FAP follows with either (A3) or (A3′), so we may use that property in both parts of the present proof.

Assume that (A1)-(A2)-(A3) hold and suppose that for all s∈S, X(s)≥ 0 and EX=0. Then by (A3) it follows that P(X=0)=1. Therefore, (A3′) holds.

Now assume that (A1)-(A2)-(A3′) hold, but not (A3). Then there exists X∈S such that: (a) EX=0, P(X=0)<1, X≥0, or (b) EX=0, P(X=0)<1, X≤0. In the case (a), using (A3′) we find that P(X=0)=1, which is a contradiction to P(X=0)<1. The case (b) can be reduced to (a) with the function Y=−X.

Due to the equivalence established in Lemma 2.3, in the rest of the article we refer to (A3) as being either (A3) or (A3′).

Remark 2.4. Suppose that we have a quintuplet (S,F,S,E,P) that satisfies assumptions (C1)-(C5) and axioms (A1)-(A2), where P is defined with (5). Let S^*∈F be a subset of S with $P(S^*)>0$such that $X\cdot I^*S\in S$. Define $F^*=\{B\cap S^* | B\in F\}$. Next, for each X∈S, we define the function $X^*:=X|S^*$ - that is, X restricted to S^*, and let S^* be the space of all those mappings. We define a linear functional E^* on S as

$$\mathrm{E}^*(X^*) = \frac{1}{P(S^*)}\mathrm{E}(X \cdot I_{S^*}), \quad X^* = X|_{S^*},$$

and the set function P^* as

$$P^*(B^*) = \mathrm{E}^*(I_{B^*}) = \frac{P(B^*)}{P(S^*)} = \frac{P(B \cap S^*)}{P(S^*)}, \quad B^* = B \cap S^* \in \mathcal{F}^*.$$

In this way we get a new quintuplet with (S^*,F^*,S^*,E^*,P^*), and it is not difficult to see that the new quintuplet inherits conditions (C1)-(C5) and axioms (A1)-(A2) as well as (A3) if it is satisfied with the original quintuplet.

In the next lemma we prove Markov's inequality from the axioms.

Lemma 2.5. Let X∈S and X(s)≥0 for all s∈S. Then

$$P(X > \varepsilon) \leq \frac{\mathrm{E}X}{\varepsilon}, \quad \varepsilon > 0. \tag{6}$$

Proof

Since X≥0, we use (C4) to conclude that

$$EX = E(X \cdot I_{0 \le X \le \varepsilon}) + E(X \cdot I_{X > \varepsilon}).$$

Further, we have

$$X(s) \cdot I_{0 \le X \le \varepsilon}(s) \ge 0, \qquad X(s) \cdot I_{X > \varepsilon}(s) \ge \varepsilon I_{X > \varepsilon}(s),$$

and then use positivity (Lemma 2.2) and (A1)-(A2) to conclude that EX≥εP(X>ε).

Lemma 2.6. Assuming (A1)-(A2), (C1)-(C5) and Markov's inequality, (A3) holds if P is countably additive probability.

Proof

Let X∈S such that X≥0 and EX=0. We need to show that P(X>0)=0. By Markov's inequality we have that P(X>ε)=0 for any ε>0, and so, using the countable additivity, we get

$$P(X>0) = P\left(\bigcup_{n=1}^{\infty}\left\{X > \frac{1}{n}\right\}\right) = \lim_{n \to +\infty} P\left(X > \frac{1}{n}\right) = 0$$

as desired.

Remark 2.7. Consider the case where P is a countably additive probability on (S,F), F is a sigma algebra, and $EX = \int_S X(s)\,dP(s)$. Axioms (A1)-(A2) and conditions (C1)-(C5) are clearly satisfied, and Markov's inequality can be proved from properties of the integral, so by Lemma 2.6, axiom (A3) also holds.

Let us now recall some facts about FAPs. A probability P which is defined on an algebra F of subsets of the set S is purely finitely additive if ν≡0 is the only countably additive measure with the property that ν(B)≤P(B) for all B∈F. A purely finitely additive probability P is strongly finitely additive-SFAP if there exist countably many disjoint sets H1,H2,…∈F such that

$$\bigcup_{i=1}^{+\infty} H_i = S \quad \text{and} \quad P(H_i) = 0 \quad \text{for all } i. \tag{7}$$

For every probability P on F there exists a countably additive probability Pc and a purely finitely additive probability Pd such that $P=\lambda P_c+(1-\lambda)P_d$ for some λ∈[0,1]. This decomposition is unique (except for λ=0 or λ=1, when it is trivially non-unique). For more details see [9].

Lemma 2.8. Assuming axioms (A1)-(A2), conditions (C1)-(C5) and positivity, if P is a SFAP, the condition of axiom (A3) is not satisfied.

Proof

Let P be a SFAP, and let Hi be a partition of S as in (7). Define X(s)=1/i if s∈Hi. Further,

$$X(s) \leq 1 \cdot I_{H_1}(s) + \frac{1}{2} \cdot I_{H_2}(s) + \cdots + \frac{1}{k} \cdot I_{\{H_k \cup H_{k+1} \cup \cdots\}}$$

and so (by positivity) 0≤EX≤1/k for every k>0, hence EX=0. This contradicts (A3).

Example 2.9

Let S=[0,+∞) and let P be the probability defined by the non-principal ultrafilter of Banach limit as s→+∞. Let X(s)=e−s. Then X≥0 and EX=0, but P(X=0)=0. In this case the convex hull K(X)=(0,1] and EX∉K(X).

Theorem 2.10. Let (S,F,S,E,P) be a quintuplet as defined above, and let $X=(X_1,\ldots,X_n)$, where Xi∈S for all i. Assuming that axioms (A1)-(A2)-(A3) and conditions (C1)-(C5) hold, EX belongs to the convex hull of the set $X(S)=\{X(t)|t\in S\}\subset R^n$.

Proof

Without loss of generality we may assume that for all i, EXi=0 (otherwise, if EXi=ci, we can observe E(Xi−ci)=0). Let K denote the convex hull of the set X(S)∈Rn. We now prove that 0∈K by induction on n. Let n=1. By (A3), EX=0 implies that either X(s)=0 for some s∈S or there are s1,s2∈S such that X(s1)> 0 and X(s2)<0X(s2)<0. In both cases it follows that 0∈K.

Now assume that the statement of the theorem is valid for all dimensions from 1 to n−1 for all quintuplets (S,F,S,E,P) that satisfy the conditions mentioned in the statement of the theorem. Let now X be a vector function with values in R^n.

If every hyperplane ϖ that contains 0 has the property that the set X(S) has a non-empty intersection with both of two open half-spaces with ϖ as boundary, then 0∈ K (see [10] for details). Otherwise, suppose that

$$L(s) := \sum_{k=1}^{n} a_k X_k(s) \geq 0 \quad \text{for every } s \in S, \tag{8}$$

with some real numbers a1,..., an such that $\sum a_k^2 > 0$. By linearity (A2) we have that EL(t)= 0, which is (by (A3')) possible together with (8) only if L(t)= 0 for all t∈S\N, where μ(N)=0. Assuming that an≠0, we find that

$$X_n(s) = -\sum_{k=1}^{n-1} \frac{a_k}{a_n} X_k(s) \quad \text{for every } s \in S \setminus N. \tag{9}$$

In other words, a separating hyperplane exists only if there exists a linear relation among n given functions with probability one. In order to eliminate X_n and to reduce the system to n−1 functions, we consider functions $X_i^*(s) = X_i(s)$ on the restricted domain S*=S\N(i=1,…,n−1) and the corresponding functional E*. Let K* be convex hull of X*(S*)∈R^{n-1}.

By hereditary property (Remark 2.4), we have that $E^*(X_i^*) = E(X_i \cdot I_{S\setminus N}) = 0$ by (C4). Note that K*⊂K. By induction assumption, the statement of the theorem holds for dimension n−1, and so

$$\sum_{i=1}^{m} \lambda_i X_k(t_i) = 0, \quad k = 1, \ldots, n-1,$$

10

with some $t_1,\ldots,t_m$∈S*⊂S (here we use the fact that $X_i^*(s) = X_i(s)$ for $s \in S^*$). Finally, using (9) and (10) we find that also

$$\sum_{i=1}^{m} \lambda_i X_n(t_i) = 0,$$

11

and so, the statement of the theorem holds for dimension n.

Remark 2.11. Theorem 2.10 provides sufficient conditions for EX to belong to the convex hull of X(S)X(S). However, by inspection of the proof, we can see that axiom (A3) is also necessary, assuming (A1)-(A2) and conditions (C1)-(C5).

Now, as a corollary to Theorem 2.10, using Carathéodory's theorem on representation of convex hull in finite dimension, we get the following result.

Theorem 2.12. Assume that axioms (A1)-(A2)-(A3) and conditions (C1)-(C5) hold on (S,F,S,E,P). Let X=(X_1,…,X_n), where Xi∈S for all i. Then there are points t_0,…,t_n and a discrete probability law given by probabilities λ_0,…, λ_n so that

$$EX_i = \sum_{j=0}^{n} \lambda_j X_i(t_j), \quad i = 1, \ldots n.$$

12

Theorem 2.12 is the most general mean value theorem for axiomatic integral. Due to Remark 2.7, the statement of this theorem applies with $EX_i = \int_S X_i(s)\,d\mu(s)$, μ is a countably additive probability measure on (S,F) where FF is a sigma algebra, and Xi are (S,F)−(R,B) measurable and integrable functions (B is the Borel sigma-field on **R**).

In the next section we consider the case of continuous functions Xi.

MEAN VALUE THEOREM FOR CONTINUOUS MULTIVARIATE MAPPINGS

Definition 3.1. A path from a point a to a point b in a topological space S is a continuous mapping $f:[0,1] \to S$ such that $f(0)= a$ and $f(1)=b$. A space S is path connected if for any two points $a,b \in S$ there exists a path that connects them.

Let us remark that any topological vector space is path connected. A path that connects points a and b is given by $f(\lambda)=\lambda a+(1-\lambda)\, b$, $\lambda \in [0,1]$. The same is true for a convex subset Sof any topological space.

The following result is a generalization of Theorem B.

Theorem 3.2. Let $X:t \mapsto X(t)$, $t \in S$, be a continuous function defined on a path-connected topological space S with values in R^n, and let K be the convex hull of the set X(S). Then each $v \in K$ can be represented as a convex combination of n or fewer points of the set X(S).

Proof

By Carathéodory's theorem, any $v \in K$ can be represented as a convex combination of at most n+1 points of the set X(S). Without loss of generality, assume that v=0. Therefore, there exist $tj \in S$ and $vj \geq 0$, $0 \leq j \leq n$, such that $ti \neq tj$ for $i \neq j$, $v_0+\cdots+v_n=1$, and

$$\nu_0 x(t_0) + \nu_1 x(t_1) + \cdots + \nu_n x(t_n) = 0. \tag{13}$$

We may also assume that all n+1 points x(tj) do not belong to one hyperplane in R^n (in particular, $x(ti) \neq x(tj)$ for $i \neq j$) and that the numbers vj are all positive; otherwise, at least one term from (13) can be eliminated. Now we apply the following reasoning.

Denote by pj(x), $0 \leq j \leq n$, the coordinates of the vector $x \in R^n$ with respect to the coordinate system with the origin at 0, and with the vector base consisting of vectors x(tj), j=1,…,n (that is, $x = \sum_{j=1}^{n} p_j(x)x(t_j)$). Then from (13) we find that $p_j(x(t_0))=-v_j/v_0<0$, j=1,…, n, i.e., the coordinates of the vector $x(t_0)$ are negative. The coordinates of vectors x(tj), j=1,2,…,n, are non-negative: $p_j(x(t_j))=1$ and $p_k(x(t_j))=0$ for $k \neq j$. Now consider a path $t=t(\lambda)$, $\lambda \in [0,1]$ which connects points t_0 and t_1, so that $t(0)=t_0$ and $t(1)=t_1$. The functions $\lambda \mapsto pj(x(t(\lambda))):=fj(\lambda)$ are continuous as mappings from [0,1] to R^n and fj(0)<0 for all j=1,…,n, whereas $f_1(1)=1$ and fj(1)=0 for j>1. Therefore, for each of functions fj there exists one or more points λ∈(0,1] such that fj(λ)=0. Since the set $N = \bigcup_{j=1}^{n} f_j^{-1}(\{0\})$ is a closed and non-empty subset of (0,1], there exists λ_0= minN, λ_0>0. Let $\bar{t} := t(\lambda_0)$. From this construction it follows that there exists (at least one) k such that $p_k(x(\bar{t})) = 0$ and $p_j(x(\bar{t})) < 0$ for j≠ k. Hence,

$$x(\bar{t}) - \sum_{1\leq j\leq n, j\neq k} p_j(x(\bar{t}))x(t_j) = 0,$$

and it follows that v=0v=0 is a convex combination of points $x(\bar{t})$ and x(tj), j=1,…, n, j≠k.

As a direct corollary to Theorems 2.10 and 3.2, we have the following mean value theorem for continuous multivariate mappings.

Theorem 3.3

Let (S,F,S,E,P) be a quintuplet as defined in Section 2, where S is a path-connected topological space. Under conditions (C1)-(C5) and axioms (A1)-(A3), let Xi, i=1,…, n, be continuous functions from S. Then there exist points $t_1,\ldots, t_n$ in S and non-negative numbers $\lambda_1,\ldots, \lambda_n$, with $\sum_{i=1}^{n} \lambda_i = 1$, such that

$$\mathrm{E}X_k = \sum_{i=1}^{n} \lambda_i X_k(t_i), \quad k = 1, \ldots, n.$$

Remark 3.4. Since Theorem 3.2 is independent of axioms and conditions of Section 2, the statement of Theorem 3.3 holds whenever Theorem 2.12 holds. In particular, it holds whenever $\mathrm{E}X_i = \int_S X_i(s)\,d\mu(s)$, where μ is a countably additive probability measure.

In fact, all what Theorem 3.3 says is that we can save one point in the representation (12) of Theorem 2.12. Although it might look not much significant, in some applications it makes difference. For example, Karamata's representation for covariance in [11] based on Kowalewski's original result for n=2 strongly depends on two points and nothing similar can be derived with three points.

SOME PARTICULAR CASES AND OPEN PROBLEMS

Riemann and Lebesgue integral on R^d

For $d = 1$, let $-\infty < a < b < +\infty$ and let X be a Riemann integrable function on [a,b]. Define

$$\mathrm{E}X = \frac{1}{b-a}\int_a^b X(s)\,\mathrm{d}s. \tag{14}$$

In terms of previous notations, S=[a,b], and S is the family of Riemann integrable functions on S. A natural choice for algebra F of sets related to Riemann integral should be the algebra of intervals in [a,b], which can be defined as the collection of all subintervals of [a,b] (including singletons and an empty set) and their finite unions. The corresponding probability is defined then as follows: If $A = \bigcup_{i=1}^{k} J_i$, where Ji are intervals,

$$P(A) = \frac{1}{b-a}\int_a^b I_A(s)\,\mathrm{d}s = \sum_{i=1}^{k}\lambda(J_i),$$

where λ(Ji) is the length of Ji. This is Jordan probability measure on [a,b], and it is well known that, even for continuous bounded functions, the set {s∈S:X(s)≤c} where c is a real number does not obligatory belong to F (this is probably first shown by an example in [12]). This fact makes it impossible to use our system of axioms directly because condition (C3) does not hold. Nevertheless, we can proceed by noticing that the algebra F as described above is a sub-algebra of the Borel sigma algebra B generated by open sets in [a,b], and since Riemann and Lebesgue integrals coincide if the integrand is Riemann integrable, we can proceed in this way. In more common notations, let us consider a general case of a functional E based on Lebesgue integral on R^d, $d \geq 1$:

$$\mathrm{E}f = \frac{1}{V(D)}\int_D f(s_1,\ldots,s_d)\,\mathrm{d}s_1\cdots\mathrm{d}s_d, \tag{15}$$

or, in shorthand,

$$\mathrm{E}f = \frac{1}{V(D)}\int_D f(s)\,\mathrm{d}\lambda(s),$$

where λ is the Lebesgue measure on R^d restricted to D, and V(D)=λ(D)>0, where D is a convex (or, in more generality, path-connected) subset of R^d. The underlying probability measure in our construction of Section 2 is

$P(\cdot) = \frac{1}{V(D)}\lambda(\cdot)$. Let $f_1,\ldots, f_n$ be continuous functions $D \mapsto R$ such that Efi is as defined in (15). Then, by Theorem 3.3, we have that there are points x1,… ,xn∈D and non-negative numbers $\lambda_1,\ldots,\lambda_n$ with $\sum_{i=1}^{n} \lambda_i = 1$ such that

$$\frac{1}{V(D)} \int_D f_1(s_1,\ldots,s_d)\,\mathrm{d}s_1 \cdots \mathrm{d}s_d = \lambda_1 f_1(x_1) + \cdots + \lambda_n f_1(x_n)$$

$$\vdots$$

$$\frac{1}{V(D)} \int_D f_n(s_1,\ldots,s_d)\,\mathrm{d}s_1 \cdots \mathrm{d}s_d = \lambda_1 f_n(x_1) + \cdots + \lambda_n f_n(x_n).$$

In words, this result shows that for an arbitrary system of *n* integrals with continuous integrands, there exists an exact quadrature rule with *n* points in *D* with coefficients λ_i which are the same for all integrals. Note that x_i are *d*-dimensional points, so in fact here we have d_n scalar parameters.

Integrals with respect to countably additive probability measure

As already noted, Theorem 3.3 holds for all integrals based on countably additive probability measure. Suppose that *S* is a path-connected topological space, and let μ be a countably additive probability measure on (S,F), where F is a sigma algebra of Borel subsets of *S*. Let $f_1,\ldots, f_n$ be continuous mappings from *S* to **R**, and suppose that

$$\mathrm{E}f_i = \int_S f(s)\,\mathrm{d}\mu(s), \quad i = 1,\ldots,n \tag{16}$$

is finite. In particular, let S=C[0,T], the space of continuous functions on the interval [0,T] with the supremum norm. Then the measure μ defines a stochastic process on [0,T], s(t), 0≤t≤ T, and fi(s) is a continuous functional of trajectories of the process. By Theorem 3.3, we have that the system of expectations (16) can be represented as

$$\mathrm{E}f_i = \sum_{j=1}^{n} \lambda_j f_i(x_j), \quad i = 1,\ldots,n$$

for some $x_j \in C[0,T]$ and $\lambda_j \geq 0$ with $\sum_j \lambda_j = 1$.

Open question

We showed in Section 2 that axiom (A3) does not hold for (integrals based on) SFAPs, so by Remark 2.11, the mean value theorem does not hold in this case. It would be of interest to describe classes of finitely additive probabilities for which axiom (A3) holds or does not hold in terms of some structural properties of measures.

COMPETING INTERESTS

The author declares that there are no competing interests.

ACKNOWLEDGEMENTS

This work is supported by grant III 44006 from the Ministry of Education, Science and Technological Development of Republic of Serbia.

REFERENCES

1. Kowalewski, G: Ein Mittelwertsatz für ein System von n Integralen. Zeitschrf. für Math. und Phys. (Schlömilch Z.) 42, 153-157 (1895)
2. Kowalewski, G: Bemerkungen zu dem Mittelwertsatze für ein System von n Integralen. Zeitschrf. für Math. und Phys. (Schlömilch Z.) 43, 118-120 (1896)
3. Jankovic, S, Merkle, M: A mean value theorem for systems of integrals. J. Math. Anal. Appl. ´ 342, 334-339 (2008)
4. Heydari, M, Avazzdeh, Z, Navabpour, H, Loghmani, GB: Numerical solution of Fredholm integral equations of the second kind by using integral mean value theorem II. High dimensional problems. Appl. Math. Model. 37, 432-442 (2013)
5. Gillman, L: An axiomatic approach to the integral. Am. Math. Mon. 100(1), 16-25 (1993)
6. Ostrowski, AM: On an integral inequality. Aequ. Math. 4, 358-373 (1970)
7. Schervish, MJ, Seidenfeld, T, Kadane, JB: On the equivalence of conglomerability and disintegrability for unbounded random variables. Technical Report 864, Carnegie Mellon University, Pittsburgh, PA (2008)
8. Schervish, MJ, Seidenfeld, T, Kadane, JB: Dominating countably many forecasts. Ann. Stat. 42, 728-756 (2014)
9. Yosida, K, Hewitt, E: Finitely additive measures. Trans. Am. Math. Soc. 72, 46-66 (1952)
10. Rockafellar, RT: Convex Analysis. Princeton University Press, Princeton (1997)
11. Karamata, J: Sur certain inégalités relatives aux quotients et à la difference de $\int fg$ et $\int f \int g$. Publ. Inst. Math. (Belgr.) **2**, 131-145 (1948)
12. Frink, O Jr: Jordan measure and Riemann integration. Ann. Math. 34, 518-526 (1933)

5

Multivariate Longitudinal Analysis with Bivariate Correlation Test

Eric Houngla Adjakossa[1,2], Ibrahim Sadissou[3,4], Mahouton Norbert Hounkonnou[2] , Gregory Nuel[1]

[1]Laboratoire de Probabilités et ModèlesAléatoires /Université Pierre et Marie Curie, Case courrier 188 - 4, Place Jussieu 75252 Paris cedex 05 France

[2]University of Abomey-Calavi, 072 B.P. 50 Cotonou, Republic of Benin

[3]Laboratoire de Biologieet de PhysiologieCellulaires /University of Abomey-Calavi, Cotonou, Republic of Benin

[4]Centre d'Etudeet de Recherchesur le PaludismeAssocié à la Grossesse et à l'Enfance (CERPAGE), Cotonou, Republic of Benin

ABSTRACT

In the context of multivariate multilevel data analysis, this paper focuses on the multivariate linear mixed-effects model, including all the correlations between the random effects when the dimensional residual terms are assumed uncorrelated. Using the EM algorithm, we suggest more general expressions of the model's parameters estimators. These estimators can be used in the framework of the multivariate longitudinal data analysis as well

Citation: Adjakossa, E. H., Sadissou, I., Hounkonnou, M. N., &Nuel, G. (2016). Multivariate longitudinal analysis with bivariate correlation test. PloS one, 11(8). (33 pages), DOI/URL: https://doi.org/10.1371/journal.pone.0159649.

Copyright: © 2016 Adjakossa et al. This is an open access article distributed under the terms of the Creative Commons Attribution 4.0 International (CC BY 4.0) License.

as in the more general context of the analysis of multivariate multilevel data. By using a likelihood ratio test, we test the significance of the correlations between the random effects of two dependent variables of the model, in order to investigate whether or not it is useful to model these dependent variables jointly. Simulation studies are done to assess both the parameter recovery performance of the EM estimators and the power of the test. Using two empirical data sets which are of longitudinal multivariate type and multivariate multilevel type, respectively, the usefulness of the test is illustrated.

INTRODUCTION

In statistical studies, one often needs to analyze data with nested sources of variability: e.g., pupils in classes, employees in companies, repeated measurements in subjects, etc. [1] referred to these type of data as grouped data which are also named multilevel data, hierarchical data or nested data in the literature [2–4]. In the analysis of such data, it is usually illuminating to take account of the variability associated with each level of nesting. There is variability, e.g., between pupils but also between classes. The measurements related to a specific subject (level of nesting) can be correlated, while observations from different subjects are usually independent, and one may draw wrong conclusions if either of these sources of variability is ignored [5]. A series of works in statistical literature focus on the analysis of univariate multilevel data (or univariate grouped data) where a single outcome of interest is analyzed [6–11]. Such analyses are generally simple to deal with due to the availability of many software packages conceived to perform them [12–14]. In practice, many scientific questions of interest require to focus on multiple outcomes, all arising from the same multilevel study, leading to the so-called multivariate multilevel data. For example, to answer some questions of interest, [15] analyzed hearing threshold data (in the Baltimore Longitudinal Study on Aging) [16] which consisted in the longitudinal recording of 22 variables. [17] also studied the joint evolution of HIV RNA and CD4+ T lymphocytes in a cohort of HIV-1 infected patients treated with highly active antiretroviral treatment, by jointly analyzing both markers. [18] used multivariate multilevel regression analysis to investigate individual level determinants of self rated health and happiness, as well as the extent of community level covariation in health and happiness. [19] also used multivariate multilevel analysis to jointly model three commonly used indicators of fear of crime which are: feeling unsafe alone at home after dark, feeling unsafe walking alone after dark and worry about becoming a victim of crime. A variety of works were devoted to joint modeling during the last few decades (see e.g., [20–24]).

These analyses often require a specification of the joint density of all outcomes or, at least, the correlation structure of the data and therefore can lead to the parsimony and/or computation (optimization) problems as well as to numerical difficulties in statistical inference, when the dimension of these outcomes increases. Many analysis strategies were proposed in the statistical literature to circumvent these problems. These strategies generally consist in reducing the dimensionality of the multivariate vector of outcomes and/or in using a small number of latent variables to model correlations within these data. Joint analysis of multivariate multilevel data then requires a trade-off between the increase of the computational complexity and the gain in information.

In this work, we focus on the multivariate linear mixed-effects model, including all the correlations between the random effects along with the independent marginal (dimensional) residuals. The correlations between two dependent variables are then those from the random effects related to these dependent variables. The class of mixed-effects models considered here assumes that both the random effects and the errors (residuals) follow Gaussian distributions. These models are intended for the analysis of multivariate multilevel data in which the dependent variables are continuous.

We use the EM algorithm to estimate the parameters of the model but here, we have two novelties: 1) we suggest a general expression of EM-based estimators which can help in analyzing multivariate longitudinal data as well as the multivariate multilevel data, not of the longitudinal type, and 2) we test the significance of the correlations between the random effects of two dependent variables, using the likelihood ratio test which allows to decide if some dependent variables are significantly correlated or not. By using this bivariate correlation test, the novelty here is the illustration, through empirical data, of some of the consequences of performing separate analyses when a joint analysis is required. Two dependent variables which are found to be uncorrelated after this test will be analyzed with two independent models (or analyzed separately). This strategy may be considered as a way toward the obtaining of a more parsimonious model in high dimension without losing much information. It may also be used in a joint modeling selection procedure.

The paper is organized as follows. In Section 2, contributions of previous works are briefly presented. We also present in this section the EM-based estimators of the parameters of the multivariate linear mixed model. Simulations studies are done in Section 3 where we also discuss the power of the likelihood ratio test which allows to test the significance of the correlation between two response variables. Two illustrations on empirical data are also done in Section 3. The first, concerning bivariate two-level data, is about a study on the effects of school differences on pupils' progress in Dutch language and arithmetics in the Netherlands. The second illustration concerns a longitudinal study on the immune response to malaria of infants in Benin.

MATERIALS AND METHODS

Previous Works

In this Section, we briefly recall the framework of the multivariate multilevel analysis (see for instance, [25, 26]). We can basically distinguish two main approaches to model such data: those which specify the joint distribution of all outcomes without the use of latent structures, and the models using latent structures. We denote by y_1, ..., y_m the m dependent vectors of interest, and $y = (y_1^T, \cdots, y_m^T)^T$.

Modeling Methods without Latent Structures

The first approach which is that of the modeling without latent structures comprises three sub-approaches consisting in a) direct specification of the correlation structure of y, b) analysis without explicit modeling of the correlation structure of y and c) conditional models.

In the case of direct specification of Cov(y), [27] and [28] factorized the covariance matrix of y by using the Kronecker product in order to have more parsimonious models in the context of fully balanced data. With the same idea of having a parsimonious structure, [29] specified the intra-outcome and inter-outcome correlations, respectively, as follows $\mathrm{Corr}(y_k(t), y_k(s)) = \exp(\alpha|s-t|^{\theta_k})$ and $\mathrm{Corr}(y_k(t), y_{k'}(s)) = \exp((\alpha|s-t|+1)^{\theta_{kk'}})$, with t and s indicating the time, and k and k' indicating the dimension. Although these models are useful, they are often too restrictive and may not be realistic in many applications, especially when the data, for example, in the longitudinal studies are unbalanced (i.e. the number of available measurements per subject and the time points at which the measurements were taken often differ from one subject to another). Another class of joint models, specifying directly the joint distribution of $\boldsymbol{y}$, and whose application is often not straightforward, due to unbalanced data structures is the so-called copula model [30, 31]. Denoting by F_i, i = 1, …, m the cumulative distribution function of the ith component of $\boldsymbol{y}$, y_i, a copula model is defined by an m-dimensional cumulative distribution function $C(u_1, \ldots, u_m)$ with uniform marginals such that $F(y_1, \ldots, y_m) = C(u_1, \ldots, u_m)$ with $(U_1, \ldots, U_m) = (F_1(y_1), \ldots, F_m(y_m))$, where F is the joint cumulative distribution of $\boldsymbol{y}$. That is, the joint distribution of $\boldsymbol{y}$ can be written in terms of its marginal distributions and a copula which describes the dependence structure between its components. While the construction of copulas is mathematically elegant, parameters estimation is often not feasible, especially in high-dimensional situations [26]. One of the rare applications of the copula-based modeling in the multivariate multilevel data analysis framework was proposed by [32], who studied the hemodynamic effect of a new antidepressant on the diastolic blood pressure, the systolic blood pressure and the heart rate of 10 healthy volunteers. They separately modeled, at first, each longitudinal series of response and used a copula to relate the marginal distributions of these responses at each observation time. In a second step, at each observation time, the conditional (on the past) distributions of each response were related using another copula describing the relationship between the corresponding variables. One of the advantages of this approach is that there is no need to use the same family of distributions for all response variables. As [33] used ARIMA process to model the error structure of earnings in a longitudinal data analysis context, time series models can also be used for modeling multivariate multilevel data in order to describe the dynamic dependence between variables and perform forecasting. The most commonly used multivariate time series model, the vector autoregressive (VAR) model which is relatively easy to estimate, is found to be similar to the multivariate multiple linear regression [34] where the errors for different response variables on the same trial are set to be correlated [35]. Other examples of VAR modeling include [36] and [37], but one drawback of the model is that the number of parameters can become very large, potentially leading to estimation problems [38].

Regarding analysis without explicit specification of $Cov(\boldsymbol{y})$, [6] proposed an extension of generalized linear models to the analysis of longitudinal data, where they introduced a class of estimating equations called generalized estimating equations (GEE). GEE estimation ensures consistent estimates of the regression parameters without specifying the joint distribution of a subject's observations. That is, GEE replaces $\mathbb{V}[y]$ by the so-called working covariance matrix W(α) which depends on an unknown vector α to estimate. The related working correlation matrix, R(α), is also considered. Incorrect choice of W(α) does not affect the consistency of the regression parameters' estimators [6]. [39] discussed the use of GEEs with multivariate discrete variables, where focus was on the modeling of the marginal (dimensional) means of these variables and their pairwise associations. The extension of the GEE method to mixed continuous-discrete responses was discussed by [40] and [41]. [42] also avoided the need of explicit modeling of the covariance structure of bivariate longitudinal responses by using SUR [5] and GEE. As pointed out by [43], ambiguities concerning the definition of the working covariance matrix can result in a breakdown of the GEE-based estimation. For example in the longitudinal data analysis, if the true structure of correlation is equicorrelation, $(R_i^*)_{jk} = \rho$, and that the working structure is autoregressive, $(R_i)_{jk} = \alpha^{|j-k|}$, there is no solution for $\hat{\alpha}$ when $-1/2 \leq \rho < -1/3$ [43]. This can be viewed as the major drawback of the GEE method since it can lead to the misspecification of within-subject associations in the context of longitudinal data analysis, for instance. Examples of procedures which bypass the need to explicitly model the underlying covariance structure of $\boldsymbol{y}$ include [42, 44, 45]. These procedures, generally, consist in regressing each component of $\boldsymbol{y}$ on relevant covariates of interest, followed by combination of these regression coefficients into a single global estimate of the covariates effect [25].

One way to avoid the direct specification of the joint distribution of $\boldsymbol{y}$ is to factorize it, leading to the so-called conditional models [46]. For two responses, the joint density $f(y_1, y_2)$ can be written as follows:

$$f(y_1, y_2) = f(y_1|y_2)f(y_2) = f(y_2|y_1)f(y_1) \tag{1}$$

The choice of the conditioning response is of course arbitrary and requires very careful reflection about plausible associations between components of $\boldsymbol{y}$. For example, in the specification of a conditional model such as $f(y_1|y_2)$, y_2 plays the role of covariate and different choices can lead to completely opposite results and conclusions [47]. In a clinical trial, for example, none of these factorizations will be of interest due to the conditioning on a post-randomization outcome which may partially attenuate the treatment effect on the other [26]. Another drawback of conditional models is that they do not directly lead to marginal inferences. Suppose that scientific interest would be in a comparison of the rate of longitudinal change in average of y_1 and y_2. The factorization $f(y_1, y_2) = f(y_1|y_2)f(y_2)$ directly allows for inferences about the marginal evolution of y_2, but the marginal expectation of y_1 requires computation of $\mathbb{E}[y_1] = \mathbb{E}[\mathbb{E}[y_1|y_2]]$, which, depending on the actual models, may be far from straightforward [26].

Modeling Methods using Latent Structures

The second approach regarding models using latent structures can also be split in two sub-approaches including the strategy based on the reduction of the dimensionality of $\boldsymbol{y}$ and the mixed-effect models. The general idea of reducing the dimension of $\boldsymbol{y}$ is to use principal-component type analysis, or a summary function, to first reduce the dimensionality of $\boldsymbol{y}$ and then, use standard univariate multilevel models for the analysis of the principal factors or the retained summaries of $\boldsymbol{y}$ [48–52]. Although it is useful, simple to understand and easy to compute, this strategy of dimension reduction has some drawbacks such as the loss of information as discussed by [25] and [26]. [25] used this approach and retained the first principal-component only which explains 31% of the total variation in their data. They found out that the summary function does not have any physical significance and the inference results cannot be interpreted in terms of the effect of the covariates on the original (response) variables. They also found that the method fails to explore the association of the components of $\boldsymbol{y}$ along time, in the case of longitudinal studies. Furthermore, the method is not applicable in situations where all the components of $\boldsymbol{y}$ are not measured at the same time point [25], although a possible extension might be the use of functional principal components [53].

Regarding the mixed-effect models, [15], [54], [55], [56] and [57] proposed the use of random-effects models for multivariate longitudinal data. They pointed out that the main disadvantage of joining separate mixed models by allowing their model-specific random effects to be correlated is the increase of the dimension of the total vector of random effects with the number of outcomes, leading to computational problems. To circumvent these problems, [15] noted that all parameters in the joint model can be estimated by fitting all the bivariate models, based on

$$f(y_s, y_t) = \int\int f(y_s|\gamma_s)f(y_t|\gamma_t)f(\gamma_s, \gamma_t)d\gamma_s d\gamma_t$$

for all $m(m - 1)/2$ pairs (y_s, y_t), $1 \leq s < t \leq m$, resulting from the main multivariate model. Estimators for the main parameters are obtained by averaging over the results from fitting the $m(m - 1)/2$ pairwise models. They then showed that the pseudo-likelihood theory can be used to derive the asymptotic distribution of these estimators, and used SAS procedures for mixed models [14] based on the Newton-Raphson algorithm to fit their models, following the approach in [17]. In some multilevel studies, focus is not to directly model $\boldsymbol{y}$, but a few number of latent variables which cannot be quantified directly (e.g., depression and anxiety), but through measurements of $\boldsymbol{y}$. In such situations, analysis may be conducted in two steps: the first produces the obtaining of the latent variables and the second proceeds to the joint analysis of these latent variables. For example, [58] proposed a latent factor linear mixed model to capture the joint trend over time of latent variables. the authors reduced, indeed, the high-dimensional responses to low-dimensional latent factors by the factor analysis model, and then used the multivariate linear mixed model to study the longitudinal trends of these latent factors, where the estimates have been done using the EM algorithm. To deal with missing values in multivariate longitudinal analysis using multivariate linear mixed-effects model, [59] proposed multiple imputations using Markov chain Monte Carlo, where they used EM algorithm for the parameters estimation. Here, the authors sped up the EM algorithm by analytically integrating the random effects out of the likelihood function, avoiding to treat them as missing data. [60] used EM based modeling to estimate the parameters of the multivariate linear mixed model under a SAS macro program encoded in IML.

Although the EM algorithm is known to be slow, one of the biggest advantages of this method is that it is not computationally expensive, even with a large number of response variables. In this context, our contribution is the writing of the EM-based estimators in a more general form than those used in [58, 59] and [60]. The expressions of the EM-based estimators used in this paper can easily perform any analysis in the framework of the multivariate multilevel data analysis using multivariate linear mixed-effects model.

Another technique somewhat close to those discussed in [58] is the structural equations-based techniques. For example, [61] developed linear structural equations with latent variables approach. Considering $y = (y_1^T, y_2^T)^T$, this approach can be expressed as follows: $y_i = \mu_i + G_i \eta_i; i = 1, 2$ and $\beta\eta_1 = \gamma\eta_2$, where η_i, $i = 1, 2$ are the latent variables, $\beta(m \times m)$ and $\gamma(m \times n)$ are coefficient matrices governing the linear relations of all variables involved in the m structural equations. G_i, $i = 1, 2$ are known matrices. The parameters of the model may be estimated by gradient and quasi-Newton methods, or a Gauss-Newton algorithm that obtains least-squares, generalized least-squares, or maximum likelihood estimates. One modeling strategy which fuses together mixed-effects model and VAR model in order to analyze multivariate multilevel data is the so-called multilevel-VAR method. For example, [62] used the multilevel-VAR model in the context of network inference in psychopathology, where they used the population standard deviation of the person-specific random effects to construct a network representing individual variability. Examples of multilevel-VAR modeling include [63] and [38].

State space models [64] which are useful to investigate the dynamical properties of latent variables can also be used to analyze multivariate multilevel data. For example, [65] introduced an extension of the basic state space model which is flexible and general in the sense of it is applicable to any time series for multiple systems.

Methods for estimating the connectivity maps containing heterogeneity may also be applied to analyze multivariate multilevel data. [66] presented the Group Iterative Multiple Model Estimation (GIMME) approach, which addresses the issue of heterogeneity (the need for individual-level maps) in effective connectivity mapping while capitalizing on shared information to arrive at group inferences. Unlike mixed-effects models, GIMME allows for the structure of the connectivity maps to be unique across individuals [66].

One can also use a nonparametric function f to handle the relationship between the components of $\boldsymbol{y}$ and the covariates [67–69]. This strategy requires also to have sufficient data per subject, in the case of multivariate longitudinal data. Other estimation strategies implemented under softwares and discussed by [70] can perhaps be extended to the multivariate analysis case, when necessary.

Let us finally point out that the software packages which can easily and accurately analyze (jointly) the data of multivariate multilevel type are extremely rare, and one arranges the data and manipulates packages primarily designed for fitting univariate models to handle their analysis. The SabreR [71] package, under the R software [72], which has been devoted to jointly fitting up to three mixed-effects models, with random intercepts only, has been recently removed from the depot. These facts prove by themselves that the analysis of multivariate multilevel data in a single framework is a challenging task. Bayesian-based approaches can be implemented using packages like R2WinBUGS [73] under the R software, and are useful but very time consuming and require a good expertise from the user who can easily be discouraged.

Model and Notations

The model discussed here is the multivariate linear mixed-effects model (or the multivariate linear multilevel model), including all the correlations between the random effects, but the marginal residual terms are assumed to be uncorrelated. For a more general multivariate linear mixed-effects model, the dependent variables are assumed to be correlated, conditional on the random-effects. That is, the marginal residual terms are correlated. In this paper, as in many other works (see for example, [59, 60, 74] and [58]), we assume that conditional on the random-effects, the dependent variables are uncorrelated. In the context of using EM algorithm in estimating the model parameters, this assumption allows to derive the EM-based estimators for the residual variance parameters. If the dimensional residual terms are assumed to be correlated, the EM-based estimators of theirs variance parameters are not easy to deal with and we don't treat this case here. This model assumes that both the random effects and the residuals follow Gaussian distribution, and is intended for the analysis of multivariate multilevel data in which the dependent variables are continuous. For the sake of simplicity we focus on the bivariate case (m = 2) in most of the paper, but the generalization to higher dimensions (m > 2) is straightforward. The model is as follows:

$$\begin{aligned} y_1 &= X_1\beta_1 + Z_1\gamma_1 + \varepsilon_1 \\ y_2 &= X_2\beta_2 + Z_2\gamma_2 + \varepsilon_2 \end{aligned} \quad (2)$$

$$\gamma = \begin{pmatrix} \gamma_1 \\ \gamma_2 \end{pmatrix} \sim \mathcal{N}\left(0; \Gamma = \begin{pmatrix} \Gamma_1 & \Gamma_{12} \\ \Gamma_{12}^T & \Gamma_2 \end{pmatrix}\right), \quad (3)$$

$$\varepsilon = \begin{pmatrix} \varepsilon_1 \\ \varepsilon_2 \end{pmatrix} \sim \mathcal{N}\left(0; \Sigma = \begin{pmatrix} \Sigma_1 & 0 \\ 0 & \Sigma_2 \end{pmatrix}\right); \quad \gamma \perp \varepsilon \quad (4)$$

For $k \in \{1, 2\}$, β_k and γ_k denote respectively the fixed effects and the random effects vector of covariates, while ε_k is the residual component. X_k is a matrix of covariates and Z_k a covariates-based design matrix. $\dim(X_k) = N_k \times p_k$ and $\dim(Z_k) = N_k \times q_k$, where N_k is the total number of observations in the dimension k of the model. p_k and q_k are, respectively, the number of fixed effect related covariates and the number of random effect related covariates in the dimension k of the model. If N_k is a constant N for any k, the index k will be removed and N will denote the total number of observations in all dimensions of the model. The bold symbols represent parameters of multiple dimensions (i.e. Σ_1 concerns dimension 1 of the model while $\mathbf{\Sigma}$ concerns both dimensions).

Another way to easily understand the model is to express it using the levels of the covariate related to the random-effects. This expression (subject-based version) of the model is, generally, used in the framework of longitudinal data analysis, and lead to EM-based estimators (expressions) which are a particular case of the estimators expressions obtained in Eqs (17), (18) and (19) (for example, see [60]). Denoting by n the total number of subjects involved in the longitudinal study, the model can be expressed as follows:

denoting by i a subject, for i = 1, …, n

$$\begin{aligned} y_{1i} &= X_{1i}\beta_1 + Z_{1i}\gamma_{1i} + \varepsilon_{1i} \\ y_{2i} &= X_{2i}\beta_2 + Z_{2i}\gamma_{2i} + \varepsilon_{2i} \end{aligned} \qquad (5)$$

with

$$\gamma_i = \begin{pmatrix} \gamma_{1i} \\ \gamma_{2i} \end{pmatrix} \sim \mathcal{N}\left(\mathbf{0}; \bar{\mathbf{\Gamma}} = \begin{pmatrix} \bar{\Gamma}_1 & \bar{\Gamma}_{12} \\ \bar{\Gamma}_{12}^{T} & \bar{\Gamma}_2 \end{pmatrix}\right) \qquad (6)$$

and

$$\varepsilon_i = \begin{pmatrix} \varepsilon_{1i} \\ \varepsilon_{2i} \end{pmatrix} \sim \mathcal{N}\left(\mathbf{0}; \bar{\Sigma} = \begin{pmatrix} \sigma_1^2 \mathrm{I}_{N_{1i}} & 0 \\ 0 & \sigma_2^2 \mathrm{I}_{N_{2i}} \end{pmatrix}\right) \qquad (7)$$

N_{1i} and N_{2i} are the dimensions of y_{1i} and y_{2i}, respectively. Here, we assume that the marginal residuals are homoscedastic $(\mathbb{V}[\varepsilon_{ki}] = \sigma_k^2 I_{N_{ki}}, k = 1, 2)$, but the residual covariance matrices can be of full form as in Eq (4). In order to make clear the relation between the model described by Eqs (2), (3) and (4), and its version expressed by Eqs (5), (6) and (7), we propose below a detailed example.

Detailed Example

We place ourselves in the case of longitudinal data where we observe two response variables y_1 and y_2 which are respectively the weight (kg) and the size (cm) of infants according to the score (V_2) of the quality of their food as well as the quality score (V_1) of their mothers' food. Infants are n = 3 girls (sex = F) and boys (sex = M) who are monitored over time. The dataset is presented by Table 1.

Table 1: Example of data

Response variables						
subject	**age**	**sex**	V_1	V_2	y_1	y_2
1	0	F	63.76	38.16	3.14	47.82
2	4	M	100.88	41.46	4.87	64.02
1	6	F	60.98	41.37	8.43	73.21
3	0	M	93.24	48.76	2.82	44.93
2	7	M	101.95	44.79	8.03	89.54
2	10	M	99.24	48.17	10.08	92.14
1	16	F	NA	44.79	13.96	86.12
3	9	M	88.38	47.91	8.47	86.42

Suppose that the model at each of two dimensions has one random intercept by subject (infant) and one random slope by subject in the direction of the infant's age (in months). For example, considering an identifiability constraint covering the sex variable whose level F is the reference, the bivariate linear mixed model can be written as follows:

$$y_1 = \underbrace{(\mathbb{1}\ V_1\ \text{sex} = M)}_{=X_1}\beta_1 + Z_1\gamma_1 + \varepsilon_1$$

$$y_2 = \underbrace{(\mathbb{1}\ V_2\ \text{sex} = M)}_{=X_2}\beta_2 + Z_2\gamma_2 + \varepsilon_2 \qquad (8)$$

where, explicitly

$$X_1 = \begin{pmatrix} 1 & 63.76 & 0 \\ 1 & 100.88 & 1 \\ 1 & 60.98 & 0 \\ 1 & 93.24 & 1 \\ 1 & 101.95 & 1 \\ 1 & 99.24 & 1 \\ 1 & 88.38 & 1 \end{pmatrix}; X_2 = \begin{pmatrix} 1 & 38.16 & 0 \\ 1 & 41.46 & 1 \\ 1 & 41.37 & 0 \\ 1 & 48.76 & 1 \\ 1 & 44.79 & 1 \\ 1 & 48.17 & 1 \\ 1 & 44.79 & 0 \\ 1 & 47.91 & 1 \end{pmatrix} \quad (9)$$

$$Z_1 = \begin{pmatrix} 1 & 0 & 0 & 0 & 0 & 0 \\ 0 & 0 & 1 & 4 & 0 & 0 \\ 1 & 6 & 0 & 0 & 0 & 0 \\ 0 & 0 & 0 & 0 & 1 & 0 \\ 0 & 0 & 1 & 7 & 0 & 0 \\ 0 & 0 & 1 & 10 & 0 & 0 \\ 0 & 0 & 0 & 0 & 1 & 9 \end{pmatrix}; Z_2 = \begin{pmatrix} 1 & 0 & 0 & 0 & 0 & 0 \\ 0 & 0 & 1 & 4 & 0 & 0 \\ 1 & 6 & 0 & 0 & 0 & 0 \\ 0 & 0 & 0 & 0 & 1 & 0 \\ 0 & 0 & 1 & 7 & 0 & 0 \\ 0 & 0 & 1 & 10 & 0 & 0 \\ 1 & 16 & 0 & 0 & 0 & 0 \\ 0 & 0 & 0 & 0 & 1 & 9 \end{pmatrix} \quad (10)$$

$\dim(X_1)=7\times3$ and $\dim(X_2)=8\times3$; $\dim(Z_1)=7\times6$ and $\dim(Z_2)=8\times6$

In the present example we have $\dim(X_1) \neq \dim(X_2)$ and $\dim(Z_1) \neq \dim(Z_2)$ due to the presence of the NA (Not Available) within the values of the variable V_1. Removing information related to this NA in the dimension 1 of the model does not affect its dimension 2.

$$\mathbf{\Gamma} = \begin{pmatrix} \Gamma_1 & \Gamma_{12} \\ \Gamma_{12}^{\top} & \Gamma_2 \end{pmatrix}$$

with,

$$\Gamma_1 = \begin{pmatrix} \eta_1^2 & \rho_\eta\eta_1\eta_2 & . & . & . & . \\ \rho_\eta\eta_1\eta_2 & \eta_2^2 & . & . & . & . \\ . & . & \eta_1^2 & \rho_\eta\eta_1\eta_2 & . & . \\ . & . & \rho_\eta\eta_1\eta_2 & \eta_2^2 & . & . \\ . & . & . & . & \eta_1^2 & \rho_\eta\eta_1\eta_2 \\ . & . & . & . & \rho_\eta\eta_1\eta_2 & \eta_2^2 \end{pmatrix} \quad (11)$$

$$\Gamma_2 = \begin{pmatrix} \tau_1^2 & \rho_\tau\tau_1\tau_2 & . & . & . & . \\ \rho_\tau\tau_1\tau_2 & \tau_2^2 & . & . & . & . \\ . & . & \tau_1^2 & \rho_\tau\tau_1\tau_2 & . & . \\ . & . & \rho_\tau\tau_1\tau_2 & \tau_2^2 & . & . \\ . & . & . & . & \tau_1^2 & \rho_\tau\tau_1\tau_2 \\ . & . & . & . & \rho_\tau\tau_1\tau_2 & \tau_2^2 \end{pmatrix} \quad (12)$$

$$\Gamma_{12} = \begin{pmatrix} \rho_{\eta_1\tau_1}\eta_1\tau_1 & \rho_{\eta_1\tau_2}\eta_1\tau_2 & \cdot & \cdot & \cdot & \cdot \\ \rho_{\eta_2\tau_1}\eta_2\tau_1 & \rho_{\eta_2\tau_2}\eta_2\tau_2 & \cdot & \cdot & \cdot & \cdot \\ \cdot & \cdot & \rho_{\eta_1\tau_1}\eta_1\tau_1 & \rho_{\eta_1\tau_2}\eta_1\tau_2 & \cdot & \cdot \\ \cdot & \cdot & \rho_{\eta_2\tau_1}\eta_2\tau_1 & \rho_{\eta_2\tau_2}\eta_2\tau_2 & \cdot & \cdot \\ \cdot & \cdot & \cdot & \cdot & \rho_{\eta_1\tau_1}\eta_1\tau_1 & \rho_{\eta_1\tau_2}\eta_1\tau_2 \\ \cdot & \cdot & \cdot & \cdot & \rho_{\eta_2\tau_1}\eta_2\tau_1 & \rho_{\eta_2\tau_2}\eta_2\tau_2 \end{pmatrix} \quad (13)$$

ρ_η, ρ_τ, $\rho_{\eta 1}\tau_1$, $\rho_{\eta 1}\tau_2$, $\rho_{\eta 2}\tau_1$, and $\rho_{\eta 2}\tau_2$ lie in [−1, 1]. All other parameters involved in Γ_1, Γ_2 and Γ_{12} are positive real numbers.

Referring to the subject-based version of the model,

$$\bar{\Gamma}_1 = \begin{pmatrix} \eta_1^2 & \rho_\eta\eta_1\eta_2 \\ \rho_\eta\eta_1\eta_2 & \eta_2^2 \end{pmatrix}, \quad \bar{\Gamma}_2 = \begin{pmatrix} \tau_1^2 & \rho_\tau\tau_1\tau_2 \\ \rho_\tau\tau_1\tau_2 & \tau_2^2 \end{pmatrix}, \quad \bar{\Gamma}_{12} = \begin{pmatrix} \rho_{\eta_1\tau_1}\eta_1\tau_1 & \rho_{\eta_1\tau_2}\eta_1\tau_2 \\ \rho_{\eta_2\tau_1}\eta_2\tau_1 & \rho_{\eta_2\tau_2}\eta_2\tau_2 \end{pmatrix},$$

$$\mathbb{V}(\gamma_1) = \bar{\Gamma}_1, \quad \mathbb{V}(\gamma_2) = \bar{\Gamma}_2, \text{ and } \mathrm{Cov}(\gamma_1, \gamma_2) = \bar{\Gamma}_{12}. \quad (14)$$

Then,

$$\Gamma_1 = \mathrm{diag}(\bar{\Gamma}_1, \cdots, \bar{\Gamma}_1), \quad \Gamma_2 = \mathrm{diag}(\bar{\Gamma}_2, \cdots, \bar{\Gamma}_2), \text{ and } \Gamma_{12} = \mathrm{diag}(\bar{\Gamma}_{12}, \cdots, \bar{\Gamma}_{12}). \quad (15)$$

EM Estimation

Let θ be the vector of unknown parameters in β_1, β_2, $\mathbf{\Gamma}$, Σ_1, Σ_2. The EM algorithm requires an initial value of θ and some expressions (estimators) to update until convergence. In the next two subsections we provide these estimators, their initial values and the stopping criterion.

EM-based Estimators of Parameters

Theorem 1. Suppose that $y = (y_1^\top, y_2^\top)^\top$ satisfies the model based on Eqs (2), (3) and (4) and θ the vector of its unknown parameters while θ_{old} is the previous value of θ provided by the EM algorithm. Let $f(\boldsymbol{y},\boldsymbol{\gamma}|\theta)$ be the joint density function of $\boldsymbol{y}$ and $\boldsymbol{\gamma}$ given θ, and $Q(\theta|\theta_{old}) = \mathbb{E}[\log f(\boldsymbol{y}, \gamma|\theta)|\boldsymbol{y}, \theta_{old}]$. Let M be the mapping $\theta_{old} \mapsto M(\theta_{old}) = \hat{\theta}$ such that:

$$M(\theta_{old}) = \arg\max_{\theta} Q(\theta|\theta_{old}) \quad (16)$$

Then, the EM-based estimator of θ, i.e. $\hat{\theta}$, is expressed through:
for k ∈ {1, …, m},

$$\hat{\beta}_k = (X_k^\top \Sigma_k^{-1} X_k)^{-1} X_k^\top \Sigma_k^{-1} (y_k - Z_k \mathbb{E}[\gamma_k|y, \theta_{old}]), \quad (17)$$

$$\hat{\mathbf{\Gamma}} = \mathbb{V}[\gamma|\boldsymbol{y}, \theta_{old}] + \mathbb{E}[\gamma|\boldsymbol{y}, \theta_{old}]\mathbb{E}[\gamma|\boldsymbol{y}, \theta_{old}]^\top, \quad (18)$$

$$\hat{\Sigma}_k = Z_k \mathbb{V}[\gamma_k|\boldsymbol{y},\theta_{old}]Z_k^\top + (y_k - X_k\beta_k - Z_k\mathbb{E}[\gamma_k|\boldsymbol{y},\theta_{old}])(y_k - X_k\beta_k - Z_k\mathbb{E}[\gamma_k|\boldsymbol{y},\theta_{old}])^\top, \tag{19}$$

where,

$$\mathbb{E}[\gamma_k|\boldsymbol{y},\theta_{old}] = Cov(\gamma_k,\boldsymbol{y}|\theta_{old})\mathbb{V}[\boldsymbol{y}|\theta_{old}]^{-1}(\boldsymbol{y} - \mathbb{E}[\boldsymbol{y}|\theta_{old}]), \tag{20}$$

$$\mathbb{V}[\gamma_k|\boldsymbol{y},\theta_{old}] = \boldsymbol{\Gamma}_k - Cov(\gamma_k,\boldsymbol{y}|\theta_{old})\mathbb{V}[\boldsymbol{y}|\theta_{old}]^{-1}Cov(\gamma_k,\boldsymbol{y}|\theta_{old})^\top \tag{21}$$

and

$$\mathbb{V}[\boldsymbol{y}|\theta_{old}] = \begin{pmatrix} Z_1\Gamma_1 Z_1^\top + \Sigma_1 & Z_1\Gamma_{12}Z_2^\top \\ Z_2\Gamma_{12}^\top Z_1^\top & Z_2\Gamma_2 Z_2^\top + \Sigma_2 \end{pmatrix}, \quad Cov(\gamma,\boldsymbol{y}|\theta_{old}) = \begin{pmatrix} Z_1\Gamma_1 & Z_1\Gamma_{12} \\ Z_2\Gamma_{12}^\top & Z_2\Gamma_2 \end{pmatrix}^\top, \tag{22}$$

$$Cov(\gamma_1,\boldsymbol{y}|\theta_{old}) = \begin{pmatrix} Z_1\Gamma_1 \\ Z_2\Gamma_{12}^\top \end{pmatrix}^\top, \quad Cov(\gamma_2,\boldsymbol{y}|\theta_{old}) = \begin{pmatrix} Z_1\Gamma_{12} \\ Z_2\Gamma_2 \end{pmatrix}^\top. \tag{23}$$

proof. For $k \in \{1,\ldots,m\}, \hat{\beta}_k, \hat{\Sigma}_k$ and $\hat{\boldsymbol{\Gamma}}$ optimize the quantity:

$$Q(\theta|\theta_{\text{old}}) = \mathbb{E}[\log f(\boldsymbol{y},\gamma|\theta)|\boldsymbol{y},\theta_{\text{old}}] \tag{24}$$

where f($\boldsymbol{y}$,γ|θ) is the joint density function of the observed data $\boldsymbol{y}$ and the random effect $\boldsymbol{\gamma}$. In the case of m = 2, we have:

$$\begin{aligned} f(\boldsymbol{y},\gamma|\theta) &= f(\boldsymbol{y}|\gamma,\theta)f(\gamma|\theta) \\ &= f(y_1|\gamma_1,\theta)f(y_2|\gamma_2,\theta)f(\gamma|\theta) \\ &= (2\pi)^{-(N_1+N_2+q)/2}|\Sigma_1|^{-1/2}|\Sigma_2|^{-1/2}|\boldsymbol{\Gamma}|^{-1/2}\exp\{-\frac{1}{2}\gamma^\top\boldsymbol{\Gamma}^{-1}\gamma \\ &\quad -\frac{1}{2}(y_1 - X_1\beta_1 - Z_1\gamma_1)^\top\Sigma_1^{-1}(y_1 - X_1\beta_1 - Z_1\gamma_1) \\ &\quad -\frac{1}{2}(y_2 - X_2\beta_2 - Z_2\gamma_2)^\top\Sigma_2^{-1}(y_2 - X_2\beta_2 - Z_2\gamma_2)\} \end{aligned} \tag{25}$$

Since f is a multivariate Gaussian, using the dominated convergence theorem and the derivative under the integral sign, the differential of $Q(\theta|\theta_{old})$ yields:

$$\begin{aligned} dQ(\theta|\theta_{\text{old}}) &= \mathbb{E}[-\frac{1}{2}\text{tr}\{\Sigma_1^{-1}d\Sigma_2 + \Sigma_2^{-1}d\Sigma_2 + \boldsymbol{\Gamma}^{-1}d\boldsymbol{\Gamma}\} \\ &\quad -\frac{1}{2}\text{tr}\{-2(y_1 - X_1\beta_1 - Z_1\gamma_1)^\top\Sigma_1^{-1}X_1 d\beta_1 \\ &\quad -\Sigma_1^{-1}(y_1 - X_1\beta_1 - Z_1\gamma_1)(y_1 - X_1\beta_1 - Z_1\gamma_1)^\top\Sigma_1^{-1}d\Sigma_1\} \\ &\quad -\frac{1}{2}\text{tr}\{-2(y_2 - X_2\beta_2 - Z_2\gamma_2)^\top\Sigma_2^{-1}X_2 d\beta_2 \\ &\quad -\Sigma_2^{-1}(y_2 - X_2\beta_2 - Z_2\gamma_2)(y_2 - X_2\beta_2 - Z_2\gamma_2)^\top\Sigma_2^{-1}d\Sigma_2\} \\ &\quad -\frac{1}{2}\text{tr}\{-\gamma^\top\boldsymbol{\Gamma}^{-1}d\boldsymbol{\Gamma}\boldsymbol{\Gamma}^{-1}\gamma\}|\boldsymbol{y},\theta_{\text{old}}] \end{aligned} \tag{26}$$

$$
\begin{aligned}
= & -\frac{1}{2}\mathrm{tr}\{\Sigma_1^{-1}\mathrm{d}\Sigma_2 + \Sigma_2^{-1}\mathrm{d}\Sigma_2 + \boldsymbol{\Gamma}^{-1}\mathrm{d}\boldsymbol{\Gamma}\} \\
& +\mathrm{tr}(y_1 - X_1\beta_1 - Z_1\mathbb{E}[\gamma_1|\boldsymbol{y},\theta_{\mathrm{old}}])^{\top}\Sigma_1^{-1}X_1\mathrm{d}\beta_1 \\
& +\frac{1}{2}\mathrm{tr}\{\Sigma_1^{-1}\mathbb{E}[(y_1 - X_1\beta_1 - Z_1\gamma_1)^{\top}(y_1 - X_1\beta_1 - Z_1\gamma_1)|\boldsymbol{y},\theta_{\mathrm{old}}]\Sigma_1^{-1}\mathrm{d}\Sigma_1\} \\
& +\mathrm{tr}(y_2 - X_2\beta_2 - Z_2\mathbb{E}[\gamma_2|\boldsymbol{y},\theta_{\mathrm{old}}])^{\top}\Sigma_2^{-1}X_2\mathrm{d}\beta_2 \\
& +\frac{1}{2}\mathrm{tr}\{\Sigma_2^{-1}\mathbb{E}[(y_2 - X_2\beta_2 - Z_2\gamma_2)^{\top}(y_2 - X_2\beta_2 - Z_2\gamma_2)|\boldsymbol{y},\theta_{\mathrm{old}}]\Sigma_2^{-1}\mathrm{d}\Sigma_2\} \\
& +\mathrm{tr}\frac{1}{2}\{\boldsymbol{\Gamma}^{-1}\mathbb{E}[\gamma\gamma^{\top}|\boldsymbol{y},\theta_{\mathrm{old}}]\boldsymbol{\Gamma}^{-1}\mathrm{d}\boldsymbol{\Gamma}\}
\end{aligned}
\tag{27}
$$

Partial derivatives of $Q(\theta|\theta_{old})$ yield:
for $k \in \{1, \ldots, m\}$,

$$\frac{\partial Q(\theta|\theta_{\mathrm{old}})}{\partial \beta_k} = (y_k - X_k\beta_k - Z_k\mathbb{E}[\gamma_k|\boldsymbol{y},\theta_{\mathrm{old}}])^{\top}\Sigma_k^{-1}X_k,$$

$$\frac{\partial Q(\theta|\theta_{\mathrm{old}})}{\partial \Sigma_k} = -\frac{1}{2}\Sigma_k^{-1} + \Sigma_k^{-1}\mathbb{E}[(y_k - X_k\beta_k - Z_k\gamma_k)^{\top}(y_k - X_k\beta_k - Z_k\gamma_k)|\boldsymbol{y},\theta_{\mathrm{old}}]\Sigma_k^{-1}$$

and

$$\frac{\partial Q(\theta|\theta_{\mathrm{old}})}{\partial \boldsymbol{\Gamma}} = \frac{1}{2}(\boldsymbol{\Gamma}^{-1}\mathbb{E}[\gamma\gamma^{\top}|\boldsymbol{y},\theta_{\mathrm{old}}]\boldsymbol{\Gamma}^{-1} - \boldsymbol{\Gamma}^{-1}).$$

We then get EM-based estimators by setting these partial derivatives equal to zero. $\mathbb{E}[\gamma_k|\boldsymbol{y},\theta_{\mathrm{old}}]$ and $\mathbb{V}[\gamma_k|\boldsymbol{y},\theta_{\mathrm{old}}]$ are straightforward to get since $(\gamma_k^{\top},\boldsymbol{y}^{\top})^{\top}$ is a multivariate Gaussian.

Initialization and Stopping Criterion of the Algorithm

Various ways exist for obtaining starting values for $\hat{\beta}_k, \hat{\boldsymbol{\Gamma}}, \hat{\Sigma}_k$, for $k = 1, \cdots, m$. Taking inspiration from [75] and [60], we have separately fitted each dimension of the model by using the lme4 package [76] under the R software and have used marginal estimated parameters to initialize $\hat{\beta}_k$ and $\hat{\Sigma}_k$. We then keep the expected random effects $\tilde{\gamma} = \hat{\mathbb{E}}[\gamma|\boldsymbol{y}]$ to initialize $\hat{\boldsymbol{\Gamma}}$ by

$$\frac{1}{n-1}\sum_{i=1}^{n}\tilde{\gamma}_i\tilde{\gamma}_i^{\top} \tag{28}$$

The stopping criterion is related to the relative error of the components of θ as follows:

$$\max_j \left| \frac{\theta_j^{(r)} - \theta_j^{(r+1)}}{\theta_j^{(r+1)}} \right| < \text{tol} \tag{29}$$

where (r) is the iteration index and θ_j the jth component of θ. tol = 10^{-5} seems to work well in practice.

Test of the Significance of $\widehat{Cor}(\gamma_1, \gamma_2)$

After the calculation of $\boldsymbol{\Gamma}$ on dataset, we sometimes need to investigate if the correlation between marginal random effects is statistically significant, by testing H_0: $Cor(\gamma_1, \gamma_2) = \mathbf{0}$ against H_1: $Cor(\gamma_1, \gamma_2) \neq \mathbf{0}$. The result of this test can help to decide if the bivariate analysis is justified or not. We perform the likelihood ratio (LR) test to choose between H_0 and H_1. We calculated S, the statistic of the likelihood ratio test.

$$S = -2\log\left(\frac{L(\theta|\text{data},\ H_0)}{L(\theta|\text{data},\ H_1)}\right) \tag{30}$$

where $L(\theta|\text{data}, H0)$ and $L(\theta|\text{data}, H1)$ are the likelihood of θ under H_0 and H_1, respectively. Under suitable and standard conditions, $S \sim \chi^2(df)$, asymptotically, under H_0 [77]. With df the difference in the number of parameters between $L(\theta|\text{data}, H0)$ and $L(\theta|\text{data}, H1)$.

RESULTS AND DISCUSSION

Simulation Studies

In this section, simulation studies are used to investigate the computational properties of the EM-based estimators. For the sake of simplicity, these simulation studies are conducted using simulated bivariate longitudinal data sets. Through these studies, we pursue two objectives: the first is to assess the accuracy of parameter estimates and the second is to analyze the power of the likelihood ratio test performed via these EM-based estimators. In the following paragraph, we explain how we choose the parameters that have been used to simulate the working longitudinal data sets.

The Working Data Sets

We suppose that we are following up a sample of subjects where the goal is to evaluate how the growth of the weight and the height of the individuals of this population are jointly explained by the sex, the score of nutrition (Nscore) and the age. We randomly choose through a uniform distribution the score of nutrition between 20 and 50, and the age between 18 and 37, using the R software. All the analysis in this paper are done using the R software. The subject's sex is also randomly chosen. The model under which we simulate the data sets is the following:

n indicating the total number of subjects, for i = 1, …, n

$$\begin{aligned} \text{weight}_i &= (\mathbb{1}_{n_i}, \text{sex}_i, \text{Nscore}_i, \text{age}_i)\beta_1 + (\mathbb{1}_{n_i}, \text{Nscore}_i)\gamma_{1i} + \varepsilon_{1i} \\ \text{height}_i &= (\mathbb{1}_{n_i}, \text{sex}_i, \text{Nscore}_i, \text{age}_i)\beta_2 + (\mathbb{1}_{n_i}, \text{Nscore}_i)\gamma_{2i} + \varepsilon_{2i} \end{aligned} \quad (31)$$

with

$$\gamma_i = \begin{pmatrix} \gamma_{1i} \\ \gamma_{2i} \end{pmatrix} \sim \mathcal{N}(\mathbf{0}, \bar{\mathbf{\Gamma}}), \varepsilon_{1i} \sim \mathcal{N}(0, \sigma_1^2 \mathrm{I}_{n_i}), \varepsilon_{2i} \sim \mathcal{N}(0, \sigma_2^2 \mathrm{I}_{n_i}), \gamma_i \perp \varepsilon_{1i} \perp \varepsilon_{2i} \quad (32)$$

The random effect related to the dependent variable 'weight' or 'height' is a vector composed by one random intercept and one random slope in the direction of the covariate 'Nscore'. The total number of observations is denoted by N.

We randomly choose β_1, β_2, σ_1 and σ_2 whose values are in the first column of Table 2. $\bar{\Gamma}$ is also randomly chosen such that it is positive definite, with the following form:

$$\bar{\mathbf{\Gamma}} = \begin{pmatrix} \eta_1^2 & \rho_\eta \eta_1 \eta_2 & \rho \eta_1 \tau_1 & \rho \eta_1 \tau_2 \\ \rho_\eta \eta_1 \eta_2 & \eta_2^2 & \rho \eta_2 \tau_1 & \rho \eta_2 \tau_2 \\ \rho \eta_1 \tau_1 & \rho \eta_2 \tau_1 & \tau_1^2 & \rho_\tau \tau_1 \tau_2 \\ \rho \eta_1 \tau_2 & \rho \eta_2 \tau_2 & \rho_\tau \tau_1 \tau_2 & \tau_2^2 \end{pmatrix} \quad (33)$$

The covariance between the random effects γ_1 and γ_2 is set, intentionally,

$$Cov(\gamma_1, \gamma_2) = \rho \begin{pmatrix} \eta_1 \tau_1 & \eta_1 \tau_2 \\ \eta_2 \tau_1 & \eta_2 \tau_2 \end{pmatrix} \quad (34)$$

in order to be able to decrease or increase the correlation between the marginal random effects γ_1 and γ_2, by changing the value of ρ, without losing the positive definiteness of $\bar{\Gamma}$. This property of $\bar{\Gamma}$ will be used to assess the power of the likelihood ratio test through simulations, by changing the value of ρ. We simulate 1000 data sets with $\rho = 0.8$, in order to assess the accuracy of estimates using the EM-based estimators. With $\rho = 0.8$, the randomly chosen $\bar{\Gamma}$ is

$$\bar{\boldsymbol{\Gamma}} = \begin{pmatrix} 27.77 & 18.80 & 41.70 & 4.93 \\ 18.80 & 36.00 & 47.47 & 5.62 \\ 41.70 & 47.47 & 97.81 & 8.91 \\ 4.93 & 5.62 & 8.91 & 1.37 \end{pmatrix} \quad (35)$$

Table 2: Comparative table of true values of parameters and estimates based on 1000 replications using true values of parameters

Parameter	Value	Empirical mean	Empirical Sd	Bias
β_1	50.67	50.669	0.763	0.000
	-4.80	-4.779	0.811	0.021
	14.00	14.012	0.345	0.012
	2.70	2.700	0.016	0.000
β_2	13.20	13.263	1.077	0.063
	-2.80	-2.796	1.186	0.003
	27.00	27.000	0.068	0.000
	1.70	1.699	0.019	0.000
σ_1	5.80	5.796	0.062	0.003
σ_2	7.60	7.602	0.082	0.002

Empirical Accuracy of the Estimates

The 1000 data sets simulated in order to assess the accuracy of the estimates performed using the EM-based estimators contain N = 5000 observations provided by n = 300 independent subjects.

The mean and the standard deviation of the 1000 estimates are presented, respectively, in the second and the third column of the Table 2. The bias of the parameter estimates, which is the absolute difference between the true value of the parameter and the mean of the 1000 estimates, is calculated as measure of performance. These bias are contained in the forth column of the Table 2.

$\hat{\mathbb{E}}[\hat{\bar{\boldsymbol{\Gamma}}}]$, $\hat{\sigma}_{\hat{\Gamma}}$ and $\text{Bias}(\hat{\bar{\boldsymbol{\Gamma}}})$ (Eqs (36), (37)) contain, respectively, the empirical mean, the empirical standard deviation and the empirical bias of $\bar{\boldsymbol{\Gamma}}$.

$$\hat{\mathbb{E}}[\hat{\bar{\boldsymbol{\Gamma}}}] = \begin{pmatrix} 27.86 & 18.71 & 41.19 & 4.93 \\ 18.71 & 35.73 & 47.09 & 5.58 \\ 41.19 & 47.09 & 95.93 & 8.87 \\ 4.93 & 5.58 & 8.87 & 1.36 \end{pmatrix}; \quad \hat{\sigma}_{\hat{\Gamma}} = \begin{pmatrix} 5.32 & 2.99 & 6.37 & 0.63 \\ 2.98 & 2.86 & 5.11 & 0.50 \\ 6.37 & 5.11 & 13.73 & 0.97 \\ 0.63 & 0.50 & 0.97 & 0.11 \end{pmatrix} \quad (36)$$

$$\text{Bias}(\hat{\boldsymbol{\Gamma}}) \begin{pmatrix} 0.085 & 0.090 & 0.515 & 0.001 \\ 0.089 & 0.265 & 0.382 & 0.045 \\ 0.514 & 0.382 & 1.881 & 0.039 \\ 0.001 & 0.045 & 0.039 & 0.007 \end{pmatrix} \quad (37)$$

The bias contained in the estimates of β_k and σ_k ranges from 0.000 to 0.063 (Table 2), and the bias contained in the estimates of the entries of $\bar{\boldsymbol{\Gamma}}$ ranges from 0.001 to 1.881 (Eq 37). These results show that $\hat{\beta}_k$ and $\hat{\sigma}_k$ (i.e. $\hat{\Sigma}_k$) seem unbiased when $\hat{\boldsymbol{\Gamma}}$ is biased.

The estimates of $\bar{\boldsymbol{\Gamma}}$ appear to be poorer than the estimates of all other parameters. In order to investigate which entries of $\bar{\boldsymbol{\Gamma}}$ are particularly poorly estimated, we calculate the coefficients of variation (CV) of these entries. The CV computed here is obtained by dividing the standard deviation of the estimates by the true value of each entry of $\bar{\boldsymbol{\Gamma}}$. The CVs give an idea of the variability of estimates around the true values and enable to compare these variabilities between them. A particularly large value of CV could lead us to suspect that the corresponding input is particularly poorly estimated. Here, the CV ranges from 0.08 to 0.19, and is represented by the Fig 1 for more visibility. Given these CV values, it seems that none of the entries of $\bar{\boldsymbol{\Gamma}}$ is particularly poorly estimated.

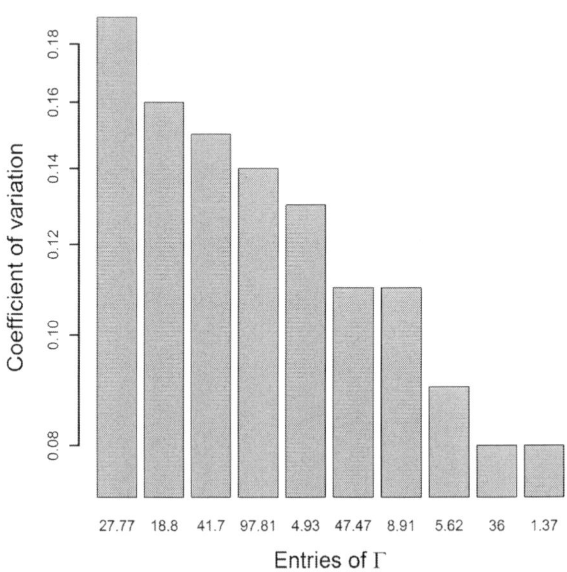

Figure 1: Coefficients of variation of entries of Γ.

N = 5000 observations and n = 300 subjects.

Deep Investigation on the Estimates' Accuracy

Here, we compute the Mean Square Error (MSE) of the EM-based estimators with N = 600,1000 and N = 3000 across n = 50, 60, 100 and 300 to investigate how both values of n and N affect the quality of the estimates. For each value of n and N, we simulate 1000 data sets on which we estimate the model parameters and compute the MSE of these estimates.

Without surprise, Table 3 shows that the quality of estimates is clearly improved when both n and N grow. Estimations performed on dataset containing N = 3000 observations are more accurate than those performed with N = 600, observing the maximum value of the MSE in each case. For N = 600, information contained in Table 3 shows that the MSE related to n = 60 (60 subjects) are better than those related to n = 300. This result suggests a good tradeoff between the number of subjects and the total number of observations in order to have accurate estimates, especially if the number of observations is not very high. Once again, it appears that $\hat{\Gamma}$ (Table 3) has the highest MSE for all values of n and N.

Table 3: Mean Square Error of EM-based estimator with 95% CI estimated on 1000 replications for various values of n **and** N

Parameter	*n*	*N* = **600**	*N* = **1000**	*N* = **3000**
β_1	50	1.27 (0.01 - 4.64)	1.10 (0.00 - 3.83)	0.71 (0.00 - 2.57)
	60	1.38 (0.00 - 5.56)	0.99 (0.00 - 3.90)	0.65 (0.00 - 2.44)
	100	1.31 (0.00 - 4.79)	0.87 (0.00 - 3.41)	0.46 (0.00 - 1.58)
	300	1.64 (0.00 - 5.93)	0.80 (0.00 - 3.23)	0.29 (0.00 - 1.13)
β_2	50	2.13 (0.01 - 7.22)	1.73 (0.00 - 6.21)	1.08 (0.00 - 4.17)
	60	2.12 (0.00 - 8.06)	1.63 (0.01 - 6.66)	0.95 (0.00 - 3.52)
	100	1.88 (0.00 - 7.14)	1.29 (0.00 - 4.76)	0.62 (0.00 - 2.24)
	300	2.30 (0.00 - 8.54)	1.23 (0.00 - 4.57)	0.43 (0.00 - 1.65)
σ_1	50	0.03 (0.00 - 0.10)	0.10 (0.00 - 0.07)	0.01 (0.00 - 0.02)
	60	0.04 (0.00 - 0.14)	0.02 (0.00 - 0.07)	0.01 (0.00 - 0.02)
	100	0.04 (0.00 - 0.14)	0.02 (0.00 - 0.08)	0.01 (0.00 - 0.02)
	300	0.06 (0.00 - 0.22)	0.03 (0.00 - 0.11)	0.01 (0.00 - 0.02)

σ_2	50	0.05 (0.00 - 0.18)	0.03 (0.00 - 0.12)	0.01 (0.00 - 0.03)
	60	0.06 (0.00 - 0.21)	0.03 (0.00 - 0.12)	0.01 (0.00 - 0.03)
	100	0.06 (0.00 - 0.24)	0.04 (0.00 - 0.15)	0.01 (0.00 - 0.03)
	300	0.09 (0.00 - 0.36)	0.04 (0.00 - 0.18)	0.01 (0.00 - 0.05)
$\bar{\Gamma}$	50	400.91 (1.52 - 1274.32)	670.46 (2.36 - 2536.73)	489.07 (2.02 - 1840.60)
	60	706.45 (2.58 - 2497.42)	620.34 (2.70 - 2293.58)	408.79 (0.85 - 1597.77)
	100	701.50 (3.79 - 2747.70)	477.55 (2.08 - 1839.26)	283.25 (1.97 - 1023.77)
	300	798.28 (3.65 - 2603.65)	547.83 (3.03 - 1721.46)	199.54 (0.81 - 736.08)

The Bivariate Likelihood Ratio Test

Considering the random effects covariance matrix $\bar{\Gamma}$ (see Eq (35)), the related correlation matrix is

$$\begin{pmatrix} 1.00 & 0.59 & 0.80 & 0.80 \\ 0.59 & 1.00 & 0.80 & 0.80 \\ 0.80 & 0.80 & 1.00 & 0.77 \\ 0.80 & 0.80 & 0.77 & 1.00 \end{pmatrix}. \tag{38}$$

That is, the matrix of the correlations between the marginal random effects (i.e., the random effects related to the two dependent variables) is

$$\mathrm{Cor}(\gamma_1, \gamma_2) = \begin{pmatrix} 0.80 & 0.80 \\ 0.80 & 0.80 \end{pmatrix} \tag{39}$$

whereas the estimate (on one of the previous simulated data) of this matrix, $\mathrm{Cor}(\gamma_1, \gamma_2)$, of the correlations between the marginal random effects, is

$$\widehat{\mathrm{Cor}}(\gamma_1, \gamma_2) = \begin{pmatrix} 0.77 & 0.78 \\ 0.90 & 0.74 \end{pmatrix} \tag{40}$$

If we decide to test H_0: $\mathrm{Cor}(\gamma_1, \gamma_2) = \mathbf{0}$ against H_1: $\mathrm{Cor}(\gamma_1, \gamma_2) \neq \mathbf{0}$ in the case of these simulated data, we must know the distribution of the LR statistic S. In order to approximate the distribution of S, under H_0, we proceed to an extensive simulation study in the next paragraph.

Empirical Distribution of S under H_0

In this paragraph, our goal is to investigate about the empirical law of the LR statistic S, under H_0, when the size N of the data set increases. The simulated data sets used in this paragraph are also of bivariate longitudinal type, with N the total number of observations coming from n subjects. We choose N as an arithmetic sequence ranging from 50 to 2000, where the common difference is 50. We choose n = N/5 as it is sufficient to have two observations per subject for fitting the model. When N/n = 1, the random-effects parameters and the residual variance are unidentifiable [1].

The expected (standard) asymptotic distribution of S, under H_0, is a $\chi^2(4)$. This may be explained by the fact that $Cov(\gamma_1, \gamma_2)$ and its transpose, $Cov(\gamma_1, \gamma_2)^T$, contain four entries, respectively, and $\bar{\Gamma}$ contains $Cov(\gamma_1, \gamma_2)$ and $Cov(\gamma_1, \gamma_2)^T$. Therefore, the difference between the number of entries of $\bar{\Gamma}$ which need to be estimated with L(θ|data,H0) and L(θ|data,H1), respectively, is df = 4. Precisely, the parameters of interest are $\rho_{\eta1}\tau_1$, $\rho_{\eta1}\tau_2$, $\rho_{\eta2}\tau_1$ and $\rho_{\eta2}\tau_2$ (see Eq (14)).

Fig 2 assumes an asymptotic distribution of $\chi^2(4)$ and plots the Kolmogorov-Smirnov test's p-value (at log10 scale) against the total number of observations of the data set that has served to compute the LR statistic S. The blue curve is obtained by applying the empirical Bartlett correction to S and the red curve is obtained without correction. The horizontal dashed line represents log10(0.05). The empirical Bartlett corrected S, say S^B, can be expressed as $\hat{S}_B = df \times S/\hat{\mathbb{E}}[S|H_0]$. This Bartlett correction is applied in order to avoid the small size distortion of the $\chi^2(df)$ distribution, when performing the LR test using a data set of small size [78]. Fig 2 thus helps to investigate how the LR distribution performs in finite and small dimension. It also helps to investigate, in the case of this bivariate correlation test, how the Bartlett correction helps to avoid the small size distortion of the chi-square approximation. As the total number of observations increases, the curves (red and blue) reach the dashed line, gradually. Assuming the $\chi^2(4)$ distribution of S, it seems important to work with a data set containing at least 500 observations coming from at least 2 subjects, and to apply the Bartlett correction in order to avoid the breakdown of the procedure.

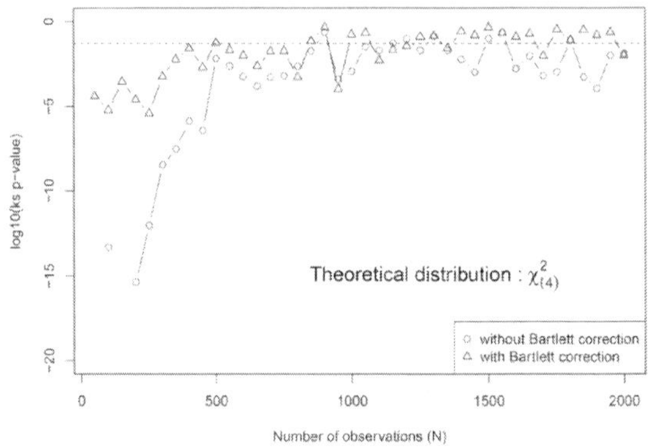

Figure 2: Empirical analysis of the asymptotic distribution of the LR statistic S **under** H_0, using longitudinal data sets (200 replications) with size N ∈ {50, 100, 150, 200, …, 2000} coming from n ∈ {10, 20, 30, …, 400} subjects.

An asymptotic distribution of $\chi^2(4)$ is assumed and the Kolmogorov-Smirnov test's p-value (at log10 scale) is ploted against the total number of observations of the data set that has served to compute the LR statistic S. The blue curve is obtained by applying the empirical Bartlett correction to S and the red curve is obtained without correction. The horizontal dashed line represents log10(0.05).

The type I error is generally controlled by the significance level of 10% (red and blue curves of Fig 3). It is clear that the control is almost full with the Bartlett correction (blue curve of Fig 3).

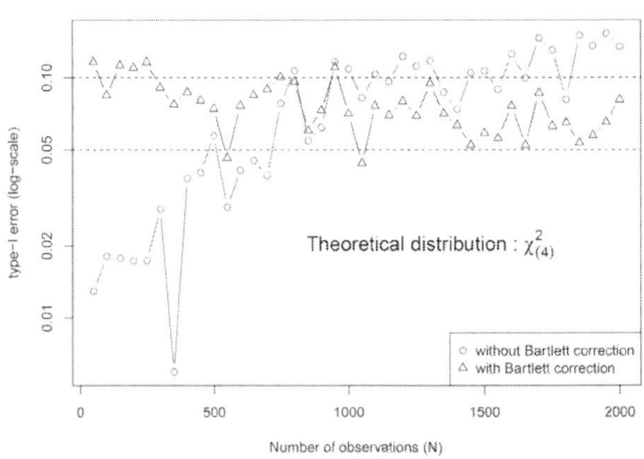

Figure 3: Empirical analysis of the asymptotic distribution of the LR statistic S **under** H_0, using longitudinal data sets (200 replications) with size N ∈ {50, 100, 150, 200, …, 2000} coming from n ∈ {10, 20, 30, …, 400} subjects.

An asymptotic distribution of $\chi^2(4)$ is assumed and the type I error (at log10 scale) is ploted against the total number of observations of the data set that has served to compute the LR statistic S. The blue curve is obtained by applying the empirical Bartlett correction to S and the red curve is obtained without correction. The horizontal dashed lines represent the significance levels of 5% and 10%, respectively.

By simulating 1000 × 3000 realizations of $\chi^2(4)$ distribution, we plot the red sheath represented in Fig 4. This sheath corresponds to the minimum and the maximum of the simulated $\chi^2(4)$ realizations. The blue curve inside the red sheath represents the empirical LR statistics obtained from the 3000 simulated data sets under H_0. This figure (Fig 4) shows that the asymptotic distribution of LR statistic related to the bivariate correlation test is not violated, since the blue curve does not go out of the red sheath.

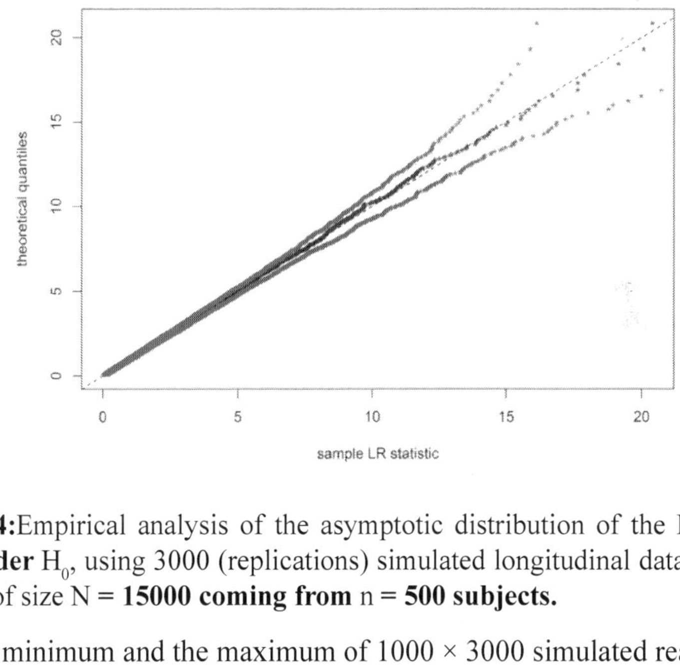

Figure 4: Empirical analysis of the asymptotic distribution of the LR statistic S **under** H_0, using 3000 (replications) simulated longitudinal data sets (under H_0) of size N = **15000 coming from** n = **500 subjects.**

The minimum and the maximum of 1000 × 3000 simulated realizations of $\chi^2(4)$ are used to construct the red sheath. The blue curve represents the LR statistics related to the bivariate correlation test.

Empirical Power of the Bivariate Correlation Test

In order to analyze the power of this likelihood ratio test performed with EM-based estimates, we calculate S on data sets which have been simulated under H_0 and H_1, respectively, leading to what we named S0 and S1 vectors containing the resulting values of S. We then plot a ROC curve with S0 and S1, where S0 is the vector of the cases while S1 contains the controls. We calculate S0 and S1 in different situations where we have changed the value of ρ in the following configuration:

$$Cov(\gamma_1, \gamma_2) = \rho \begin{pmatrix} \eta_1\tau_1 & \eta_1\tau_2 \\ \eta_2\tau_1 & \eta_2\tau_2 \end{pmatrix} \quad (41)$$

We maintain fixed $\eta_1 = 5.27$, $\eta_2 = 6.00$, $\tau_1 = 9.89$, $\tau_2 = 1.17$ and change ρ (∈ {0.1, 0.2, 0.3, …, 0.9}). The number of subjects (n) and the total number of observations (N) have also been modified throughout these simulation studies. In each case, the estimated Area Under Curve (AUC) of the ROC curve with its confidence interval have been recorded to produce Fig 5.

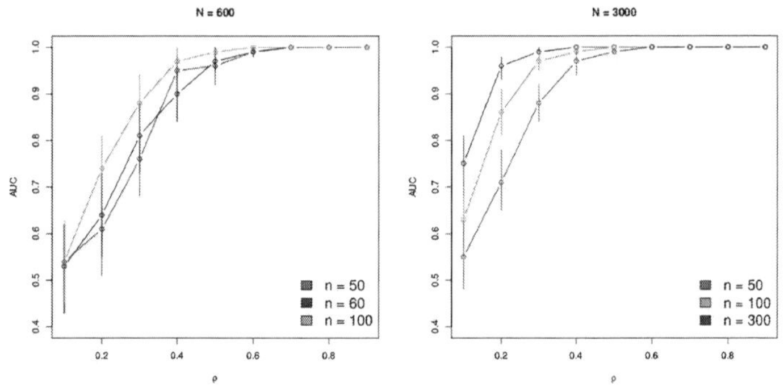

Figure 5: Empirical analysis of the power of the correlation test.

AUC values of ROC curves with their confidence interval computed for different ρ, number of subjects (n) and observations (N). Left panel for N = 600, n = 50, 60, 100. Right panel for N = 3000, n = 50, 100, 300.

With n = 50 subjects, we detect, indeed, a correlation of 0.6 when the total number of observations is N = 3000; in contrast, if the total number of observations is N = 600, we perfectly detect a correlation of 0.7.

Unsurprisingly, confidence intervals of AUC are also more accurate with N = 3000 than with N = 600. With a sufficient number of observations and subjects, weak correlations are easily detected. For example, we perfectly detect a correlation of 0.2 with N = 3000 and n = 300 where AUC = 0.96(0.93 − 0.98) according to Fig 5. However, we detect quite well a correlation of 0.3 with N = 600 and n = 60 where AUC = 0.81(0.73 − 0.88).

In the case where estimates are of a higher quality (because they are performed on data sets having a sufficient number of observations N = 5000 and subjects n = 300), we plot ROC curves with low values of ρ (0.1, 0.2 and 0.3). We then show in Fig 6, the estimated AUC and its 95% confidence interval.

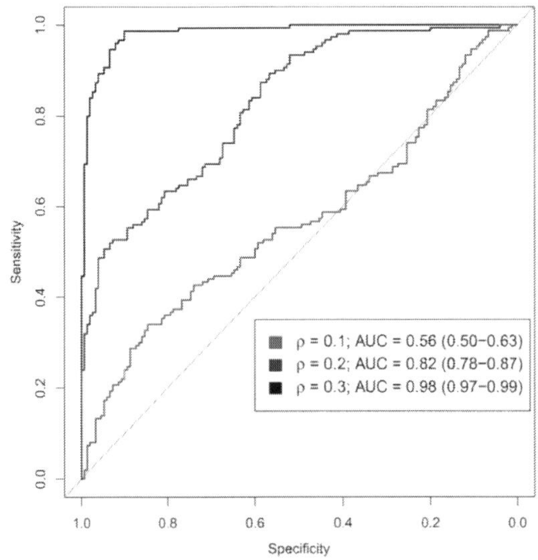

Figure 6: Analysis of the power of the likelihood ratio test performed under EM-estimators.

ROC curves with ρ ∈ {0.1, 0.2, 0.3}. N = 5000 observations, n = 300 subjects, 95% CI on AUC.

Fig 6 shows that EM-based estimators lead to a good power of the bivariate correlation test, when we have a sufficient number of observations and subjects in the longitudinal study case. This goodness of the power of the bivariate correlation test persists when the correlation between marginal random effects is low (about 0.2).

Applications on Real Data Sets

In this section we analyze two data sets by using the likelihood ratio test through the EM-based estimators presented above. The first dataset is of multivariate multilevel type and the second is, specifically, of longitudinal multivariate type.

Application to Education Data in the Netherlands

The data used here are named 'bdf' under the package nlme [13] of the R software. These data contain N = 3776 Grade eight students (aged about eleven years) in n = 208 elementary schools in the Netherlands [79]. These pupils were tested twice (with an interval of one year between grade seven and grade eight) on their proficiency in Dutch language and arithmetics, where the goal was to investigate which characteristics of schools can account for the differences in the effectiveness of schools with regard to pupils' progress in language and arithmetics. Most of the previous analyses of this dataset were concerned with investigating how the language test score depends on the pupil's intelligence, his family's socio-economic status and on related class or school variables. By fitting two independent (separate) models, [79] found that variables in Table 4 have a significant effect on post-test scores (language post-test and arithmetic post-test). These variables are: socio-economic status, intelligence score, age, gender and nationality. They also found a significant random slope related to the language pre-test and to the gender in the language post-test model.

Table 4: Modeling of covariates on post-test achievement in language and arithmetic from [79]

Model	**Language (post)**	**Arithmetic (post)**
Language pre-test	0.567	-
Arithmetic pre-test	-	0.413
Socio-economic status	0.143	0.132
Intelligence score	0.124	0.270
Age	−0.069	−0.081
Gender (female)	0.187	−0.103
Ethnicity (foreign)	n.s.	−0.105

Based on these results from [79] and some of their data (n = 131 schools, N = 2287 pupils; age and ethnicity are not present), we have fitted the bivariate linear mixed-effect model where post-test scores are the response variables and covariates are the pre-test scores, socio-economic status, intelligence score, gender and minority (a factor indicating if the pupil is a member of a minority group). Random intercepts and random slopes related to pre-test scores are integrated to the model on the school level in the configuration shown by the Table 5.

Table 5: Model configuration.

	Variables	
Model parts	**Language (post)**	**Arithmetic (post)**
Fixed effects	Language (pre)	Arithmetic (pre)
	Socio-eco. status	Socio-eco. status
	Intelligence score	Intelligence score
	Gender	Gender
	Minority	Minority
Random effects (school level)	1	1
	Language (pre)	Arithmetic (pre)

Table 6 contains estimated fixed effects and residual standard deviations of the model.

Table 6: Estimated fixed effects and residual standard deviations in the joint bivariate model fitted to school data

	Response variables			
	Language (post)		**Arithmetic (post)**	
Covariates	**Estimate**	**p-value**	**Estimate**	**p-value**
Intercept	4.690		−1.456	
Language (pre)	0.795	0.000	-	-
Arithmetic (pre)	-	-	0.807	0.000
Socio-eco. status	0.101	0.000	0.089	0.000
Intelligence score	0.474	0.000	0.810	0.000
Gender (female)	1.778	0.000	−0.522	0.003
Minority (yes)	−0.302	0.589	−0.545	0.194
σ_1 and σ_2	5.356	4.066		

The estimated covariance matrix Γ^ of the random effects is:

$$\hat{\Gamma} = \begin{pmatrix} 15.15 & -0.54 & 18.20 & -0.27 \\ -0.54 & 0.02 & -0.65 & 0.01 \\ 18.20 & -0.65 & 30.69 & -0.48 \\ -0.27 & 0.01 & -0.48 & 0.01 \end{pmatrix}$$

The null hypothesis, H_0, that the arithmetic post-test score and the language post-test score are independent, is rejected with a p-value of 1.436×10^{-7}. This result justifies a joint analysis of post-scores conditionally on the covariates present in the model and is therefore a supplementary information obtained from the data. The estimated correlation matrix of the random effects is:

$$\hat{\rho} = \begin{pmatrix} 1.00 & -1.00 & 0.84 & -0.82 \\ -1.00 & 1.00 & -0.84 & 0.82 \\ 0.84 & -0.84 & 1.00 & -1.00 \\ -0.82 & 0.82 & -1.00 & 1.00 \end{pmatrix}$$

Table 6 shows that covariates which are significant in the independent models are also significant in the joint model. The Minority covariate is neither significant in the joint model, nor significant in the independent models. $\hat{\Gamma}_{ij}$ and $\hat{\rho}_{ij}$ identify the item which is at the intersection of the row i and the column j of the matrices $\hat{\Gamma}$ and $\hat{\rho}$, respectively. These matrices are filled from the top to the bottom in the order of (Intercept$_{y1}$, Slope$_{y1}$, Intercept$_{y2}$, Slope$_{y2}$).

There is a clear inter-school variability with respect to the post-test scores ($\hat{\Gamma}_{11}$=15.15 and Γ^33=30.69). Everything else being equal, schools that have good scores in arithmetics also have good scores in language ($\hat{\rho}_{31}$=0.84). The schools in which the differential effect of the pre-test score in arithmetics on the post-test score is strongly negative are in average above the average post-test score in language ($\hat{\rho}_{41}$=−0.82); same as with the language pre-test score ($\hat{\rho}_{32}$=−0.84). This confirms that the scores in language and in arithmetics vary in the same direction in schools. The schools in which the differential effect of the pre-test score (in arithmetics or language) on the post-test score is strongly negative are in average above the average post score $(\hat{\rho}_{21} = -1 \text{ and } \hat{\rho}_{43} = -1)$, and vice versa. These schools have strived to bring the level of all pupils above the average. In contrast, pupils with a good initial level maintain their level without becoming excellent. The differential effect of the pre-test score has a very weak variability (arithmetics score: $\hat{\Gamma}_{22} = 0.02$; language score: $\hat{\Gamma}_{44} = 0.01$) and this implies that the pre-test score explains about 0.15% (in arithmetics) and 0.03% (in language) of inter-school variability of post-test scores.

We have fitted the bivariate model without random slopes (with random intercept only) to investigate if it fits more to the data than the model with random slopes, due to the weak variability of these random slopes. The results are presented in Table 7, where the estimated fixed effects and their significance generally remain the same.

Table 7: Results of the model with random intercepts only

	Response variables			
	Language (post)		Arithmetic (post)	
Covariates	**Estimate**	**p-value**	**Estimate**	**p-value**
Intercept	4.698		−1.446	
Language (pre)	0.789	0.000	-	-
Arithmetic (pre)	-	-	0.789	0.000
Socio-eco. status	0.103	0.000	0.093	0.000
Intelligence score	0.480	0.000	0.809	0.000
Gender (female)	1.788	0.000	−0.526	0.002
Minority (yes)	−0.391	0.489	−0.498	0.249
σ_1 and σ_2	5.384	4.091		

The estimated covariance matrix, related to results contained in Table 7 of random effects is

$$\hat{\Gamma} = \begin{pmatrix} 5.15 & 5.41 \\ 5.41 & 7.07 \end{pmatrix} \tag{42}$$

which indicates a correlation of $\hat{\rho}$=0.895 between the random marginal intercepts, confirming a strong positive correlation between post-test scores in arithmetics and language. With a p-value of 8.505×10^{-5}, the likelihood ratio test indicates that the data are more compatible with the model incorporating random intercepts and random slopes at a time.

Fixed effects seem very strong and do not significantly change between the independent and bivariate models. In contrast, a posterior distribution of random effects changes significantly between the independent model and the joint bivariate model. For example, we plot the joint distribution of random effects conditional on the data concerning School 47 in the education dataset under the independent model and the joint bivariate model. Fig 7a shows the joint posterior distribution of random intercepts under the independent model whereas Fig 7b presents the same posterior distribution under the joint bivariate model. A clear difference appears between these two distributions. We notice the same difference between distributions of random intercepts and slopes as shown in Fig 7c and 7d as well as the joint distribution of random slopes in Fig 7e and 7f. The joint bivariate model seems to fit more to the present data and we retain it for their analysis.

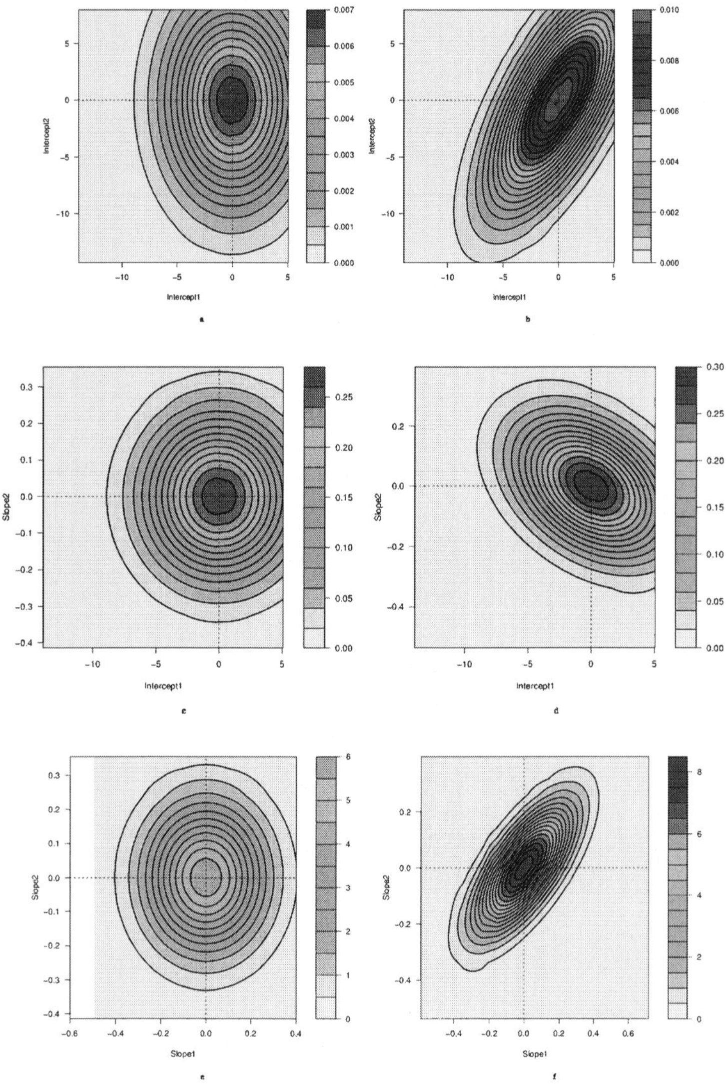

Figure 7: Posterior distributions of random intercepts conditional on the data related to School 47 in the education dataset.

Left panels assume independence across the two dimensions while right panels assume dependence. Top panels for the joint distribution of the random intercepts, middle panels for the joint distribution of random intercept in first dimension and random slope in the second dimension, bottom panels for the joint distribution of the random slopes.

Application to Malaria Immune Response Data in Benin

The data come from a study which was conducted in nine villages (Avamécentre, Gbédjougo, Houngo, Anavié, Dohinoko, Gbétaga, Tori Cada Centre, Zébè and Zoungoudo) of Tori Bossito area (Southern Benin), where P. falciparum is the most common species in the study area (95%) [80] from June 2007 to January 2010. The aim of this study was to evaluate the determinants of malaria incidence in the first months of life of child in Benin. Details of the follow-up procedures have been published elsewhere [81].

Data Description

Mothers (n = 620) were enrolled at delivery and their newborns were actively followed-up during the first year of life. One questionnaire was conducted to gather information on women's characteristics (age, parity, use of Intermittent Preventive Treatment during pregnancy (IPTp) and bed net possession) and on the course of their current pregnancy. Maternal peripheral blood as well as cord blood were collected into Vacutainer® EDTA (Ethylene diaminetetraacetic acid) tubes. At birth, newborn's weight and length were measured by midwives and gestational age was estimated using the Ballard method [82].

During the follow-up of newborns, axillary temperature was measured weekly. Symptomatic malaria cases, defined as fever (>37.5°C) with TBS and/or RDT positive, were treated with an artemisinin-based combination therapy as recommended by the Benin National Malaria Control Program. Systematically, TBS were made every month to detect asymptomatic infections. Every three months, venous blood was sampled to quantify the level of antibody against malaria promised candidate vaccine antigens. The environmental risk of exposure to malaria was modeled for each child, derived from a statistical predictive model based on climatic, entomological parameters, and characteristics of children's immediate surroundings as reported by [83].

Concerning the antibody quantification, two recombinant P. falciparum antigens were used to perform IgG subclass (IgG1 and IgG3) antibody quantification by Enzyme-Linked ImmunoSorbent Assay (ELISA) standard methods developed for evaluating malaria vaccines by the African Malaria Network Trust (AMANET [www.amanet148trust.org]). Protocol was described in detail [84].

Data Analysis

For our analysis, we use some of the data and we rename the proteins used in the study described above, for confidentiality reasons (some important findings are yet to be published). Thus, the proteins we use here, are named A1, A2, B and C, and are related to the antigens IgG1 and IgG3 as mentioned above in the description of the study. A1 and A2 are different domains of the same protein A, and C and D are two different proteins. Information contained in the multivariate longitudinal dataset of malaria are described in the Table 8, where Y denotes a protein which is one of the following:

$$\text{IgG1\_A1, IgG3\_A1, IgG1\_A2, IgG3\_A2, IgG1\_B, IgG3\_B, IgG1\_C, IgG3\_C} \tag{43}$$

Table 8: Variables present in the analyzed dataset

Variable	Description
id	Child ID
conc.Y	concentration of Y
conc_CO.Y	Measured concentration of Y in the umbilical cord blood
conc_M3.Y	Predicted concentration of Y in the child's peripheral blood at 3 months
ap	Placental apposition
hb	Hemoglobin level
inf_trim	Number of malaria infections in the previous 3 months
pred_trim	Quarterly average number of mosquitoes child is exposed to
nutri_trim	Quarterly average nutrition scores

The aim of the analysis of these data is to evaluate the effect of the malaria infection on the child's immune (against malaria). Since the antigens which characterize the child's immune status interact together in the human body, we analyze the characteristics of the joint distribution of these antigens, conditional on the malaria infection and other factors of interest. The dependent variables are then provided by conc.Y (Table 8) which describes the level of the protein Y in the children at 3, 6, 9, 12, 15 and 18 months. All other variables in the Table 8 are covariates. We then have eight dependent variables which describe the longitudinal profile (in the child) of the proteins listed in Eq (43).

In the models that we fit to these data, we specify one random intercept by child and one random slope by child in the direction of the malaria infection. The illustration we do here is to jointly analyze each of the 28 pairs of proteins, in order to investigate if some profiles of proteins are independent, conditional on the configuration of the fitted model. After performing the bivariate correlation test on all 28 bivariate models, the obtained p-values, with a Bonferroni correction, range from 4.16×10^{-33} to 0.932. The p-value 0.932 is the only one which is not significant. This p-value corresponds to the pair of proteins (IgG3_A1, IgG1_B).

To investigate the general configuration of these proteins, in terms of correlations, we build their hierarchical cluster tree using −log(p-value) as dissimilarity. This hierarchical cluster tree is presented by the Fig 8.

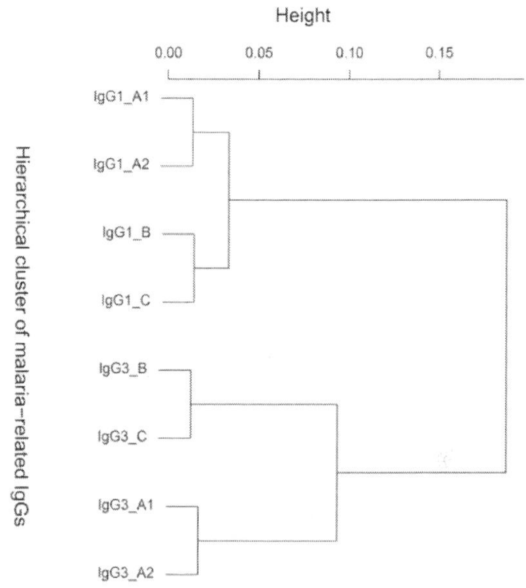

Figure 8: Hierarchical cluster tree on malaria-related proteins.

The branch related to the IgG1 is different from the branch related to the IgG3. In other words, IgG1_A1, IgG1_A2, IgG1_B and IgG1_C are on the same branch which is different from the branch containing IgG3_A1, IgG3_A2, IgG3_B and IgG3_C (Fig 8). Relatively to both IgG1 and IgG3, A1 and A2 go together, and B and C also go together. These results are biologically very consistent, since A1 and A2 are domains of the same protein, and B and C are two different proteins. On the cluster (Fig 8), it also appears that the proteins IgG3_A1 and IgG1_B which are not significantly correlated (according to our bivariate test) are distant. Statistically, the model which may be used to jointly analyze these 8 protein profiles is not probably the model which contains all the 27 significant correlations, avoiding overfitting problems. Based on the results provided by the bivariate correlation test, it may be useful to perform a regularization procedure in the fitting of the full eight-variate model.

CONCLUSION

In the context of the multivariate linear mixed-effects model, we have suggested the more general expressions of the EM-based estimators than those used in the literature to analyze multivariate longitudinal data. These estimators fit the framework of the multivariate multilevel data analysis which, obviously, englobes the multivariate longitudinal data analysis framework. We also have built a likelihood ratio test based on these EM estimators to test the independence of two dimensions of the model. Furthermore, the simulation studies have validated the power of this test and have shown that this is an extremely sensitive test. In the context of longitudinal data, it allows to detect a modest correlation signal with a very small sample ($\rho = 0.3$, AUC = 0.81, with $n = 60$ subjects and $N = 600$ observations). In the simulation studies, the empirical distribution of the likelihood ratio statistic fits the $\chi^2(4)$. The asymptotic properties of likelihood ratio statistics, under nonstandard conditions, have been shown by [85] and [86]. These works have been generalized by [87] to cover a large class of estimation problems which allow sampling from non identically distributed random variables. The asymptotic distribution of the LR statistic derived by [87] is a mixture of chi-squared distributions. In the context of likelihood ratio tests for variance components in linear mixed-effects models, [88] used the results of [87] to prove that the proposed mixture of chi-squared distributions is the actual asymptotic distribution of such LR used as test statistics for null variance components with one or two random effects. Based on these works, Further theoretical investigations may be done to properly find out the asymptotic distribution of the likelihood ratio statistic in the case of this bivariate correlation test. Finally, we have illustrated the usefulness of the test on two different real-life data. The first dataset, which is of multivariate multilevel type, concerns the effects of school and classroom characteristics on pupils' progress in Dutch language and arithmetics, where the scores in language and arithmetics are the two response variables which have been considered. Our method has yielded results that are consistent both with information in existing publications and with a conceptual understanding of the phenomenon. On this dataset, we have highlighted a joint effect between the scores in arithmetics and language within schools in the Netherlands. The second dataset, which is of longitudinal multivariate type, concerns a study of the effect of the malaria infection on the child's immune response in Benin. By jointly analyzing all the pairs of protein profiles of interest, we have plotted a hierarchical cluster tree of these proteins, using the bivariate correlation test. Information contained in this hierarchical cluster tree is consistent with the biological literature related to this issue.

The model as it is written is easily extendable to more dimensions despite a sparsity problem in choosing the parameterization of the covariance matrix or the precision matrix. Probably we could use this two-dimensional dependence test to structure a larger covariance matrix. The bivariate correlation test can help to construct iteratively, using a stepwise procedure, a parsimonious joint model containing all the components of $\boldsymbol{y}$. This stepwise procedure may consist in adding to the constructing model, at each step, the significant correlation between two dependent variables. Using a model selection strategy, the model which fits more to the data will be retained. It could possibly be advantageous to turn to graphical LASSO type approaches to make a penalized estimation of this covariance (or precision) matrix. We could also resort to the rapid optimization methods such as that implemented in the lme4 [76] package, given the slow pace of the EM algorithm. It would be useful to assess the interest of this method compared to some heuristics such as the one which consists in setting one marginal response variable as a covariate of the other(s).

ACKNOWLEDGMENTS

We warmly thank the SCAC (Service de Coopérationetd'ActionsCulturell es) of the France Embassy in Benin, as well as the IRD (Institut de Recherche pour le Développement) for their financial support in the realization of this work.

REFERENCES

1. Pinheiro J, Bates D. Mixed-effects models in S and S-PLUS. Springer Science & Business Media; 2006.
2. Snijders TA. Multilevel analysis. Springer; 2011.
3. Gelman A, Hill J. Data analysis using regression and multilevel/hierarchical models. Cambridge University Press; 2006.
4. Zuur A, Ieno EN, Walker N, Saveliev AA, Smith GM. Mixed effects models and extensions in ecology with R. Springer Science & Business Media; 2009.
5. Zellner A. An efficient method of estimating seemingly unrelated regressions and tests for aggregation bias. Journal of the American statistical Association. 1962;57(298):348–368. 10.1080/01621459.1962.10480664
6. Liang KY, Zeger SL. Longitudinal data analysis using generalized linear models. Biometrika. 1986;73(1):13–22. 10.1093/biomet/73.1.13
7. Lindstrom MJ, Bates DM. Newton Raphson and EM algorithms for linear mixed-effects models for repeated-measures data. Journal of the American Statistical Association. 1988;83(404):1014–1022. 10.1080/01621459.1988.10478693
8. Zeger SL, Liang KY, Albert PS. Models for longitudinal data: a generalized estimating equation approach. Biometrics. 1988; p. 1049–1060. 10.2307/2531734
9. Molenberghs G, Verbeke G. Models for discrete longitudinal data. 2005;.
10. Verbeke G, Molenberghs G. Linear mixed models for longitudinal data. Springer; 2009.
11. Diggle P, Heagerty P, Liang KY, Zeger S. Analysis of longitudinal data. 25. Oxford University Press; 2013.
12. Bates D, Maechler M, Bolker B, Walker S. lme4: Linear mixed-effects models using Eigen and S4. R package version 1.1-7. This is computer program (R package) The URL of the package is: http://CRANR-projectorg/package=lme4 2014;.
13. Pinheiro J, Bates D, DebRoy S, Sarkar D. R Core Team (2014). nlme: linear and nonlinear mixed effects models. R package version 3.1–117. URL: http://cranr-projectorg/web/packages/nlme/index html 2014;.
14. Littell RC, Milliken GA, Stroup WW, Wolfinger RD, Schabenberger

O. SAS system for mixed models Cary. Nc: sas institute. 1996;.

15. Fieuws S, Verbeke G. Pairwise fitting of mixed models for the joint modeling of multivariate longitudinal profiles. Biometrics. 2006;62(2):424–431. 10.1111/j.1541-0420.2006.00507.x
16. Shock NW, GreulichrC, Costa PT, Andres R, Lakatta EG, Arenberg D, et al. Normal human aging: The Baltimore longitudinal study of aging. 1984;.
17. Thiébaut R, Jacqmin-Gadda H, Chêne G, Leport C, Commenges D. Bivariate linear mixed models using SAS proc MIXED. Computer methods and programs in biomedicine. 2002;69(3):249–256. 10.1016/S0169-2607(02)00017-2
18. Subramanian S, Kim D, Kawachi I. Covariation in the socioeconomic determinants of self rated health and happiness: a multivariate multilevel analysis of individuals and communities in the USA. Journal of Epidemiology and Community Health. 2005;59(8):664–669. 10.1136/jech.2004.025742 [PMC free article]
19. Tseloni A, Zarafonitou C. Fear of crime and victimization a multivariate multilevel analysis of competing measurements. European Journal of Criminology. 2008;5(4):387–409. 10.1177/1477370808095123
20. Sy J, Taylor J, Cumberland W. A stochastic model for the analysis of bivariate longitudinal AIDS data. Biometrics. 1997; p. 542–555. 10.2307/2533956
21. Fieuws S, Verbeke G, Maes B, Vanrenterghem Y. Predicting renal graft failure using multivariate longitudinal profiles. Biostatistics. 2008;9(3):419–431. 10.1093/biostatistics/kxm041
22. Charnigo R, Kryscio R, Bardo MT, Lynam D, Zimmerman RS. Joint modeling of longitudinal data in multiple behavioral change. Evaluation & the health professions. 2011;34(2):181–200. 10.1177/0163278710392982 [PMC free article]
23. Wang XF. Joint generalized models for multidimensional outcomes: A case study of neuroscience data from multimodalities. Biometrical Journal. 2012;54(2):264–280. 10.1002/bimj.201100041 [PMC free article]
24. Brombin C, Di Serio C, Rancoita PM. Joint modeling of HIV data in multicenter observational studies: A comparison among different approaches. Statistical methods in medical research. 2014; p. 0962280214526192.

25. Bandyopadhyay S, Ganguli B, Chatterjee A. A review of multivariate longitudinal data analysis. Statistical methods in medical research. 2011;20(4):299–330. 10.1177/0962280209340191
26. Verbeke G, Fieuws S, Molenberghs G, Davidian M. The analysis of multivariate longitudinal data: A review. Statistical methods in medical research. 2014;23(1):42–59. 10.1177/0962280212445834 [PMC free article]
27. Galecki AT. General class of covariance structures for two or more repeated factors in longitudinal data analysis. Communications in Statistics-Theory and Methods. 1994;23(11):3105–3119. 10.1080/03610929408831436
28. O'Brien LM, Fitzmaurice GM. Analysis of longitudinal multiple-source binary data using generalized estimating equations. Journal of the Royal Statistical Society: Series C (Applied Statistics). 2004;53(1):177–193. 10.1046/j.0035-9254.2003.05296.x
29. Carey VJ, Rosner BA. Analysis of longitudinally observed irregularly timed multivariate outcomes: regression with focus on cross-component correlation. Statistics in medicine. 2001;20(1):21–31. 10.1002/1097-0258(20010115)20:1<21::AID-SIM639>3.0.CO;2-5
30. Sklar M. Fonctions de répartition à n dimensions etleursmarges. Université Paris 8; 1959.
31. Nelsen RB. An introduction to copulas. Springer; 1999.
32. Lambert P, Vandenhende F. A copula-based model for multivariate non-normal longitudinal data: analysis of a dose titration safety study on a new antidepressant. Statistics in medicine. 2002;21(21):3197–3217. 10.1002/sim.1249
33. MaCurdy TE. The use of time series processes to model the error structure of earnings in a longitudinal data analysis. Journal of econometrics. 1982;18(1):83–114. 10.1016/0304-4076(82)90096-3
34. Tsay RS. Multivariate Time Series Analysis: With R and Financial Applications. John Wiley & Sons; 2013.
35. Johnson RA, Wichern DW, Education P. Applied multivariate statistical analysis. vol. 4. Prentice hall Englewood Cliffs, NJ; 2007.
36. Tschacher W, Ramseyer F. Modeling psychotherapy process by time-series panel analysis (TSPA). Psychotherapy Research. 2009;19(4-5):469–481. 10.1080/10503300802654496
37. Tschacher W, Zorn P, Ramseyer F. Change mechanisms of schema-

centered group psychotherapy with personality disorder patients. PloS one. 2012;7(6):e39687 10.1371/journal.pone.0039687 [PMC free article]

38. Horváth C, Wieringa JE. Pooling data for the analysis of dynamic marketing systems. StatisticaNeerlandica. 2008;62(2):208–229. 10.1111/j.1467-9574.2007.00382.x
39. Liang KY, Zeger SL, Qaqish B. Multivariate regression analyses for categorical data. Journal of the Royal Statistical Society Series B (Methodological). 1992; p. 3–40.
40. Zeger SL, Liang KY. Longitudinal data analysis for discrete and continuous outcomes. Biometrics. 1986; p. 121–130. 10.2307/2531248
41. Prentice RL, Zhao LP. Estimating equations for parameters in means and covariances of multivariate discrete and continuous responses. Biometrics. 1991; p. 825–839. 10.2307/2532642
42. Rochon J. Analyzing bivariate repeated measures for discrete and continuous outcome variables. Biometrics. 1996; p. 740–750. 10.2307/2532914
43. Crowder M. On the use of a working correlation matrix in using generalized linear models for repeated measures. Biometrika. 1995;82(2):407–410. 10.1093/biomet/82.2.407
44. Gray SM, Brookmeyer R. Estimating a treatment effect from multidimensional longitudinal data. Biometrics. 1998; p. 976–988. 10.2307/2533850
45. Gray SM, Brookmeyer R. Multidimensional longitudinal data: estimating a treatment effect from continuous, discrete, or time-to-event response variables. Journal of the American Statistical Association. 2000;95(450):396–406. 10.1080/01621459.2000.10474209
46. Geys H, Molenberghs G, Ryan LM. Pseudolikelihood modeling of multivariate outcomes in developmental toxicology. Journal of the American Statistical Association. 1999;94(447):734–745. 10.1080/01621459.1999.10474176
47. Zhang M, Tsiatis AA, Davidian M, Pieper KS, Mahaffey KW. Inference on treatment effects from a randomized clinical trial in the presence of premature treatment discontinuation: the SYNERGY trial. Biostatistics. 2011;12(2):258–269. 10.1093/biostatistics/kxq054 [PMC free article]
48. McArdle JJ. Dynamic but structural equation modeling of repeated measures data In: Handbook of multivariate experimental psychology.

Springer; 1988. p. 561–614.

49. Duncan SC, Duncan TE. A multivariate latent growth curve analysis of adolescent substance use. Structural Equation Modeling: A Multidisciplinary Journal. 1996;3(4):323–347. 10.1080/10705519609540050
50. Oort FJ. Three-mode models for multivariate longitudinal data. British Journal of Mathematical and Statistical Psychology. 2001;54(1):49–78. 10.1348/000711001159429
51. Hancock GR, Kuo WL, Lawrence FR. An illustration of second-order latent growth models. Structural Equation Modeling. 2001;8(3):470–489. 10.1207/S15328007SEM0803_7
52. Fieuws S, Verbeke G. Joint models for high-dimensional longitudinal data. Longitudinal data analysis. 2009; p. 367–391.
53. Ramsay J, Silverman B. Functional Data Analysis. 1997; 1997.
54. Reinsel G. Estimation and prediction in a multivariate random effects generalized linear model. Journal of the American Statistical Association. 1984;79(386):406–414. 10.1080/01621459.1984.10478064
55. MacCallum RC, Kim C, Malarkey WB, Kiecolt-Glaser JK. Studying multivariate change using multilevel models and latent curve models. Multivariate Behavioral Research. 1997;32(3):215–253. 10.1207/s15327906mbr3203_1
56. Ribaudo H, Thompson S. The analysis of repeated multivariate binary quality of life data: a hierarchical model approach. Statistical methods in medical research. 2002;11(1):69–83. 10.1191/0962280202sm272ra
57. Beckett L, Tancredi D, Wilson R. Multivariate longitudinal models for complex change processes. Statistics in medicine. 2004;23(2):231–239. 10.1002/sim.1712
58. An X, Yang Q, Bentler PM. A latent factor linear mixed model for high-dimensional longitudinal data analysis. Statistics in medicine. 2013;32(24):4229–4239. 10.1002/sim.5825 [PMC free article]
59. Schafer JL, Yucel RM. Computational strategies for multivariate linear mixed-effects models with missing values. Journal of Computational and Graphical Statistics. 2002;11(2):437–457. 10.1198/106186002760180608
60. Shah A, Laird N, Schoenfeld D. A random-effects model for multiple characteristics with possibly missing data. Journal of the American

Statistical Association. 1997;92(438):775–779. 10.1080/01621459.1997.10474030

61. Bentler PM, Weeks DG. Linear structural equations with latent variables. Psychometrika. 1980;45(3):289–308. 10.1007/BF02293905
62. Bringmann LF, Vissers N, Wichers M, Geschwind N, Kuppens P, Peeters F, et al. A network approach to psychopathology: new insights into clinical longitudinal data. PloS one. 2013;8(4):e60188 10.1371/journal.pone.0060188 [PMC free article]
63. Funatogawa I, Funatogawa T, Ohashi Y. An autoregressive linear mixed effects model for the analysis of longitudinal data which show profiles approaching asymptotes. Statistics in medicine. 2007;26(9):2113–2130. 10.1002/sim.2670
64. Hamilton JD. State-space models. Handbook of econometrics. 1994;4:3039–3080. 10.1016/S1573-4412(05)80019-4
65. Lodewyckx T, Tuerlinckx F, Kuppens P, Allen NB, Sheeber L. A hierarchical state space approach to affective dynamics. Journal of mathematical psychology. 2011;55(1):68–83. 10.1016/j.jmp.2010.08.004 [PMC free article]
66. Gates KM, Molenaar PC. Group search algorithm recovers effective connectivity maps for individuals in homogeneous and heterogeneous samples. Neuroimage. 2012;63(1):310–319. 10.1016/j.neuroimage.2012.06.026
67. Rice JA, Wu CO. Nonparametric mixed effects models for unequally sampled noisy curves. Biometrics. 2001; p. 253–259. 10.1111/j.0006-341X.2001.00253.x
68. Faraway JJ. Extending the linear model with R: generalized linear, mixed effects and nonparametric regression models. CRC press; 2005.
69. Wu H, Zhang JT. Nonparametric regression methods for longitudinal data analysis: mixed-effects modeling approaches. vol. 515 John Wiley & Sons; 2006.
70. Davidian M, Giltinan DM. Nonlinear models for repeated measurement data: an overview and update. Journal of Agricultural, Biological, and Environmental Statistics. 2003;8(4):387–419. 10.1198/1085711032697
71. Crouchley R, Stott D, Pritchard J, Grose D. Multivariate Generalised Linear Mixed Models via sabreR (Sabre in R). 2010;.
72. R Core Team. R: A Language and Environment for Statistical Computing; 2014. Available from: http://www.R-project.org/

73. Sturtz S, Ligges U, Gelman A. R2WinBUGS: A Package for Running WinBUGS from R. Journal of Statistical Software. 2005;12(3):1–16. 10.18637/jss.v012.i03

74. Fieuws S, Verbeke G. Joint modelling of multivariate longitudinal profiles: pitfalls of the random-effects approach. Statistics in Medicine. 2004;23(20):3093–3104. 10.1002/sim.1885

75. Laird N, Lange N, Stram D. Maximum likelihood computations with repeated measures: application of the EM algorithm. Journal of the American Statistical Association. 1987;82(397):97–105. 10.1080/01621459.1987.10478395

76. Bates D, Maechler M, Bolker B, Walker S. lme4: Linear mixed-effects models using Eigen and S4; 2013. Available from: http://CRAN.R-project.org/package=lme4

77. Wilks SS. The large-sample distribution of the likelihood ratio for testing composite hypotheses. The Annals of Mathematical Statistics. 1938;9(1):60–62. 10.1214/aoms/1177732360

78. Bartlett MS. Properties of sufficiency and statistical tests. Proceedings of the Royal Society of London Series A, Mathematical and Physical Sciences. 1937; p. 268–282. 10.1098/rspa.1937.0109

79. Brandsma H, Knuver J. Effects of school and classroom characteristics on pupil progress in language and arithmetic. International Journal of Educational Research. 1989;13(7):777–788. 10.1016/0883-0355(89)90028-1

80. Djènontin A, Bio-Bangana S, Moiroux N, Henry MC, Bousari O, Chabi J, et al. Culicidae diversity, malaria transmission and insecticide resistance alleles in malaria vectors in Ouidah-Kpomasse-Tori district from Benin (West Africa): A pre-intervention study. Parasit Vectors. 2010;3:83 10.1186/1756-3305-3-83 [PMC free article]

81. Le Port A, Cottrell G, Martin-Prevel Y, Migot-Nabias F, Cot M, Garcia A. First malaria infections in a cohort of infants in Benin: biological, environmental and genetic determinants. Description of the study site, population methods and preliminary results. BMJ open. 2012;2(2):e000342 10.1136/bmjopen-2011-000342 [PMC free article]

82. Ballard J, Khoury J, Wedig K, Wang L, Eilers-Walsman B, Lipp R. New Ballard Score, expanded to include extremely premature infants. The Journal of pediatrics. 1991;119(3):417–423. 10.1016/S0022-3476(05)82056-6

83. Cottrell G, Kouwaye B, Pierrat C, Le Port A, Bouraïma A, Fonton N, et al. Modeling the influence of local environmental factors on malaria transmission in Benin and its implications for cohort study. 2012;. [PMC free article]
84. Courtin D, Oesterholt M, Huismans H, Kusi K, Milet J, Badaut C, et al. The quantity and quality of African children's IgG responses to merozoite surface antigens reflect protection against Plasmodium falciparum malaria. PloS one. 2009;4(10):e7590 10.1371/journal.pone.0007590 [PMC free article]
85. Chant D. On asymptotic tests of composite hypotheses in nonstandard conditions. Biometrika. 1974;61(2):291–298. 10.1093/biomet/61.2.291
86. Self SG, Liang KY. Asymptotic properties of maximum likelihood estimators and likelihood ratio tests under nonstandard conditions. Journal of the American Statistical Association. 1987;82(398):605–610. 10.1080/01621459.1987.10478472
87. Vu H, Zhou S, et al. Generalization of likelihood ratio tests under nonstandard conditions. The Annals of Statistics. 1997;25(2):897–916. 10.1214/aos/1031833677
88. Giampaoli V, Singer JM. Likelihood ratio tests for variance components in linear mixed models. Journal of Statistical Planning and Inference. 2009;139(4):1435–1448. 10.1016/j.jspi.2008.06.016

6

Generalized Inferences about the Mean Vector of Several Multivariate Gaussian Processes

Pilar Ibarrola[1] **and Ricardo Vélez**[2]

[1]Statistics Department, Universidad Complutense de Madrid, 28040 Madrid, Spain
[2]Statistics Department, UNED, 28040 Madrid, Spain

ABSTRACT

We consider in this paper the problem of comparing the means of several multivariate Gaussian processes. It is assumed that the means depend linearly on an unknown vector parameter θ and that nuisance parameters appear in the covariance matrices. More precisely, we deal with the problem of testing hypotheses, as well as obtaining confidence regions for θ. Both methods will be based on the concepts of generalized p value and generalized confidence region adapted to our context.

INTRODUCTION

The generalized p values to test statistical hypotheses in the presence of nuisance parameters are introduced by Tsui and Weerahandi (1989) [1],

Citation: Ibarrola, P., &Vélez, R. (2015). Generalized inferences about the mean vector of several multivariate Gaussian processes. Journal of Probability and Statistics, 2015. (10 pages), DOI/URL: https://doi.org/10.1155/2015/479762

Copyright: © 2015 PilarIbarrola and Ricardo Vélez. This is an open access article distributed under the Creative Commons Attribution 3.0 Unported (CC BY 3.0) License.

where the univariate Behrens-Fisher problem, as well as other examples, is considered in order to illustrate the usefulness of this approach. Afterwards Weerahandi (1993) [2] introduces the generalized confidence intervals.

In 2004, Gamage et al. [3] developed a procedure based on the generalized p values to test the equality of the mean vectors of two multivariate normal populations with different covariance matrices. They also construct a confidence region for the means difference, using the concept of generalized confidence regions. Finally, by means of the generalized p value approach, a solution is obtained for the heteroscedastic MANOVA problem, but without reaching the desirable invariance property.

In 2007, Lin et al. [4] considered the generalized inferences on the common mean vector of several multivariate normal populations. They obtained a confidence region for the common mean vector and simultaneous confidence intervals for its components. Their method is numerically compared with other existing methods, with respect to the expected area and coverage probabilities.

In 2008, Xu and Wang [5] considered the problem of comparing the means of K populations with heteroscedastic variances. They provided a new generalized p value procedure for testing the equality of means, assuming that the variables are univariate and normally distributed. Numerical results show that their generalized p value test works better than a generalized F-test. We will set out our MANOVA problem as a generalization of their framework.

In 2012, Zhang [6] considered the general linear hypothesis testing (GLHT) in anheteroscedastic one-way MANOVA. The multivariate Behrens-Fisher problem is a special case of GLHT.

In this paper we first consider the generalized inference for the case of two continuous time Gaussian processes. Later, the results will be extended for such N processes. In both cases, for the testing problem, the main step is constructing a generalized test process and analyzing the associated generalized p value, proving some linear invariance properties.

With respect to the construction of generalized confidence regions, one should use a generalized pivotal quantity and use the approach of multiple comparisons as in [4].

Finally, in the same line of Zhang [6], we consider the general linear hypothesis testing (GLHT) as a generalization of the MANOVA, adapting the setting and method of this paper.

It must be emphasized that all the references above develop these techniques for discrete univariate or multivariate models, whereas here we are concerned with a continuous time model. It is well known that when the underlying phenomenon is in essence continuous, even if it is observed at a sequence of epochs$n\Delta t$, different models may be necessary for distinct values of Δt. On the contrary a continuous time model embodies simultaneously all the statistical properties of the time series obtained for each value of Δt.

CONTINUOUS TIME GENERALIZED TESTS AND CONFIDENCE REGIONS

Let $\{X_t\}_{t \in T}$ be a k-dimensional stochastic process with distribution depending on the unknown parameter $\xi = (\theta, \zeta)$, θ being the vector of parameters of interest and ζ a nuisance parameter vector. For any random vector Y, $\tilde{Y}$ will denote its observed value.

For the problem of testing a null hypothesis H_0 :$\theta \leq \theta_0$ against the alternative H_1 :$\theta > \theta_0$, where θ_0 is a given vector (the inequalities like $\theta_1 \leq \theta_2$ should be understood componentwise.), a generalized test process is defined, following [3], as follows.

Definition 1. A generalized test process $W_t(X, \tilde{X}, \theta, \zeta)$ is, for each $t \in T$, a one-dimensional function depending on $\{Xs\}s \leq t$ and its observed value $\{\tilde{X}_s\}_{s \leq t}$, as well as the parameter value $\xi = (\theta, \zeta)$, satisfying the following: (1)the distribution of $W_t(X, \tilde{X}, \theta, \zeta)$ does not depend on ζ, for any fixed $\tilde{X}$,,(2)the observed value $\tilde{W}_t = W_t(\tilde{X}, \tilde{X}, \theta_0, \zeta)$ does not depend on ζ, (3) $P_\theta\{W_t(X, \tilde{X}, \theta, \zeta) \geq w\}$ is nondecreasing in every component of θ, for any $w \in \mathbb{R}$ and any fixed $\tilde{X}$ and ζ.

Under the above conditions, the generalized p value is defined as

$$p_t = \sup_{\theta \leq \theta_0} P_\theta \left\{ W_t \left(X, \tilde{X}, \theta, \zeta \right) \geq \tilde{W}_t \right\}$$

$$= P_{\theta_0} \left\{ W_t \left(X, \tilde{X}, \theta_0, \zeta \right) \geq \tilde{W}_t \right\}. \qquad (1)$$

When testing H_0 :$\theta = \theta_0$ versus H_1 : $\theta \neq \theta_0$, condition (3) must be replaced by(3') $W_t(X, \tilde{X}, \theta, \zeta)$ is stochastically larger under H_1 than under H_0, for any fixed $\tilde{X}$ and ζ.

In this case, the generalized p value is given by

$$p_t = P_{\theta_0}\left\{W_t\left(X, \widetilde{X}, \theta_0, \zeta\right) \geq \widetilde{W_t}\right\}. \tag{2}$$

Towards the confidential estimation of θ, we give the following definition.

Definition 2. A generalized pivotal quantity $W_t(X, \widetilde{X}, \theta, \zeta)$ is, for each $t \in T$, a one-dimensional function satisfying the following:

(1) the distribution of $W_t(X, \widetilde{X}, \theta, \zeta)$ does not depend on θ nor ζ,

(2) the observed value $\widetilde{W_t} = W_t(\widetilde{X}, \widetilde{X}, \theta, \zeta)$ does not depend on ζ.

Then, if $c_1 < c_2$ are such that

$$P_\theta\left\{c_1 \leq W_t\left(X, \widetilde{X}, \theta, \zeta\right) \leq c_2\right\} = 1 - \gamma, \tag{3}$$

$\{\theta \mid c_1 \leq \widetilde{W_t} \leq c_2\}$ is a $100(1-\gamma)\%$ generalized confidence region for θ.

ESTIMATION METHOD

In the previous paper [7] we have considered the confidence estimation of a k-dimensional parameter θ, when observing a continuous time p-dimensional Gaussian process Xt, with covariances function $B(t, t') = \sigma^2 K(t, t')$ and mean function $\mu(t) = A(t)\theta$, where $K(t, t')$ and $A(t)$ are known matrices, but σ^2 and θ are unknown parameters. More concrete assumptions were specified by Ibarrola and Velez [7] and ´ we will here suppose that they hold for all the considered processes.

The estimation method of θ, described in [7], is based on the estimator

$$\theta_t = \Sigma_t^{-1}\widehat{\theta}_t \quad \text{with } \widehat{\theta}_t = \int_0^t \widehat{f}_t'(u)\, X_u\, du, \tag{4}$$

where $\widehat{f}_t'$ is the $k \times p$-matrix with columns in $L^{2,p}[0, t]$ satisfying the equation

$$A(u) = \int_0^t \left[K\left(u, u'\right) + A(u)\, A'\left(u'\right)\right] \widehat{f}_t\left(u'\right) du', \quad (\forall u \leq t) \tag{5}$$

and Σ_t is given by

$$\Sigma_t = \int_0^t \widehat{f}_t'(u)\, A(u)\, du. \tag{6}$$

As proved in [7], $\{\theta_t\}$ does not depend on σ^2 and constitutes a Gaussian process such that

$$E_\theta\left[\theta_t\right] = \theta,$$

$$\mathrm{Cov}_\theta\left(\theta_t, \theta_{t'}\right) = \sigma^2 \sigma_{t \vee t'} \quad \text{where } \sigma_t = \Sigma_t^{-1} - I. \tag{7}$$

Consequently, θ_t and $\theta_{t_2'} - \theta_{t_1'}$ are independent if $t_1' < t_2' \le t$, while

$$E_\theta\left[\left(\theta_{t'} - \theta\right)\left(\theta_{t_2} - \theta_{t_1}\right)'\right] = \sigma^2\left(\sigma_{t_2} - \sigma_{t_1}\right) \quad \text{if } t' \le t_1 < t_2. \tag{8}$$

Moreover, θ_t is a mean square consistent estimator of θ, since all the eigenvalues of the covariance matrix σ_t converge to 0.

In order to estimate σ^2, if $t_1 < t_2$ are arbitrarily chosen so that $\sigma_{t_1} - \sigma_{t_2}$ is nonsingular, we can consider the random variable

$$\eta = \left(\theta_{t_2} - \theta_{t_1}\right)'\left(\sigma_{t_1} - \sigma_{t_2}\right)^{-1}\left(\theta_{t_2} - \theta_{t_1}\right) \tag{9}$$

such that η/σ^2 has a χ_k^2 distribution.

THE BEHRENS-FISHER PROBLEM

We first consider the case of two independent Gaussian stochastic processes, $\{X_t = (X_t^1, \ldots, X_t^p)'\}$ and $\{Y_s = (Y_s^1, \ldots, Y_s^q)'\}$, of dimensions p and q, respectively, and with similar characteristics. More precisely,

$$E_{\theta_1}\left[X_t\right] = A_1(t)\,\theta_1,$$

$$\mathrm{Cov}\left(X_t, X_{t'}\right) = \sigma_1^2 K_1\left(t, t'\right),$$

$$E_{\theta_2}\left[Y_s\right] = A_2(s)\,\theta_2,$$

$$\mathrm{Cov}\left(Y_s, Y_{s'}\right) = \sigma_2^2 K_2\left(s, s'\right), \tag{10}$$

where $\theta_1, \theta_2 \in \mathbb{R}^k$ and σ_1^2, σ_2^2 are unknown parameters, while

$A_1(t)$, $A_2(s)$ and $K_1(t,t')$, $K_2(s,s')$ are given matrices of appropriate dimensions.

We will focus on the Behrens-Fisher type problem of comparing the parameters θ_1 and θ_2 and, more concretely, our aim is to make inferences about $\theta_1 - \theta_2$, based on the progressive observation of both processes X_t and Y_s.

The progressive estimators $\theta_t^{(1)}$ and $\theta_t^{(2)}$ of θ_1 and θ_2 can be constructed according to (4), (5), and (6). For$i$ = 1, 2, $\Sigma_t^{(i)}$ and $\sigma_t^{(i)}$ will denote the characteristics defined in (6), (7) for $\{X_t\}$ and $\{Y_s\}$, respectively. We thus obtain the unbiased estimator of $\delta=\theta_1 - \theta_2$:

$$T_{t,s} = \theta_t^{(1)} - \theta_s^{(2)} \tag{11}$$

which is normally distributed, satisfies E[$T_{t,s]}=\delta$, and, since $\{\theta_t^{(1)}\}$ and $\{\theta_s^{(2)}\}$ are independent, has covariance

$$\operatorname{Cov}\left(T_{t,s}, T_{t',s'}\right) = \sigma_1^2\sigma_{t\vee t'}^{(1)} + \sigma_2^2\sigma_{s\vee s'}^{(2)}. \tag{12}$$

In order to estimate σ_1^2 and σ_2^2, we can take

$$\eta_1 = \left(\theta_{t_2}^{(1)} - \theta_{t_1}^{(1)}\right)'\left(\sigma_{t_1}^{(1)} - \sigma_{t_2}^{(1)}\right)^{-1}\left(\theta_{t_2}^{(1)} - \theta_{t_1}^{(1)}\right),$$

$$\eta_2 = \left(\theta_{s_2}^{(2)} - \theta_{s_1}^{(2)}\right)'\left(\sigma_{s_1}^{(2)} - \sigma_{s_2}^{(2)}\right)^{-1}\left(\theta_{s_2}^{(2)} - \theta_{s_1}^{(2)}\right), \tag{13}$$

where $t_1 < t_2$ and $s_1 < s_2$ are such that $\sigma_{t_1}^{(1)} - \sigma_{t_2}^{(1)}$ and $\sigma_{s_1}^{(2)} - \sigma_{s_2}^{(2)}$ are nonsingular. In this way η_1/σ_1^2 and η_2/σ_2^2 are independent, χ_k^2-distributed random variables, independent of $T_{t,\cdot}$.

Generalized Value for the Behrens-Fisher Problem

We consider the problem of testing H_0 :δ=0 against the alternative H_1 : $\delta \neq$ 0 and we look for a generalized test process $W_{t,s}(\theta_t^{(1)}, \theta_s^{(2)}, \tilde{\theta}_t^{(1)}, \tilde{\theta}_s^{(2)}, \delta, \sigma_1^2, \sigma_2^2)$.

The covariance matrix of $T_{t,}$ is

$$U_{t,s} = \sigma_1^2\sigma_t^{(1)} + \sigma_2^2\sigma_s^{(2)} \tag{14}$$

and can be estimated by means of

$$V_{t,s} = \eta_1\sigma_t^{(1)} + \eta_2\sigma_s^{(2)}. \tag{15}$$

Recall that $\widetilde{T}_{t,s}$, $\widetilde{\eta}_i$, and $\widetilde{V}_{t,s}$ represent the observed values, obtained when $\theta_t^{(1)}$ and $\theta_s^{(2)}$ s are replaced with $\widetilde{\theta}_t^{(1)}$ and $\widetilde{\theta}_s^{(2)}$, and let us define

$$d_{t,s} = \widetilde{V}_{t,s}^{-1/2} T_{t,s},$$

$$Z_{t,s} = \left(\widetilde{V}_{t,s}^{-1/2} U_{t,s} \widetilde{V}_{t,s}^{-1/2}\right)^{-1/2} d_{t,s}. \tag{16}$$

Under H_0, $d_{t,}$ has distribution $N_k(0, \widetilde{V}_{t,s}^{-1/2} U_{t,s} \widetilde{V}_{t,s}^{-1/2})$, so that the distribution of $Z_{t,s}$ is $N_k(0, I)$. Moreover

$$R_{t,s} = \widetilde{V}_{t,s}^{-1/2} \left(\sigma_1^2 \sigma_t^{(1)} \frac{\widetilde{\eta}_1}{\eta_1} + \sigma_2^2 \sigma_s^{(2)} \frac{\widetilde{\eta}_2}{\eta_2} \right) \widetilde{V}_{t,s}^{-1/2} \tag{17}$$

does not depend on δ and its distribution is independent of the parameters, since $u_1 = \eta_1 / \sigma_1^2$ and $u_2 = \eta_2 / \sigma_2^2$ are independent, χ_k^2 distributed random variables, independent of $Z_{t,s}$. The observed value of $R_{t,}$ is $\widetilde{R}_{t,s} = \widetilde{V}_{t,s}^{-1/2} U_{t,s} \widetilde{V}_{t,s}^{-1/2}$. Finally,

$$W_{t,s} = Z_{t,s}' R_{t,s} Z_{t,s}$$

$$= Z_{t,s}' \widetilde{V}_{t,s}^{-1/2} \left(\frac{\widetilde{\eta}_1}{u_1} \sigma_t^{(1)} + \frac{\widetilde{\eta}_2}{u_2} \sigma_s^{(2)} \right) \widetilde{V}_{t,s}^{-1/2} Z_{t,s} \tag{18}$$

is a one-dimensional random variable with distribution not depending on (σ_1^2, σ_2^2), whose observed value is $\widetilde{W}_{t,s} = \widetilde{T}_{t,s}' \widetilde{V}_{t,s}^{-1} \widetilde{T}_{t,s}$, that neither depends on (σ_1^2, σ_2^2).

Under H_1, when $\delta \neq 0$, as $T_{t,s}$ has distribution $N_k(\delta, U_{t,s})$, the distribution of $Z_{t,s}$ is $N_k(\delta, I)$, with $\bar{\delta} = (\widetilde{V}_{t,s}^{-1/2} U_{t,s} \widetilde{V}_{t,s}^{-1/2})^{-1/2} \widetilde{V}_{t,s}^{-1/2} \delta$. Note that, given u_1 and u_2, since $R_{t,}$ is a positive definite matrix, $W_{t,s}$ is stochastically larger under H_1 than under H_0. So, since the conditions(1), (2), and (3) of Definition 1 are verified, we have proved the following result.

Proposition 3. $W_{t.}$ *is a generalized test process for testing* $H_0 : \delta = 0$ *against* $H_1 : \delta \neq 0$.

In order to simplify the expression of $W_{t,}$, we will put

$$M_1 = M_1(t, s) = \widetilde{V}_{t,s}^{-1/2} \widetilde{\eta}_1 \sigma_t^{(1)} \widetilde{V}_{t,s}^{-1/2},$$

$$M_2 = M_2(t, s) = \widetilde{V}_{t,s}^{-1/2} \widetilde{\eta}_2 \sigma_s^{(2)} \widetilde{V}_{t,s}^{-1/2} \tag{19}$$

which are positive definite matrices such that $M_1 + M_2 = I$. Then

$$W_{t,s} = Z'_{t,s}\left(\frac{M_1}{u_1} + \frac{M_2}{u_2}\right)Z_{t,s}$$

$$= \frac{1}{2k}\overline{Z}'_{t,s}\left(\frac{M_1}{\beta} + \frac{I - M_1}{1-\beta}\right)\overline{Z}_{t,s}, \tag{20}$$

where, under H_0, $\overline{Z}_{t,s} = \sqrt{2k}Z_{t,s}/(u_1 + u_2)^{1/2}$ has a k-multivariate Student's t-distribution with $2k$ degrees of freedom and $\beta = u_1/(u_1 + u_2)$ is a random variable with distribution $Beta(k/2, k/2)$, which is independent of $\overline{Z}_{t,s}$.

Thus the generalized p value of the given test:

$$p_{t,s}(M_1) = P_{\delta=0}\left\{W_{t,s} \geq \widetilde{W}_{t,s}\right\}$$

$$= P_{\delta=0}\left\{\overline{Z}'_{t,s}\left(\frac{M_1}{\beta} + \frac{I - M_1}{1-\beta}\right)\overline{Z}_{t,s} \geq 2k\widetilde{W}_{t,s}\right\} \tag{21}$$

may be calculated once M_1 is observed.

Invariance Properties of the Generalized p Value

Let us assume that the basic processes are transformed by means of

$$X_t^{\star} = Q_1 X_t,$$

$$Y_s^{\star} = Q_2 Y_s, \tag{22}$$

whereQ_1, Q_2 are nonsingular square matrices of order p and q, respectively. The characteristics of the transformed processes are

$$E_{\theta_1}[X_t^{\star}] = A_1^{\star}(t)\theta_1,$$

$$E_{\theta_2}[Y_t^{\star}] = A_2^{\star}(s)\theta_2,$$

$$\mathrm{Cov}(X_t^{\star}, X_{t'}^{\star}) = \sigma_1^2 K_1^{\star}(t, t'),$$

$$\mathrm{Cov}(Y_s^{\star}, Y_{s'}^{\star}) = \sigma_2^2 K_2^{\star}(s, s'), \tag{23}$$

where $A_i^{\star}(\cdot) = Q_i A_i(\cdot)$ and $K_i^{\star}(\cdot,\cdot) = Q_i K_i(\cdot,\cdot) Q_i'$.

For $X_t^{\star}$, the solution $\widehat{f}_t^{\star}(u)$ of (5) with the new characteristics will be related to $\widehat{f}_t(u)$ by means of $\widehat{f}_t^{\star}(u) = (Q_1')^{-1}\widehat{f}_t(u)$. Therefore, according to (4) and (6), we get

$$\widehat{\theta}_t^{(1)\star} = \widehat{\theta}_t^{(1)},$$

$$\Sigma_t^{(1)\star} = \Sigma_t^{(1)}, \tag{24}$$

so that $\theta_t^{(1)\star} = \theta_t^{(1)}, \sigma_t^{(1)\star} = \sigma_t^{(1)}$, and $\eta_1^\star = \eta_1$.

The same results hold for $Y_s^\star$ yielding $T_{t,s}^\star = T_{t,s}$ as $W_{t,s}^\star = W_{t,s}$. In this sense the generalized test process and the corresponding generalized p value are invariant under the proposed transformations.

Generalized Confidence Region for

For any value of the unknown parameter $T_{t,s}^\delta = T_{t,s} - \delta$, the difference $N_k(0, U_{t,s})$ and $Z_{t,s}^\delta = (\widetilde{V}_{t,s}^{-1/2} U_{t,s} \widetilde{V}_{t,s}^{-1/2})^{-1/2} \widetilde{V}_{t,s}^{-1/2} T_{t,s}^\delta$ is $N_k(0, I)$. Hence

$$W_{t,s}^\delta = \left(Z_{t,s}^\delta\right)' R_{t,s} Z_{t,s}^\delta \tag{25}$$

has the same distribution as $W_{t,s}$ under H_0 :δ=0, which is independent of all the parameters, while its observed value $\widetilde{W}_{t,s}^\delta$ does not depend on the nuisance parameters (σ_1^2, σ_2^2). Since conditions(1) and (2) of Definition 2 are accomplished, the following result is established.

Proposition 4. *$W_{t,s}^\delta$ is a generalized pivotal quantity and $\{\delta \in \mathbb{R}^k \mid \widetilde{W}_{t,s}^\delta \le c_\gamma\}$ is a generalized confidence region for δ, whenever $P\{W_{t,s}^\delta \le c_\gamma\} = 1 - \gamma$.*

According to (20), once M_1 is observed, the constant c_γ may be determined and the confidence region for δ is the ellipsoid in $\mathbb{R}^k$:

$$\mathscr{R} = \left\{\delta \in \mathbb{R}^k \mid \left(\delta - \widetilde{T}_{t,s}\right)' \widetilde{V}_{t,s}^{-1} \left(\delta - \widetilde{T}_{t,s}\right) \le c_\gamma\right\} \tag{26}$$

which is centered at $\widetilde{T}_{t,s}$ and with axes in the direction of each eigenvector of $\widetilde{V}_{t,s}$ of length $\pm\sqrt{c_\gamma \lambda}$, where λ is the corresponding eigenvalue. Thus the k-dimensional volume of the confidence region results in

$$\frac{\left(\pi c_\gamma\right)^{k/2} \left|\widetilde{V}_{t,s}\right|^{1/2}}{\Gamma(1 + k/2)}. \tag{27}$$

Simultaneous confidence intervals for the k components of δ can be obtained from the following consequence of the Cauchy-Schwarz inequality:

$x \in \mathbb{R}^k$ *satisfies* $x'x \le c^2$ if and only if $|y'x| < c|y'y|^{1/2}$ *for all* $y \ne 0$. Thus, with $x = \widetilde{V}_{t,s}^{-1/2}(\delta - \widetilde{T}_{t,s})$ and $y = \widetilde{V}_{t,s}^{1/2}a$, the above equivalence allows to express

$$\mathscr{R} = \left\{\delta \in \mathbb{R}^k \mid \left|a'\left(\delta - \widetilde{T}_{t,s}\right)\right| \le c_\gamma^{1/2}\left(a'\widetilde{V}_{t,s}a\right)^{1/2} \ \forall a \ne 0\right\} \tag{28}$$

and therefore

$$\mathscr{R} \subset \left\{\delta \in \mathbb{R}^k \mid \left|\delta_i - \left(\widetilde{T}_{t,s}\right)_i\right| \le c_\gamma^{1/2}\left(\widetilde{V}_{t,s}\right)_{i,i}^{1/2} \ \forall i = 1,\ldots,k\right\}. \tag{29}$$

This last set provides simultaneous confidence intervals for the components of δ with confidence level greater than $1-\gamma$.

An Example with Simulation Results

In order to evaluate the performance of the proposed confidence region and confidence intervals, we will analyze a particular situation that allows accomplishing simulation studies.

The considered problem is the case of two Wiener processes, X_t and Y_s, with

$$\begin{aligned} \mathrm{E}\left[\mathbf{X}_t\right] &= A_1 t^a \theta_1, \\ \mathrm{Cov}\left(\mathbf{X}_t, \mathbf{X}_{t'}\right) &= \sigma_1^2 K_1\left(t \wedge t'\right), \\ \mathrm{E}\left[\mathbf{Y}_s\right] &= A_2 s^b \theta_2, \\ \mathrm{Cov}\left(\mathbf{Y}_s, \mathbf{Y}_{s'}\right) &= \sigma_2^2 K_2\left(s \wedge s'\right). \end{aligned} \tag{30}$$

The estimators $\theta_t^{(i)}$ defined in (4) together with their covariance matrices $\sigma_t^{(i)}$ have been determined in [8]:

$$\theta_t^{(1)} = \frac{2a-1}{at^{2a-1}}\left(A_1' K_1 A_1\right)^{-1} A_1 K_1^{-1} \int_0^t u^{a-1}\, dX_u \tag{31}$$

while

$$\left(\sigma_t^{(1)}\right)^{-1} = \frac{a^2 t^{2a-1}}{2a-1} A_1' K_1^{-1} A_1. \tag{32}$$

Similar results hold for $\theta_s^{(2)}$ and $\sigma_s^{(2)}$. We will take $a=b=1$, so that

$$\sigma_t^{(1)} = t^{-1}\left(A_1' K_1^{-1} A_1\right)^{-1},$$

$$\sigma_t^{(2)} = s^{-1}\left(A_2' K_2^{-1} A_2\right)^{-1}. \tag{33}$$

Since $(A_1' K_1^{-1} A_1)^{-1}$ and $(A_2' K_2^{-1} A_2)^{-1}$ are positive definite symmetric matrices, there exists a nonsingular $k \times k$ matrix L that simultaneously diagonalizes them:

$$L'\left(A_1' K_1^{-1} A_1\right)^{-1} L = I,$$

$$L'\left(A_2' K_2^{-1} A_2\right)^{-1} L = \Lambda = \operatorname{diag}(\lambda_1, \ldots, \lambda_k) \tag{34}$$

and therefore $L'\widetilde{V}_{t,s} L = \widetilde{\eta}_1 t^{-1} I + \widetilde{\eta}_2 s^{-1} \Lambda$.

The random variable $T_{t,s}^{\star} = L' T_{t,s}$ has distribution $N_k(\delta^{\star}, \sigma_1^2 t^{-1} I + \sigma_2^2 s^{-1} \Lambda)$ with $\delta^{\star} = L'\delta$ and the confidence region (26) may be written as

$$\mathscr{R} = \left\{\delta \in \mathbb{R}^k \mid \left(\delta^* - \widetilde{T}_{t,s}^{\star}\right)' \left(\widetilde{\eta}_1 t^{-1} I + \widetilde{\eta}_2 s^{-1} \Lambda\right)^{-1} \left(\delta^* - \widetilde{T}_{t,s}^{\star}\right) \le c_\gamma\right\} = L'^{-1}\left\{\delta^{\star} \in \mathbb{R}^k \mid \left(\delta^* - \widetilde{T}_{t,s}^{\star}\right)' \left(\widetilde{\eta}_1 t^{-1} I + \widetilde{\eta}_2 s^{-1} \Lambda\right)^{-1} \left(\delta^* - \widetilde{T}_{t,s}^{\star}\right) \le c_\gamma\right\}. \tag{35}$$

Thus, the basic case corresponds to $\sigma_t^{(1)} = t^{-1} I$ and $\sigma_s^{(2)} = s^{-1}\Lambda$, from which the confidence region for more general cases may be constructed. With this simple choice, the matrix M_1 in (19) is the diagonal matrix

$$M_1 = \text{diag}\left(\frac{s\tilde{\eta}_1}{s\tilde{\eta}_1 + t\tilde{\eta}_2\lambda_1}, \ldots, \frac{s\tilde{\eta}_1}{s\tilde{\eta}_1 + t\tilde{\eta}_2\lambda_k}\right) \tag{36}$$

and (20) may be written as

$$W_{t,s} = \frac{1}{2k}\sum_{i=1}^{k} Z_i^2 \left(\frac{1}{\beta}\frac{s\tilde{\eta}_1}{s\tilde{\eta}_1 + t\tilde{\eta}_2\lambda_i} + \frac{1}{1-\beta}\frac{t\tilde{\eta}_2\lambda_i}{s\tilde{\eta}_1 + t\tilde{\eta}_2\lambda_i}\right). \tag{37}$$

In the following algorithms we suppose that the values of the dimension k and the variance parameters: σ_1^2, σ_2^2 , and $\lambda_1,\ldots,\lambda_k$ are given. The values of t, s, and γ are also fixed.

Algorithm 5. (1) Generate the independent variables u_1, u_2 with a χ_k^2 distribution.

(2) Compute $\tilde{\eta}_1 = \sigma_1^2 u_1$ and $\tilde{\eta}_2 = \sigma_2^2 u_2$, and subsequently, for each $i = 1, \ldots, k$,

$$C_i = \frac{s\tilde{\eta}_1}{s\tilde{\eta}_1 + t\tilde{\eta}_2\lambda_i},$$

$$D_i = \frac{t\tilde{\eta}_2\lambda_i}{s\tilde{\eta}_1 + t\tilde{\eta}_2\lambda_i}. \tag{38}$$

(3) Fixing a large number of iterations m, for each $j = 1, \ldots, m$,

(a) generate Z_j with a k-variate t-distribution with $2k$ degrees of freedom;

(b) generate β_j with a Beta(k/2, k/2) distribution;

(c) compute

$$W_j = \frac{1}{2k}\sum_{i=1}^{k} Z_{j,i}^2 \left(\frac{C_i}{\beta_j} + \frac{D_i}{1-\beta_j}\right). \tag{39}$$

(4) Determine c_γ as the $1-\gamma$ percentile of the sample $(W_j)_{j=1}^m$.

The obtained value of c_γ may be validated by generating a new sample $(W'_j)_{j=1}^{m'}$ in which the proportion $1-\gamma'$ of (W'_j) less than c_γ may be estimated. Table 1 shows the results obtained by means of a matlab program, using the data $k = 10$, $\sigma_1^2 = 2$, $\sigma_2^2 = 4$, $\lambda_i = 1/i$ for $i = 1, \ldots, 10$, $t = 25$, $s = 50$ and $m=m' = 1000$.

Table 1: Sample percentiles and its validation

$1-\gamma$	c_γ	$1-\gamma'$
0.95	2.7182	0.97
0.90	2.3249	0.935
0.85	1.8182	0.848

The next algorithm is designed to obtain the expected volume of the generalized confidence region and the coverage probability of a given vector $\delta = (\delta_1,...,)$.

Algorithm 6. Given a large number of iterations m, for each $j = 1, \ldots,$,

(1) generate a k-dimensional vector $\widetilde{T}_j$ with independent components, such that $\widetilde{T}_{j,i}$ has a distribution $N(\delta_i, \sigma_1^2/t + \sigma_2^2\lambda_i/s)$ for each $i = 1, \ldots, k$;

(2) generate $\widetilde{\eta}_{1,j}$ and $\widetilde{\eta}_{2,j}$ as in Algorithm 5;

(3) compute

$$W_j = \sum_{i=1}^{k} \left(\delta_i - \widetilde{T}_{j,i}\right)^2 \frac{ts}{\widetilde{\eta}_{1,j}s + \widetilde{\eta}_{2,j}\lambda_i t}, \tag{40}$$

where the coefficients are the terms of the diagonal matrix $(\widetilde{\eta}_1 t^{-1} I + \widetilde{\eta}_2 s^{-1} \Lambda)^{-1}$;

(4) use the value c_γ given in Algorithm 5 to compute the estimated coverage probability of δ as the proportion $p_c(\delta)$ of W_j less than c_γ;

(5) according to (27) compute the volume

$$v_j = \frac{\pi^{k/2} c_\gamma^{k/2}}{\Gamma(1 + k/2)} \left(\prod_{i=1}^{k} \left(\widetilde{\eta}_{1,j} t^{-1} + \widetilde{\eta}_{2,j} s^{-1} \lambda_i\right)\right)^{1/2}; \tag{41}$$

and the estimated expected volume $\overline{v} = (1/m) \sum_{j=1}^{m} v_j$.

Taking for instance $\delta = (1, -1, 2, 0, 3, 4, -2., 5, -3, 1)$ and with the remaining data as before, the obtained coverage probability and estimated expected volume are shown in Table 2.

Finally an algorithm may be designed in order to simulate the simultaneous confidence intervals for the components of δ and to estimate the joint coverage probability.

Table 2: Coverage probability and expected volume of the confidence region

$1-\gamma$	$p_c(\delta)$	$\bar{v}$
0.95	0.982	271.85
0.90	0.947	130.45
0.85	0.891	37.30

Algorithm 7. Given a large number of iterations m, for each $j = 1, \ldots,$ the following hold.

(1) For each $i = 1, \ldots,$, compute, according to the results of the previous algorithms, the extremes of the intervals $I_{j,i} = (l_{j,i}, L_{j,i})$, where

$$l_{j,i} = \tilde{T}_{j,i} - c_\gamma^{1/2}\left(\frac{\tilde{\eta}_{1,j}}{t} + \frac{\tilde{\eta}_{2,j}\lambda_i}{s}\right)^{1/2},$$

$$L_{j,i} = \tilde{T}_{j,i} + c_\gamma^{1/2}\left(\frac{\tilde{\eta}_{1,j}}{t} + \frac{\tilde{\eta}_{2,j}\lambda_i}{s}\right)^{1/2}. \quad (42)$$

(2) Compute the estimated joint coverage probability of δ as the proportion $p(\delta)$ of iterations such that $\delta_i \in I_{j,i}$ for all $i = 1, \ldots, k$.

(3) Compute

$$l_i = \frac{1}{m}\sum_{j=1}^{m} l_{j,i},$$

$$L_i = \frac{1}{m}\sum_{j=1}^{m} L_{j,i} \quad (43)$$

that gives the expected length $L_i - l_i$ of the interval $\bar{I}_i = (l_i, L_i)$.
With the same data as before and the same vector δ, the obtained results are shown in Table 3. Let us observe that the coverage probability of the generalized confidenceregion (δ) always exceeds the confidence level and that the coverage probability of the confidence intervals is always very close to 1. Taking alternative parameter values, some other simulations have been made with similar results.

Table 3: Confidence intervals and joint coverage probabilities

$1-\gamma=0.95$		$1-\gamma=0.90$		$1-\gamma=0.85$	
$\bar{I}_i$	L_i-l_i	$\bar{I}_i$	L_i-l_i	$\bar{I}_i$	L_i-l_i
(−1.29, 3.32)	4.61	(−1.09, 3.14)	4.23	(−0.88, 2.89)	3.73
(−2.79, 0.79)	3.58	(−2.64, 0.65)	3.29	(−2.46, 0.44)	2.90
(0.41, 3.57)	3.16	(0.56, 3.46)	2.90	(0.72, 3.28)	2.56
(−1.47, 1.45)	2.92	(−1.34, 1.35)	2.69	(−1.19, 1.18)	2.37
(1.60, 4.36)	2.76	(1.73, 4.28)	2.55	(1.87, 4.13)	2.25
(2.67, 5.33)	2.66	(2.77, 5.22)	2.45	(2.93, 5.09)	2.16
(−3.29, −0.71)	2.58	(−3.19, −0.81)	2.38	(−3.05, −0.95)	2.10
(3.73, 6.25)	2.52	(3.83, 6.16)	2.33	(3.99, 6.04)	2.05
(−4.24, −1.77)	2.47	(−4.14, −1.86)	2.28	(−4.01, −1.99)	2.01
(−0.21, 2.22)	2.43	(−0.12, 2.12)	2.24	(0.01, 1.99)	1.98
$p_c(\boldsymbol{\delta})=1$		$p_c(\boldsymbol{\delta})=0.9997$		$p_c(\boldsymbol{\delta})=0.9997$	

INFERENCES ABOUT THE VECTOR MEANS OF SEVERAL INDEPENDENT GAUSSIAN PROCESSES

As a generalization we now consider the case of N independent multivariate Gaussian processes $X_{t_1}^{(1)}, X_{t_2}^{(2)}, \ldots, X_{t_N}^{(N)}$ with dimensions p_1, p_2,..., and time parameters t_1, t_2, ...,t_N varying in some interval T. We assume, for each i = 1, . . . ,, the following:

$$\mathrm{E}_{\theta_i}\left[X_{t_i}^{(i)}\right] = A_i\left(t_i\right)\theta_i,$$

$$\mathrm{Cov}\left(X_{t_i}^{(i)}, X_{t_i'}^{(i)}\right) = \sigma_i^2 K\left(t_i, t_i'\right), \tag{44}$$

where $\theta_1, \ldots, \theta_N \in \mathbb{R}^k$ and $\sigma_1^2, \ldots, \sigma_N^2 \in \mathbb{R}$ are unknown parameters.

Testing the Equality of Means

We first consider a test of the null hypothesis $H_0 : \theta_1 = \theta_2 = \cdots = \theta_N$, or $H_0 : \boldsymbol{\delta} = \mathbf{0}$, with $\delta_i = \theta_i - \theta_N$ for $i = 1, 2, \ldots, N-1$.

According to the method of Section 3, we may consider the estimators $\theta_{t_i}^{(i)}$ of the parameters θ_i, whose distribution is $N_k(\theta_i, \sigma_i^2\sigma_{t_i}^{(i)})$, and also the differences $T_{t_i,t_N}^{(i)} = \theta_{t_i}^{(i)} - \theta_{t_N}^{(N)}$ for i = 1, 2, . . . , N − 1 with distributions $N_k(\delta_i, \sigma_i^2\sigma_{t_i}^{(i)} + \sigma_N^2\sigma_{t_N}^{(N)})$.

The (N − 1)--dimensional vector

$$T'_{\mathbf{t}} = T'_{t_1,\ldots,t_N} = \left(T^{(1)}_{t_1,t_N}, T^{(2)}_{t_2,t_N}, \ldots, T^{(N-1)}_{t_{N-1},t_N}\right) \tag{45}$$

will have a multivariate normal distribution with vector mean δ and covariance matrix

$$U_{\mathbf{t}} = \begin{pmatrix} \sigma_1^2\sigma^{(1)}_{t_1} & & & \\ & \sigma_2^2\sigma^{(2)}_{t_2} & & \\ & & \ddots & \\ & & & \sigma_{N-1}^2\sigma^{(N-1)}_{t_{N-1}} \end{pmatrix} \tag{46}$$

$$+\begin{pmatrix} \sigma_N^2\sigma^{(N)}_{t_N} & \sigma_N^2\sigma^{(N)}_{t_N} & \cdots & \sigma_N^2\sigma^{(N)}_{t_N} \\ \sigma_N^2\sigma^{(N)}_{t_N} & \sigma_N^2\sigma^{(N)}_{t_N} & \cdots & \sigma_N^2\sigma^{(N)}_{t_N} \\ \vdots & \vdots & \ddots & \vdots \\ \sigma_N^2\sigma^{(N)}_{t_N} & \sigma_N^2\sigma^{(N)}_{t_N} & \cdots & \sigma_N^2\sigma^{(N)}_{t_N} \end{pmatrix} \tag{47}$$

$$= \operatorname{diag}\left(\sigma_1^2\sigma^{(1)}_{t_1}, \ldots, \sigma_{N-1}^2\sigma^{(N-1)}_{t_{N-1}}\right) + \mathbf{J}\sigma_N^2\sigma^{(N)}_{t_N}\mathbf{J}', \tag{48}$$

where J is the Kronecker product $1_{N-1} \otimes I_k$ (composed by a column of $N-1$ copies of the identity matrix I_k).

For each $i = 1, \ldots,$, given $t_{i,1} < t_{i,2} < t_i$ such that $\sigma^{(i)}_{t_{i,1}} - \sigma^{(i)}_{t_i,2}$ is nonsingular, let

$$\eta_i = \left(\theta^{(i)}_{t_{i,2}} - \theta^{(i)}_{t_{i,1}}\right)' \left(\sigma^{(i)}_{t_{i,1}} - \sigma^{(i)}_{t_{i,2}}\right)^{-1} \left(\theta^{(i)}_{t_{i,2}} - \theta^{(i)}_{t_{i,1}}\right) \tag{49}$$

which are independent random variables, independent also of T_{t} and such that $u_i = \eta_i/\sigma_i^2$ has χ_k^2 distribution.

The covariance matrix U_{t} may be estimated by means of

$$V_{\mathbf{t}} = \operatorname{diag}\left(\eta_1\sigma^{(1)}_{t_1}, \ldots, \eta_{N-1}\sigma^{N-1}_{t_{N-1}}\right) + \mathbf{J}\eta_N\sigma^{(N)}_{t_N}\mathbf{J}' \tag{50}$$

with observed value

$$\widetilde{V}_{\mathbf{t}} = \operatorname{diag}\left(\widetilde{\eta}_1\sigma^{(1)}_{t_1}, \ldots, \widetilde{\eta}_{N-1}\sigma^{(1)}_{t_{N-1}}\right) + \mathbf{J}\widetilde{\eta}_N\sigma^{(N)}_{t_N}\mathbf{J}'. \tag{51}$$

As in (17), the matrix

$$R_t = \widetilde{V}_t^{-1/2}\left[\operatorname{diag}\left(\sigma_1^2\sigma_{t_1}^{(1)}\frac{\widetilde{\eta}_1}{\eta_1},\ldots,\sigma_{N-1}^2\sigma_{t_{N-1}}^{(N-1)}\frac{\widetilde{\eta}_{N-1}}{\eta_{N-1}}\right)\right.$$
$$\left.+\,\mathbf{J}\sigma_N^2\sigma_{t_N}^{(N)}\frac{\widetilde{\eta}_N}{\eta_N}\mathbf{J}'\right]\widetilde{V}_t^{-1/2}$$
$$= \widetilde{V}_t^{-1/2}\left[\operatorname{diag}\left(\sigma_{t_1}^{(1)}\frac{\widetilde{\eta}_1}{u_1},\ldots,\sigma_{t_{N-1}}^{(N-1)}\frac{\widetilde{\eta}_{N-1}}{u_{N-1}}\right)\right.$$
$$\left.+\,\mathbf{J}\sigma_{t_N}^{(N)}\frac{\widetilde{\eta}_N}{u_N}\mathbf{J}'\right]\widetilde{V}_t^{-1/2} \tag{52}$$

does not depend on any parameter and is independent of T_t; moreover its observed value is

$$\widetilde{R}_t = \widetilde{V}_t^{-1/2}U_t\widetilde{V}_t^{-1/2}. \tag{53}$$

Let us define

$$d_t = \widetilde{V}_t^{-1/2}T_t,$$
$$Z_t = \widetilde{R}_t^{-1/2}d_t \tag{54}$$

that have, under H_0, distributions $N_{(N-1)k}(0, \widetilde{R}_t)$ and $N_{(N-1)k}(0, I)$, respectively. So that, finally

$$W_t = Z_t'R_tZ_t \text{ and its observed value } \widetilde{W}_t = \widetilde{T}_t'\widetilde{V}_t^{-1}\widetilde{T}_t \tag{55}$$

are scalars not depending on the parameters and the distribution of W_t neither depends on $(\sigma_1^2,\ldots,\sigma_N^2)$. Moreover, W_t is a positive definite quadratic form in Z_t whose distribution under H_1 is $N_{(N-1)k}(\bar{\boldsymbol{\delta}}, I)$ with $\bar{\boldsymbol{\delta}} = (\widetilde{V}_t^{-1/2}U_t\widetilde{V}_t^{-1/2})^{-1/2}\widetilde{V}_t^{-1/2}\boldsymbol{\delta}$. Hence W_t is stochastically larger under H_1 than under H_0 and we get the following.

Proposition 8. *W_t is a generalized test process for testing $H_0: \delta = 0$ against H_1: $\boldsymbol{\delta} \neq \mathbf{0}$.*

An alternative expression will be more suitable for the calculation of the p values. For each $i = 1, \ldots, N-1$, let Gi be the $(N-1)k \times (N-1)k$ square matrix obtained by replacing in the matrix (46) the i block with $\widetilde{\eta}_i\sigma_{t_i}^{(i)}$ while the remaining are replaced with 0 (of order $k\times k$); that means

$$G_i = G_i(t_i) = \operatorname{diag}\left(\mathbf{0},\ldots,\mathbf{0},\widetilde{\eta}_i\sigma_{t_i}^{(i)},\mathbf{0},\ldots,\mathbf{0}\right). \tag{56}$$

Now

$$M_i = \widetilde{V}_{\mathbf{t}}^{-1/2} G_i \widetilde{V}_{\mathbf{t}}^{-1/2},$$

$$M_N = \widetilde{V}_{\mathbf{t}}^{-1/2} \mathbf{J} \widetilde{\eta_N} \sigma_{t_N}^{(N)} \mathbf{J}' \widetilde{V}_{\mathbf{t}}^{-1/2} \tag{57}$$

satisfy $\sum_{i=1}^{N-1} M_i + M_N = I$ and allow expressing

$$R_{\mathbf{t}} = \sum_{i=1}^{N-1} \frac{M_i}{u_i} + \frac{M_N}{u_N} = \sum_{i=1}^{N-1} \frac{M_i}{u_i} + \frac{I - \sum_{i=1}^{N-1} M_i}{u_N}. \tag{58}$$

Consequently

$$W_{\mathbf{t}} = Z_{\mathbf{t}}' \left(\sum_{i=1}^{N-1} \frac{M_i}{u_i} + \frac{I - \sum_{i=1}^{N-1} M_i}{u_N} \right) Z_{\mathbf{t}}$$

$$= \frac{1}{Nk} \overline{Z}_{\mathbf{t}}' \left(\sum_{i=1}^{N-1} \frac{M_i}{\beta_i} + \frac{I - \sum_{i=1}^{N-1} M_i}{1 - \sum_{i=1}^{N-1} \beta_i} \right) \overline{Z}_{\mathbf{t}}, \tag{59}$$

where

$$\overline{Z}_{\mathbf{t}} = \frac{\sqrt{Nk}}{\left(\sum_{i=1}^{N} u_i\right)^{1/2}} Z_{\mathbf{t}} \tag{60}$$

has a $(N - 1)k$-multivariate Student t-distribution, with Nk degrees of freedom, while $\beta_i = u_i / (\sum_{i=1}^{N-1} u_i)$ is such that $(\beta_1, \dots, \beta_{N-1})$ has a Dirichlet distribution with parameters $(k/2, \dots, k/2)$ and is independent of $\overline{Z}_{\mathbf{t}}$.

After the observation of $M_1, \dots, _{-1}$, the generalized p value of the given test is

$$p_{\mathbf{t}}(M_1, \dots, M_{N-1})$$

$$= P_{\delta=0} \left\{ \frac{1}{Nk} \overline{Z}_{\mathbf{t}}' \left(\sum_{i=1}^{N-1} \frac{M_i}{\beta_i} + \frac{I - \sum_{i=1}^{N-1} M_i}{1 - \sum_{i=1}^{N-1} \beta_i} \right) \overline{Z}_{\mathbf{t}} \right.$$

$$\left. \geq \overline{W}_{\mathbf{t}} \right\}. \tag{61}$$

As in Section 4.2, an analogous property of invariance of the test statistic $W_{\mathbf{t}}$ may be stated with respect to the transformations of each process

$$X_{t_i}^{(i)*} = Q_i X_{t_i}^{(i)} \tag{62}$$

for nonsingular matrices Q_i.

Confidence Region forδ

For any value of the unknown parameter $\boldsymbol{\delta}$, the difference $T_{\mathrm{t}}^{\boldsymbol{\delta}} = T_{\mathrm{t}} - \boldsymbol{\delta}$ has distribution $N_{(N-1)k}(0, U_{\mathrm{t}})$ and $Z_{\mathrm{t}}^{\boldsymbol{\delta}} = (\widetilde{V}_{\mathrm{t}}^{-1/2} U_{\mathrm{t}} \widetilde{V}_{\mathrm{t}}^{-1/2})^{-1/2} \widetilde{V}_{\mathrm{t}}^{-1/2} T_{\mathrm{t}}^{\boldsymbol{\delta}}$ is $N_{(N-1)k}(0, I)$. Hence

$$W_{\mathrm{t}}^{\boldsymbol{\delta}} = \left(Z_{\mathrm{t}}^{\boldsymbol{\delta}}\right)' R_{\mathrm{t}} Z_{\mathrm{t}}^{\boldsymbol{\delta}} \tag{63}$$

has the same distribution as W_{t} under $H_0 : \delta = 0$, which is independent of all the parameters, while its observed value $\widetilde{W_{\mathrm{t}}^{\delta}}$ does not depend on the nuisance parameters. The following proposition is thus established.

Proposition 9. $W_{\mathrm{t}}^{\boldsymbol{\delta}}$ *is a generalized pivotal quantity and a generalized confidence region for* δ *is given by* $\{\boldsymbol{\delta} \in \mathbb{R}^{(N-1)k} \mid \widetilde{W}_{\mathrm{t}}^{\boldsymbol{\delta}} \leq c_{\gamma}\}$, *if* $P\{W_{\mathrm{t}}^{\boldsymbol{\delta}} \leq c_{\gamma}\} = 1 - \gamma$.

As observed in Section 4.3, the just obtained confidence region may be written as

$$\mathscr{R} = \Big\{\boldsymbol{\delta} \in \mathbb{R}^{(N-1)k} \mid \left|a'\left(\widetilde{T}_{\mathrm{t}} - \boldsymbol{\delta}\right)\right| \leq c_{\gamma}^{1/2}\left(a'\widetilde{V}_{\mathrm{t}} a\right)^{1/2} \ \forall a \in \mathbb{R}^{(N-1)k},\ a \neq 0\Big\}. \tag{64}$$

For $j \leq N-1$ and $i \leq k$, let $a_{j,i}$ represent the vector in $\mathbb{R}^{(N-1)k}$ whose only nonzero term is 1 in the $(j-1)k+i$ position, whereas $a_{j,m,i} = a_{j,i} - a_{m,i}$ for $j < m \leq N-1$. Then

$$\begin{aligned}
a_{j,i}'\boldsymbol{\delta} &= \theta_i^{(j)} - \theta_i^{(N)},\\
a_{j,i}'\widetilde{T}_{\mathrm{t}} &= \left(\widetilde{\theta}_{t_j}^{(j)} - \widetilde{\theta}_{t_N}^{(N)}\right)_i,\\
a_{j,m,i}'\boldsymbol{\delta} &= \theta_i^{(j)} - \theta_i^{(m)},\\
a_{j,m,i}'\widetilde{T}_{\mathrm{t}} &= \left(\widetilde{\theta}_{t_j}^{(j)} - \widetilde{\theta}_{t_m}^{(m)}\right)_i.
\end{aligned} \tag{65}$$

And from (51) we get

$$\begin{aligned}
a_{j,i}'\widetilde{V}_{\mathrm{t}} a_{j,i} &= \widetilde{\eta}_j\left(\sigma_{t_j}^{(j)}\right)_{i,i} + \widetilde{\eta}_N\left(\sigma_{t_N}^{(N)}\right)_{i,i},\\
a_{j,m,i}'\widetilde{V}_{\mathrm{t}} a_{j,m,i} &= \widetilde{\eta}_j\left(\sigma_{t_j}^{(j)}\right)_{i,i} + \widetilde{\eta}_m\left(\sigma_{t_m}^{(m)}\right)_{i,i}.
\end{aligned} \tag{66}$$

Therefore, as in Section 4.3, $\mathscr{R}$ is contained in the intersection of all the intervals:

$$\left\{\left|\left(\theta_j - \theta_m\right)_i - \left(\widetilde{\theta}_{t_j}^{(j)} - \widetilde{\theta}_{t_m}^{(m)}\right)_i\right| \right.$$

$$\left. \leq c_\gamma^{1/2}\left[\widetilde{\eta}_j \sigma_{t_j}^{(j)} + \widetilde{\eta}_m \sigma_{t_m}^{(m)}\right]_{i,i}\right\} \tag{67}$$

for $1 \leq j < m \leq N$ and $1 \leq i \leq k$. This gives a set of simultaneous confidence intervals for all the components of the differences between the mean parameters of the N processes.

General Linear Hypotheses

When the null hypothesis $H_0 : \theta_1 = \theta_2 = \cdots = \theta_N$ is rejected, one can focus on testing a linear relation between the mean parameters. That is, if $\theta' = (\theta_1', \ldots, \theta_N')$, L is an $r \times Nk$ matrix, and l is a vector in $\mathbb{R}^r$, the objective is to test the null hypothesis $H_0' : \mathbf{L}\boldsymbol{\theta} = \mathbf{l}$ against the alternative $H_1' : \mathbf{L}\boldsymbol{\theta} \neq \mathbf{l}$.

We denote $\boldsymbol{\theta}_\mathbf{t}' = (\theta_{t_1}^{(1)'}, \ldots, \theta_{t_N}^{(N)'})$ which has distribution $N_{Nk}(\boldsymbol{\theta}, \mathbf{U_t})$, where $\mathbf{U_t} = \mathrm{diag}(\sigma_1^2 \sigma_{t_1}^{(1)}, \ldots, \sigma_N^2 \sigma_{t_N}^{(N)})$. Now, $T_\mathrm{t} = \mathbf{L}\boldsymbol{\theta}_\mathbf{t} - \mathbf{l}$ has $N_r(\mathbf{L}\boldsymbol{\theta} - \mathbf{l}, \mathbf{L}\mathbf{U_t}\mathbf{L}')$ distribution, whose mean vector is 0 under H_0'. Moreover $\mathrm{U_t}$ may be estimated by

$$\mathbf{V_t} = \mathbf{L}\,\mathrm{diag}\left(\eta_1 \sigma_{t_1}^{(1)}, \ldots, \eta_N \sigma_{t_N}^{(N)}\right)\mathbf{L}' \tag{68}$$

which has observed value $\widetilde{\mathbf{V}}_\mathbf{t} = \mathbf{L}\,\mathrm{diag}(\widetilde{\eta}_1 \sigma_{t_1}^{(1)}, \ldots, \widetilde{\eta}_N \sigma_{t_N}^{(N)})\mathbf{L}'$. Observing that, under H_0',

$$D_\mathbf{t} = \widetilde{\mathbf{V}}_\mathbf{t}^{-1/2} T_\mathbf{t} \tag{69}$$

has distribution $N_r(\mathbf{0}, \widetilde{\mathbf{V}}_\mathbf{t}^{-1/2}\mathbf{L}\mathbf{U_t}\mathbf{L}'\widetilde{\mathbf{V}}_\mathbf{t}^{-1/2})$, we get that

$$Z_\mathbf{t} = \left(\widetilde{\mathbf{V}}_\mathbf{t}^{-1/2}\mathbf{L}\mathbf{U_t}\mathbf{L}'\widetilde{\mathbf{V}}_\mathbf{t}^{-1/2}\right)^{-1/2} D_\mathbf{t} \tag{70}$$

has $N_r(\mathbf{0}, I)$ distribution. Therefore, if

$$R_\mathbf{t} = \widetilde{\mathbf{V}}_\mathbf{t}^{-1/2}\mathbf{L}\,\mathrm{diag}\left(\sigma_1^2 \sigma_{t_1}^{(1)} \frac{\widetilde{\eta}_1}{\eta_1}, \ldots, \sigma_N^2 \sigma_{t_N}^{(N)} \frac{\widetilde{\eta}_N}{\eta_N}\right)\mathbf{L}'\widetilde{\mathbf{V}}_\mathbf{t}^{-1/2}, \tag{71}$$

then

$$W_\mathbf{t} = Z_\mathbf{t}' R_\mathbf{t} Z_\mathbf{t} \tag{72}$$

is a one-dimensional random variable with distribution, under H_0', independent of the parameters and whose observed value, given by $\widetilde{W}_t = T_t'\widetilde{V}_t T_t$, does not depend on the parameters. Since W_t is a positive quadratic form in the normal vector Z_t whose mean under H_1 is different from 0, W_t is stochastically larger under H_1 than under H_0, so that the following proposition is proved.

Proposition 10. *W_t is a generalized test process for the contrast of $H_0' : \mathbf{L}\boldsymbol{\theta} = \mathbf{l}$ against $H_1' : \mathbf{L}\boldsymbol{\theta} \neq \mathbf{l}$.*

With the same matrix G_i defined in (56), one can put

$$M_i = M_i(t_i) = \widetilde{V}_t^{-1/2}\mathbf{L}G_i\mathbf{L}'\widetilde{V}_t^{-1/2} \tag{73}$$

that satisfy $\sum_{i=1}^{N} M_i = I$. Since

$$R_t = \sum_{i=1}^{N}\frac{M_i}{u_i}, \tag{74}$$

where $u_1,\dots,$ are mutually independent, χ_k^2 distributed and independent of Z_t, we get

$$W_t = \frac{1}{Nk}\overline{Z}_t'\left(\sum_{i=1}^{N-1}\frac{M_i}{\beta_i} + \frac{I - \sum_{i=1}^{N-1} M_i}{1 - \sum_{i=1}^{N-1}\beta_i}\right)\overline{Z}_t \tag{75}$$

with $\overline{Z}_t$, given by (60), distributed according to a $(N - 1)k$ multivariate Student t distribution and $\beta_i = u_i/(\sum_{i=1}^{N-1} u_i)$ such that $(\beta_1,\dots,\beta_{N-1})$ has a Dirichlet distribution with parameters $(k/2, \dots, k/2)$ and is independent of $\overline{Z}_t$. Thus, the generalized p value of the given test can be calculated as in (61), although the expressions of M_i are now different.

REFERENCES

1. K. Tsui and S. Weerahandi, "Generalized *p*-values in significance testing of hypotheses in the presence of nuisance parameters," *Journal of the American Statistical Association*, vol. 84, no. 406, pp. 602–607, 1989.
2. S. Weerahandi, "Generalized confidence intervals," *Journal of the American Statistical Association*, vol. 88, no. 423, pp. 899–905, 1993.
3. J. Gamage, T. Mathew, and S. Weerahandi, "Generalized p-values and generalized confidence regions for the multivariate Behrens-Fisher problem and MANOVA," *Journal of Multivariate Analysis*, vol. 88, no. 1, pp. 177–189, 2004.
4. S. H. Lin, J. C. Lee, and R. S. Wang, "Generalized inferences on the common mean vector of several multivariate normal populations," *Journal of Statistical Planning and Inference*, vol. 137, no. 7, pp. 2240–2249, 2007.
5. L.-W. Xu and S.-G. Wang, "A new generalized *p*-value for ANOVA under heteroscedasticity," *Statistics and Probability Letters*, vol. 78, no. 8, pp. 963–969, 2008.
6. J. T. Zhang, "An approximate hotelling T^2-test for heteroscedastic one-way MANOVA," *Open Journal of Statistics*, vol. 2, no. 1, pp. 1–11, 2012.
7. P. Ibarrola and R. Vélez, "Testing and confidence estimation of the mean of a multidimensional Gaussian process," *Statistics*, vol. 36, no. 4, pp. 317–327, 2002.
8. P. Ibarrola and R. Vélez, "On Behrens-Fisher problem for continuous time Gaussian processes," *Linear Algebra and Its Applications*, vol. 389, no. 1–3, pp. 63–76, 2004.

7

A New Test of Multivariate Nonlinear Causality

ZhidongBai[1] ,Yongchang Hui[2], Dandan Jiang[3] , Zhihui Lv[1] , Wing-Keung Wong[4,5], Shurong Zheng[1]

[1] KLASMOE and School of Mathematics and Statistics, Northeast Normal University, Changchun, China
[2] School of Mathematics and Statistics, Xi'an Jiaotong University, Xi'an, China
[3] School of Mathematics, Jilin University, Changchun, China
[4] Department of Finance, Asia University, Taichung, Taiwan
[5]Department of Economics, Lingnan University, Hong Kong, China

ABSTRACT

The multivariate nonlinear Granger causality developed by Bai et al. (2010) (Mathematics and Computers in simulation. 2010; 81: 5-17) plays an important role in detecting the dynamic interrelationships between two groups of variables. Following the idea of Hiemstra-Jones (HJ) test proposed by Hiemstra and Jones (1994) (Journal of Finance. 1994; 49(5): 1639-1664), they attempt to establish a central limit theorem (CLT) of their test statistic by applying the asymptotical property of multivariate *U*-statistic. However, Bai et al. (2016) (2016; arXiv: 1701.03992) revisit the HJ test and find that

Citation: Bai, Z., Hui, Y., Jiang, D., Lv, Z., Wong, W. K., & Zheng, S. (2018). A new test of multivariate nonlinear causality. PloS one, 13(1).DOI: https://doi.org/10.1371/journal.pone.0185155.

Copyright: © 2018 Bai et al. This is an open access article distributed under the terms of the Creative Commons Attribution 4.0 International (CC BY 4.0) License.

the test statistic given by HJ is NOT a function of *U*-statistics which implies that the CLT neither proposed by Hiemstra and Jones (1994) nor the one extended by Bai et al. (2010) is valid for statistical inference. In this paper, we re-estimate the probabilities and reestablish the CLT of the new test statistic. Numerical simulation shows that our new estimates are consistent and our new test performs decent size and power.

INTRODUCTION

After the pioneering work of Granger [1], Granger causality tests have been developed into a set of useful methods to detect causal relations between time series in economics and finance. Linear Granger causality tests within the linear autoregressive model class have been developed in many directions, e.g., Hurlin et al. [2] proposed a procedure for causality tests with panel data, Ghysels et al. [3] test for Granger causality with mixed frequency data based on the multiple-horizon framework established by Dufour and Renault [4] and Dufour et al. [5]. Though linear tests of Granger causality have been investigated very deeply, they are limited in their capability to detect nonlinear causality.

The real world is "almost certainly nonlinear" as Granger [6] notes, so it is more important to test the nonlinear causality. Baek and Brock [7] develop a nonlinear Granger causality test which is modified by Hiemstra and Jones [8] later on to study the bivariate nonlinear causal relationship between two series. Among the various tests of nonlinear Granger causality, the Hiemstra-Jones test (hereafter, the HJ test) proposed by Hiemstra and Jones [8] is the most cited by scholars and the most frequently applied by practitioners in economics and finance. There were over 1100 Google Scholar hits by September 2016, which illustrates its significance in the economics and finance literatures.

Bai et al. [9] extend the HJ test from bivariate setting to multivariate setting catering to the practical needs that economic and financial factors usually move together and influence others in groups. This extension encourages a large amount of applications. For example, Lam et al. [10] suggest to use such technics to make better investment decisions. Zheng and Chen [11] proposed a complete double selection method in identifying external influential factors for a particular stock market. Choudhry et al. [12] investigates the nonlinear dynamic co-movements between gold returns, stock market returns and stock market volatility during the recent global financial crisis. Choudhry et al. [13] investigate the relationship between stock market volatility and the business cycle in four major economies US, Canada, Japan and the UK.

However, several works note that counterintuitive results are obtained from the HJ test, Diks and Panchenko ([14, 15]) find that the HJ test is seriously over-rejecting in simulation studies. In accordance with the evidence presented by Diks and Panchenko ([14, 15]), Bai et al. [16] reinvestigate the HJ test and reveal some of the underlying reasons for the questionable performance of HJ test. They find that the estimators of the probabilities in the definition are not *U*-statistics as Hiemstra and Jones [8] claimed and the central limit theorem of the test statistics is not valid. Bai et al. [16] propose a set of consistent estimators of the probabilities in the definition of Hiemstra and Jones [8] and provide a new test statistic with its asymptotic distribution.

Considering the significant importance of the multivariate nonlinear Granger causality test, there is an urgent need to reinvestigate Bai et al. [9] and extend Bai et al. [16] to multivariate setting. The remainder of this paper is organized as follows. In Section 2, we simply review the procedure of the multivariate nonlinear Granger causality test (here after BWZ test) extended by Bai et al. [9]. In Section 3, we re-estimate the probabilities in the definition of Bai et al. [9] and establish the asymptotic distribution of the new test statistics. Simulation results are presented in Section 4. Finally, we provide some concluding remarks in Section 5.

THE MULTIVARIATE NONLINEAR CAUSALITY TEST EXTENDED FROM HJ TEST

Bai et al. [9] consider two strictly stationary and weakly dependent vector time series processes $X_t = (X_{1,t}, X_{2,t}, \cdots, X_{n1},t)'$, $Y_t = (Y_{1,t}, Y_{2,t}, \cdots, Y_{n2},t)'$. The m_{xi}-length lead vector of $X_{i,t}$ is defined as $X_{i,t}^{m_{x_i}} \equiv (X_{i,t}, X_{i,t+1}, \cdots, X_{i,t+m_{x_i}-1}), m_{x_i} = 1, 2, \cdots, t = 1, 2, \cdots$, similarly L_{xi}-length lag vector of $X_{i,t}$, and L_{yi}-length lag vector of $Y_{i,t}$ are defined as $X_{i,t-L_{x_i}}^{L_{x_i}} \equiv (X_{i,t-L_{x_i}}, X_{i,t-L_{x_i}+1}, \cdots, X_{i,t-1}), L_{x_i} = 1, 2, \cdots, t = L_{x_i}+1, L_{x_i}+2, \cdots, Y_{i,t-L_{y_i}}^{L_{y_i}} \equiv (Y_{i,t-L_{y_i}}, Y_{i,t-L_{y_i}+1}, \cdots, Y_{i,t-1}), L_{y_i} = 1, 2, \cdots, t = L_{y_i}+1, L_{y_i}+2, \cdots$. Denote $M_x = (m_{x1}, \cdots, m_{xn1})$, $m_x = \max(m_{x1}, \cdots, m_{xn1})$, $L_x = (L_{x1}, \cdots, L_{xn1})$, $l_x = \max(L_{x1}, \cdots, L_{xn1})$, $L_y = (L_{y1}, \cdots, L_{yn2})$, $l_y = \max(L_{y1}, \cdots, L_{yn2})$. For given M_x, L_x, L_y, e, Bai et al. [9] define that

$$
\begin{aligned}
\{\| X_t^{M_x} - X_s^{M_x} \| < e\} &\equiv \{\| X_{i,t}^{m_{x_i}} - X_{i,s}^{m_{x_i}} \| < e, \quad for \quad all \quad i = 1, 2, \cdots, n_1\} \\
\{\| X_{t-L_x}^{L_x} - X_{s-L_x}^{L_x} \| < e\} &\equiv \{\| X_{i,t-L_{x_i}}^{L_{x_i}} - X_{i,s-L_{x_i}}^{L_{x_i}} \| < e, \quad for \quad all \quad i = 1, 2, \cdots, n_1\} \\
\{\| Y_{t-L_y}^{L_y} - Y_{s-L_y}^{L_y} \| < e\} &\equiv \{\| Y_{i,t-L_{y_i}}^{L_{y_i}} - Y_{i,s-L_{y_i}}^{L_{y_i}} \| < e, \quad for \quad all \quad i = 1, 2, \cdots, n_2\},
\end{aligned}
$$

where$\|\cdot\|$ denotes the maximum norm defined as $\|X - Y\| = \max(|x_1 - y_1|, |x_2 - y_2|, \cdots, |x_n - y_n|)$ for any two vectors $X = (x_1, \cdots, x_n)$ and $Y = (y_1, \cdots, y_n)$.

Definition 1 *The vector time series $\{Y_t\}$ does not strictly Granger cause another vector time series $\{X_t\}$ if*

$$
\begin{aligned}
&P(\| X_t^{M_x} - X_s^{M_x} \| < e \,|\, \| X_{t-L_x}^{L_x} - X_{s-L_x}^{L_x} \| < e, \| Y_{t-L_y}^{L_y} - Y_{s-L_y}^{L_y} \| < e) \\
&= P(\| X_t^{M_x} - X_s^{M_x} \| < e \,|\, \| X_{t-L_x}^{L_x} - X_{s-L_x}^{L_x} \| < e), \qquad (1)
\end{aligned}
$$

where $P(\cdot|\cdot)$ denotes conditional probability.

Using the notation

$$
\begin{aligned}
C_1(M_x + L_x, L_y, e) &\equiv P(\| X_{t-L_x}^{M_x+L_x} - X_{s-L_x}^{M_x+L_x} \| < e, \| Y_{t-L_y}^{L_y} - Y_{s-L_y}^{L_y} \| < e) \\
C_2(L_x, L_y, e) &\equiv P(\| X_{t-L_x}^{L_x} - X_{s-L_x}^{L_x} \| < e, \| Y_{t-L_y}^{L_y} - Y_{s-L_y}^{L_y} \| < e) \\
C_3(M_x + L_x, e) &\equiv P(\| X_{t-L_x}^{M_x+L_x} - X_{s-L_x}^{M_x+L_x} \| < e) \\
C_4(L_x, e) &\equiv P(\| X_{t-L_x}^{L_x} - X_{s-L_x}^{L_x} \| < e),
\end{aligned}
$$

Bai, et al. [9] re-express Eq (1) as

$$
\frac{C_1(M_x + L_x, L_y, e)}{C_2(L_x, L_y, e)} = \frac{C_3(M_x + L_x, e)}{C_4(L_x, e)}. \qquad (2)
$$

For two sets of simultaneous samples $\{x_{i,t}, i = 1, \cdots, n_1, t = 1, \cdots, T\}$ and $\{y_{i,t}, i = 1, \cdots, n_2, t = 1, \cdots, T\}$, they propose the following test statistic

$$
\sqrt{n}\left(\frac{C_1(M_x + L_x, L_y, e, n)}{C_2(L_x, L_y, e, n)} - \frac{C_3(M_x + L_x, e, n)}{C_4(L_x, e, n)}\right), \qquad (3)
$$

where

$$
\begin{aligned}
&C_1(M_x + L_x, L_y, e, n) \\
&\equiv \frac{2}{n(n-1)} \sum\sum_{t<s} \prod_{i=1}^{n_1} I(x_{i,t-L_{x_i}}^{m_{x_i}+L_{x_i}}, x_{i,s-L_{x_i}}^{m_{x_i}+L_{x_i}}, e) \cdot \prod_{i=1}^{n_2} I(y_{i,t-L_{y_i}}^{L_{y_i}}, y_{i,s-L_{y_i}}^{L_{y_i}}, e), \\
&C_2(L_x, L_y, e, n) \equiv \frac{2}{n(n-1)} \sum\sum_{t<s} \prod_{i=1}^{n_1} I(x_{i,t-L_{x_i}}^{L_{x_i}}, x_{i,s-L_{x_i}}^{L_{x_i}}, e) \cdot \prod_{i=1}^{n_2} I(y_{i,t-L_{y_i}}^{L_{y_i}}, y_{i,s-L_{y_i}}^{L_{y_i}}, e), \\
&C_3(M_x + L_x, e, n) \equiv \frac{2}{n(n-1)} \sum\sum_{t<s} \prod_{i=1}^{n_1} I(x_{i,t-L_{x_i}}^{m_{x_i}+L_{x_i}}, x_{i,s-L_{x_i}}^{m_{x_i}+L_{x_i}}, e), \\
&C_4(L_x, e, n) \equiv \frac{2}{n(n-1)} \sum\sum_{t<s} \prod_{i=1}^{n_1} I(x_{i,t-L_{x_i}}^{L_{x_i}}, x_{i,s-L_{x_i}}^{L_{x_i}}, e),
\end{aligned}
$$

and

$$I(x,y,e)=\begin{cases}0, & \text{if } \| x-y \| > e \\ 1, & \text{if } \| x-y \| \le e\end{cases}.$$

Remark: Following the instruction of Hiemstra and Jones [8], Bai et al. [9] take $C_j(*, n)$s as multivariate U-statistic estimators of their counterparts $C_j(*)$s and apply the asymptotic property of U-statistic to show the limiting results for the test statistics (3). However the $C_j(*, n)$s are not U-statistics, because the expectations of the general terms are not the same. Moreover, the $C_j(*)$s are related to the indices t and s (in fact, related to $|t - s|$ for strongly stationary processes), while the $C_j(*, n)$s were independent of t and s for summing up over them. Therefore, the $C_j(*, n)$ estimators are neither consistent nor asymptotic normal estimators of their counterparts $C_j(*)$.

A NEW MULTIVARIATE NONLINEAR CAUSALITY TEST

We first remind the reader that the pair (s, t) (in fact, $|t - s|$ for strongly stationary processes) in Eq (1) of Definition 1 is a key parameter of the probabilities $C_j(*)$. In fact, both Hiemstra and Jones [8] and Bai et al. [9] note this, and there is no problem in Eq (1) of Definition 1. However, it seems that they overlooked this fact in their proposed estimation of $C_j(*)$. The improper estimators $C_j(*, n)$ thus lead to an invalid asymptotic distribution of the test statistic.

We now begin to state the procedure for our new test. If $\{X_{i,1}\}$ and $\{Y_{j,1}\}$, $i \in \{1, 2, \cdots, n_1\}, j \in \{1, 2, \cdots, n_2\}$ share a common standard deviation 1, and thereby share a common scale parameter. For any given pair (s, t), we denote

$$C_1(M_x + L_x, L_y, e; t, s) \equiv P(\| X_{t-L_x}^{M_x+L_x} - X_{s-L_x}^{M_x+L_x} \| < e, \| Y_{t-L_y}^{L_y} - Y_{s-L_y}^{L_y} \| < e)$$

$$C_2(L_x, L_y, e; t, s) \equiv P(\| X_{t-L_x}^{L_x} - X_{s-L_x}^{L_x} \| < e, \| Y_{t-L_y}^{L_y} - Y_{s-L_y}^{L_y} \| < e)$$

$$C_3(M_x + L_x, e; t, s) \equiv P(\| X_{t-L_x}^{M_x+L_x} - X_{s-L_x}^{M_x+L_x} \| < e)$$

$$C_4(L_x, e; t, s) \equiv P(\| X_{t-L_x}^{L_x} - X_{s-L_x}^{L_x} \| < e)$$

Under the assumption of the stationary, for the given pair (s, t), if $|t - s| = l$, we denote $C_1(M_x + L_x, L_y, e; t, s) \equiv C_1(M_x + L_x, L_y, e; t, l)$, which does not depend on t, so we can write $C_1(M_x + L_x, L_y, e; l)$ instead of $C_1(M_x + L_x, L_y, e, t; l)$, the same to the others. So under the assumption of strictly stationary, for each $l > 0$, we examine whether there is nonlinear Granger causality from $\{Y_t\}$ to $\{X_t\}$ by testing the following hypothesis

$$H_0 \ : \quad \frac{C_1(M_x + L_x, L_y, e; l)}{C_2(L_x, L_y, e; l)} = \frac{C_3(M_x + L_x, e; l)}{C_4(L_x, e; l)} . \tag{4}$$

If we consider two sets of simultaneous samples (To implement the test, each series is standardized at first so that all series share a common standard deviation, and thereby share a common scale parameter.) $\{x_{i,t}, i = 1, \cdots, n_1, t = 1, \cdots, T\}$ and $\{y_{j,t}, j = 1, \cdots, n_2, t = 1, \cdots, T\}$, we first provide the consistent estimators of $C_1(M_x + L_x, L_y, e; l)$, $C_2(L_x, L_y, e; l)$, $C_3(M_x + L_x, e; l)$ and $C_4(L_x, e; l)$ are

$$\hat{C}_1(M_x + L_x, L_y, e; l) \equiv \frac{1}{n} \sum_{t=L_{xy}+1}^{T-l-m_x+1} \prod_{i=1}^{n_1} I(x_{i,t-L_{x_i}}^{m_{x_i}+L_{x_i}}, x_{i,t+l-L_{x_i}}^{m_{x_i}+L_{x_i}}, e) \cdot \prod_{j=1}^{n_2} I(y_{j,t-L_{y_j}}^{L_{y_j}}, y_{j,t+l-L_{y_j}}^{L_{y_j}}, e),$$

$$\hat{C}_2(L_x, L_y, e; l) \equiv \frac{1}{n} \sum_{t=L_{xy}+1}^{T-l-m_x+1} \prod_{i=1}^{n_1} I(x_{i,t-L_{x_i}}^{L_{x_i}}, x_{i,t+l-L_{x_i}}^{L_{x_i}}, e) \cdot \prod_{j=1}^{n_2} I(y_{j,t-L_{y_j}}^{L_{y_j}}, y_{j,t+l-L_{y_j}}^{L_{y_j}}, e),$$

$$\hat{C}_3(M_x + L_y, e; l) \equiv \frac{1}{n} \sum_{t=L_{xy}+1}^{T-l-m_x+1} \prod_{i=1}^{n_1} I(x_{i,t-L_{x_i}}^{m_{x_i}+L_{x_i}}, x_{i,t+l-L_{x_i}}^{m_{x_i}+L_{x_i}}, e),$$

$$\hat{C}_4(L_x, e; l) \equiv \frac{1}{n} \sum_{t=L_{xy}+1}^{T-l-m_x+1} \prod_{i=1}^{n_1} I(x_{i,t-L_{x_i}}^{L_{x_i}}, x_{i,t+l-L_{x_i}}^{L_{x_i}}, e),$$

where $L_{xy} = \max(L_x, L_y)$, $n = T - L_{xy} - l - m_x + 1$.

The consistency of our proposed estimators can be shown straightforwardly and the detail of the proof is omitted. We use a simple numerical study to show that our estimators are consistent whereas those of Bai et al. [9] are not. Let

$$\begin{pmatrix} X_{1,t} \\ X_{2,t} \end{pmatrix} = \begin{pmatrix} 1 & 0 \\ 0 & 1 \end{pmatrix} \begin{pmatrix} a_{1,t-1} \\ a_{2,t-1} \end{pmatrix} + \begin{pmatrix} a_{1,t} \\ a_{2,t} \end{pmatrix},$$

where $\{a_{1,t}\}$, $\{a_{2,t}\}$ are i.i.d. and mutually independent random variables generated from $N(0, 1)$, while $\{Y_t\}$ could be any stationary sequence. Let $l = 1$, $L_x = L_y = M_x = 1$. We can calculate the exact values of $C_4(L_x, e; l)$, which are 0.2709 and 0.5057, respectively, when $e = 1$ and $e = 1.5$. For simplicity, we denote the true value of $C_4(L_x, e; l)$, the estimate proposed by Bai et al. [9] and our new estimate as C_4, C^BWZ4 and C^4, respectively, in Table 1. Additionally, Table 1 provides the estimated values with their corresponding relative estimation errors in brackets when $T = 1000$, 2000 and 4000. It is obvious that C^BWZ4 is not consistent.

Table 1: $C_4(L_x, e; l)$ and estimated values

T=	e = 1			e = 1.5		
	C_4	C^4	C^BWZ4	C_4	C^4	C^BWZ4
1000	0.2709	0.2497(7.83%)	0.0154(94.32%)	0.5057	0.4774(5.60%)	0.0732(85.53%)
2000	0.2709	0.2639(2.58%)	0.0176(93.50%)	0.5057	0.4847(4.15%)	0.0795(84.28%)
4000	0.2709	0.2692(0.63%)	0.0193(92.88%)	0.5057	0.4909(2.93%)	0.0820 (83.78%)

Note: The true value of $C_4(L_x, e; l)$ is denoted C_4, the BWZ estimate and our new estimate are denoted C^BWZ4 and C^4, respectively. The relative estimation errors are in the accompanying brackets.

Now, we propose

$$T_n = \sqrt{n}\left(\frac{\hat{C}_1(M_x + L_x, L_y, e, l)}{\hat{C}_2(L_x, L_y, e, l)} - \frac{\hat{C}_3(M_x + L_x, e, l)}{\hat{C}_4(L_x, e, l)}\right) \tag{5}$$

as the test statistic, and we derive the following asymptotic distribution of T_n for the Granger causality test.

Theorem 1 *Stationary sequences* $\{x_{i,t}, i = 1, \cdots, n_1, t = 1, \cdots, T\}$ *and* $\{y_{j,t}, j = 1, \cdots, n_2, t = 1, \cdots, T\}$ *are strong mixing, with mixing coefficients satisfying the conditions of Lemma 1 presented in Appendix, for given values of* l, L_x, L_y, M_x *and* $e > 0$, *under the null hypothesis that* $\{Y_t\}$ *does not strictly Granger cause* $\{X_t\}$, *then the test statistic is defined in* (5)

$$\sqrt{n}\left(\frac{\hat{C}_1(M_x + L_x, L_y, e, l)}{\hat{C}_2(L_x, L_y, e, l)} - \frac{\hat{C}_3(M_x + L_x, e, l)}{\hat{C}_4(L_x, e, l)}\right) \xrightarrow{d} N(0, \sigma^2(M_x, L_x, L_y, e, l)).$$

The asymptotic variance $\sigma^2(M_x, L_x, L_y, e, l)$ with its consistent estimator $\hat{\sigma}^2$(Mx,Lx,Ly,e,l) and the proof of theorem 1 are given in the Appendix (With some basic definitions from [17]). The hypothesis H_0 defined in (4) is rejected at α if

$$|T_n| / \hat{\sigma}(M_x, L_x, L_y, e, l) > z_{\alpha/2},$$

where $z_{\alpha/2}$ is the up $\alpha/2$ quantile of the standard normal distribution. In this situation, we will conclude that there exists nonlinear Granger causality from $\{Y_t\}$ to $\{X_t\}$.

There are several possible methods to estimate the asymptotic variance $\sigma^2(M_x, L_x, L_y, e, l)$. A model-based approach uses known laws of $\{X_t\}$ and $\{Y_t\}$ to calculate the expectations in the formula given in the Appendix and simply substitutes $C_j(*)$, $j = 1, 2, 3, 4$ with their corresponding estimates. However, in practice, we can hardly avoid model misspecification and may obtain improper laws of $\{X_t\}$ and $\{Y_t\}$. We suggest the use of bootstrap methods as in the simulation studies we use to test hypothesis H_0.

NUMERICAL STUDY

Simulaiton

In this subsection, we perform numerical studies using simulations to illustrate the applicability and superiority of the new multivariate nonlinear Granger causality test developed in Section 3. Let R be the times of rejecting the null hypothesis that Y_t does not strictly Granger cause X_t nonlinearly in 10,000 replications at the α level, and thus, the empirical power is $R/10,000$. In our simulation, the level $\alpha = 0.05$, we standardized the series and chose the same lag length and lead length: $L_x = L_y = M_x = 1$. We set three situations of l and two situations of e: $l = 1$, $l = 2$, $l = 3$ and $e = 1$, $e = 1.5$.

Consider the following model:

$$X_t = \beta Y_{1,t-1} Y_{2,t-1} + \varepsilon_t, \qquad (6)$$

where $\{(Y_{1,t}, Y_{2,t})'\}$ are i.i.d. and mutually independent random variables generated from standard normal distribution $N(0, 1)$, $\{\varepsilon_t\}$ is Gaussian white noise generated from $N(0, 0.1)$ and independent of $\{Y_{1,t}\}$, $\{Y_{2,t}\}$. There is no nonlinear Granger causality from Y_t to X_t when $\beta = 0$, and causality strengthens when β increases.

From the results displayed in Table 2, we conclude first that our test possesses decent size, as we can see when $\beta = 0$ the empirical size are all closed to the test level 0.05 for different settings of parameters and sample size. Second, our test possesses very appropriate power, as we see that empirical power increases as β increases, especially when sample size is 500 the empirical power sharply increase to 1. Further, we find that different settings of e may influence the test results. Though the influence is little in our simulation, we still suggest that practitioners choose a couple of different values of e.

Table 2: Test multivariate nonlinear Granger causality form Y_t **to** X_t

$T = 200$	$e = 1$			$e = 1.5$		
	$l = 1$	$l = 2$	$l = 3$	$l = 1$	$l = 2$	$l = 3$
$\beta = 0$	0.0419	0.0441	0.0432	0.0444	0.0506	0.0438
$\beta = 0.1$	0.1509	0.2341	0.2184	0.3647	0.5209	0.5121

$\beta = 0.2$	0.5629	0.7425	0.7284	0.8352	0.9416	0.9416
$\beta = 0.3$	0.8178	0.9323	0.9267	0.929	0.9810	0.9825
$\beta = 0.4$	0.8994	0.9712	0.9719	0.9512	0.9878	0.9889
$\beta = 0.5$	0.9366	0.9812	0.9808	0.9586	0.9914	0.9915
$\beta = 0.6$	0.9475	0.9870	0.9862	0.9640	0.9918	0.9940
$\beta = 0.7$	0.9574	0.9875	0.9874	0.9664	0.9921	0.9942
$\beta = 0.8$	0.9615	0.9888	0.9882	0.9688	0.9926	0.9953
$\beta = 0.9$	0.9633	0.9896	0.9901	0.9711	0.9923	0.994
$T = 500$	$e = 1$			$e = 1.5$		
	$l = 1$	$l = 2$	$l = 3$	$l = 1$	$l = 2$	$l = 3$
$\beta = 0$	0.0560	0.0560	0.0478	0.0501	0.0529	0.0423
$\beta = 0.1$	0.3969	0.5081	0.5344	0.7835	0.9043	0.9017
$\beta = 0.2$	0.9622	0.9908	0.9900	0.9994	1.0000	1.0000
$\beta = 0.3$	0.9989	1.0000	0.9997	0.9999	1.0000	1.0000
$\beta = 0.4$	0.9998	1.0000	1.0000	1.0000	1.0000	1.0000
$\beta = 0.5$	1.0000	1.0000	1.0000	1.0000	1.0000	1.0000
$\beta = 0.6$	1.0000	1.0000	1.0000	1.0000	1.0000	1.0000
$\beta = 0.7$	1.0000	1.0000	1.0000	1.0000	1.0000	1.0000
$\beta = 0.8$	1.0000	1.0000	1.0000	1.0000	1.0000	1.0000
$\beta = 0.9$	1.0000	1.0000	1.0000	1.0000	1.0000	1.0000

Note: Y_t and X_t are from the model present in Eq (6). $L_x = L_y = M_x = 1$ in our test. Simulation is conducted with the test level $\alpha = 5\%$, and 10,000 replications.

Application

Now we apply our new method to detect the nonlinear causality from returns to exchanging volumes on China's stock market. To eliminate the potential influence of asymmetry information caused by language barriers, different accounting standards and foreign currencies exchanging, we consider China's A shares which are denominated in the local currency the Yuan (also called RMB) and traded among China's citizens. A shares listed on the Shanghai Stock Exchange are named SHA while the ones listed on the Shenzhen Stock Exchange are named SZA.

We denote the prices of SHA and SZA at t as P_t^{SH} and P_t^{SZ} and denote the according trading volumes as V_t^{SH} and V_t^{SZ}. The daily stock returns of SHA and SZA (from Jan. 1, 2010 to Dec. 30, 2016) are expressed as $Y_{1t} = \ln(P_t^{SH}/P_{t-1}^{SH})$ and $Y_{2t} = \ln(P_t^{SZ}/P_{t-1}^{SZ})$ respectively, the percentage changes in trading volume are expressed as $X_{1t} = \ln(V_t^{SH}/V_{t-1}^{SH})$ and $X_{2t} = \ln(V_t^{SZ}/V_{t-1}^{SZ})$. We apply our new multivariate causality test proposed in section 2.2 to detect the causalities between $(Y_{1t}, Y_{2t})'$ and $(X_{1t}, X_{2t})'$, we let $L_{x1} = L_{x2} = L_{y1} = L_{y2} = m_{x1} = m_{x2} = 1$ and consider $l = 1$, $l = 2$ and $l = 3$, $e = 1$ and $e = 1.5$.

Table 3 presents some descriptive statistics of Y_{1t}, Y_{2t}, X_{1t} and X_{2t}. It is worth noting that the standard deviations of Y_{1t}, Y_{2t}, X_{1t} and X_{2t} are 0.014843, 0.017552, 0.195853 and 0.171018 respectively. To implement our proposed test, Y_{1t}, Y_{2t}, X_{1t} and X_{2t} need to be standardized at first so that all series share a common standard deviation 1. To avoid tedious notation we also denote the standardized sequences $(Y_{1t} - Mean(Y_{1t}))/SD(Y_{1t})$, $(Y_{2t} - Mean(Y_{2t}))/SD(Y_{2t})$, $(X_{1t} - Mean(X_{1t}))/SD(X_{1t})$ and $(X_{2t} - Mean(X_{2t}))/SD(X_{2t})$ as Y_{1t}, Y_{2t}, X_{1t} and X_{2t}. Table 4 reports the results of detecting causality from $(Y_{1t}, Y_{2t})'$ to $(X_{1t}, X_{2t})'$ as well as two pairs of subset, Y_{1t} to X_{1t} and Y_{2t} to X_{2t}. Generally speaking, there exists causality from stock returns to volume changes, as we can see on the row which testing no causality from $(Y_{1t}, Y_{2t})' \rightarrow (X_{1t}, X_{2t})'$, p values are all much smaller than level 0.05. Further, we try to dig which subset possess the dynamic causal explanatory ability. Results on the last row of Table 3 strongly suggest that the returns of SZA nonlinearly cause the trading volume of SZA. However, the causal explanatory ability is weak from the returns of SHA to the trading volume of SHA. We think these results coincide with practical situation, noticing that companies listed on the Shanghai Stock Exchange are likely to be big state-owned companies, while those listed on the Shenzhen Stock Exchange tend to be smaller export-oriented companies. Investors who invest on big state-owned companies are tend to be value investment and would like to hold their stocks with in a long time but pay less attention on the price changing comparing to these invest on Shenzhen A shares.

Table 3: Descriptive statistics of the series Y_{1t}, Y_{2t}, X_{1t} **and** X_{2t}

Series	Min	Max	Mean	Median	Standard Deviation
Y_{1t}	-0.088694	0.055990	-0.000027	0.000482	0.014843
Y_{2t}	-0.086205	0.063390	0.000290	0.001650	0.017552
X_{1t}	-1.219501	1.400890	0.000118	-0.010721	0.195853
X_{2t}	-1.263281	1.380028	0.000490	-0.004409	0.171018

Note: It is worth noting that the standard deviations of the four series are not the same. To implement the test, all series need to be standardized at first so that all series share a common standard deviation 1.

Table 4: Testing nonlinear causality from prices to trading volumes of China's A shares

H_0: No causality from left to right	$e = 1$			$e = 1.5$		
	$l = 1$	$l = 2$	$l = 3$	$l = 1$	$l = 2$	$l = 3$
$(Y_{1t}, Y_{2t})' \rightarrow (X_{1t}, X_{2t})'$	0.0209*	0.0722	0.0096**	0.0077**	0.0089**	0.0013**
$Y_{1t} \rightarrow X_{1t}$	0.0548	0.0296*	0.0251*	0.0052**	0.0609	0.0977
$Y_{2t} \rightarrow X_{2t}$	0.0065**	0.0020**	0.0005**	0.0007**	0.0005**	0.0001**

Note: "*" and "**" mean significant at level 0.05 and 0.01 respectively.

CONCLUSION AND REMARKS

In this paper, we reinvestigate the multivariate nonlinear Granger causality test extended by Bai et al. [9] which attempt to uncover significant nonlinearities in the dynamic interrelationships between two groups of variables. We find that Bai et al. [9] as well as Hiemstra and Jones [8] take the estimators of the probabilities in their definition as U-statistics and establish a CLT of the test statistic by applying the asymptotic property of U-statistics. After revealing that the estimators proposed by Bai et al. [9] is not U- statistics, we show that their estimators are also not consistent.

The procedure of our new test begins with presenting consistent estimators of probabilities in the definition. Numerical study supports that our estimators are consistent, further our new test possesses admirable properties both in size and power.

There are still amounts of appealing aspects in nonlinear Granger causality test. It is worth noting that Diks and Wolski [18] extend the test in Diks and Panchenko [15] which highlight a need for substitutions for the relationship tested in the HJ test.

APPENDIX

A1: Central Limit Theorems for Strong Mixing Stationary Sequence

$\{(Z_t, F_t), -\infty < t < \infty\}$ is a stochastic process defined on the probability space (Ω, F, P). The history and the future of Z_t are σ-algebras M∞t={Fs,s>t} and σ-algebras $\mathfrak{M}_t^{\infty} = \{\mathcal{F}_s, s > t\}$ respectively.

Let $\{(Z_i, F_i)\}$ be a stationary sequence with $E(Z_i) = 0$, $E(Z_i^2) < 0$, and set $S_n^m = \sum_{i=m+1}^{n+m} Z_i$, $\sigma_n^2 = Var(S_n^m)$. We shall say that the sequence satisfies the central limit theorem if

$$\lim_{n\to\infty} P\left\{\frac{S_n^m}{\sigma_n} < z\right\} = (2\pi)^{-\frac{1}{2}} \int_{-\infty}^{z} e^{-\frac{1}{2}u^2} du = \Phi(z) .$$

Definition A1: A stationary process $\{Z_t\}$ is said to be strongly mixing (completely regular) if $\alpha(\tau) = \sup_{A\in \mathcal{M}_{-\infty}^0, B\in \mathcal{M}_\tau^\infty} |P(AB) - P(A)P(B)| \to 0$ as $\tau \to \infty$ through positive values.

Lemma A1: Let the stationary sequence $\{Z_i\}$ satisfy the strong mixing condition with mixing coefficient $\alpha(n)$, and let $E|Z_i|^{2+\delta} < \infty$ for some $\delta > 0$. If $\sum_{n=1}^{\infty} \alpha(n)^{\delta/(2+\delta)} < \infty$, then $\sigma^2 = E(Z_0^2) + 2\sum_{j=1}^{\infty} E(Z_0 Z_j) < \infty$, and if $\sigma \neq 0$, then $\lim_{n\to\infty} P\{\sigma^{-1} n^{-\frac{1}{2}} \sum_{i=1}^{n} Z_i < z\} = \Phi(z)$.

Readers can be referred to Ibragimov [17] for a proof and detailed discussion.

A2: Proof of Theorem 1

Assume $\{x_{i,1}, x_{i,2}, \cdots, x_{i,T}\}$ and $\{y_{j,1}, y_{j,2}, \cdots, y_{j,T}\}$, $i \in \{1, 2, \cdots, n_1\}$, $j \in \{1, 2, \cdots, n_2\}$ are both strong mixing stationary sequences whose mixing coefficient satisfying the conditions in Lemma 1. Then the following four sequences

$$\left\{Z_{1t} = n^{-1}\left(\prod_{i=1}^{n_1} I(x_{i,t-L_{x_i}}^{m_{x_i}+L_{x_i}}, x_{i,t+l-L_{x_i}}^{m_{x_i}+L_{x_i}}, e)\right) \cdot \prod_{j=1}^{n_2} I(y_{j,t-L_{y_j}}^{L_{y_j}}, y_{j,t+l-L_{y_j}}^{L_{y_j}}, e) - C_1(M_x + L_x, L_y, e; l)\right\},$$
$$\left\{Z_{2t} = n^{-1}\left(\prod_{i=1}^{n_1} I(x_{i,t-L_{x_i}}^{L_{x_i}}, x_{i,t+l-L_{x_i}}^{L_{x_i}}, e)\right) \cdot \prod_{j=1}^{n_2} I(y_{j,t-L_{y_j}}^{L_{y_j}}, y_{j,t+l-L_{y_j}}^{L_{y_j}}, e) - C_2(L_x, L_y, e; l)\right\},$$
$$\left\{Z_{3t} = n^{-1}\left(\prod_{i=1}^{n_1} I(x_{i,t-L_{x_i}}^{m_{x_i}+L_{x_i}}, x_{i,t+l-L_{x_i}}^{m_{x_i}+L_{x_i}}, e)\right) - C_3(L_x, L_y, e; l)\right\},$$
$$\left\{Z_{4t} = n^{-1}\left(\prod_{i=1}^{n_1} I(x_{i,t-L_{x_i}}^{L_{x_i}}, x_{i,t+l-L_{x_i}}^{L_{x_i}}, e)\right) - C_4(L_x, e; l)\right\},$$
$$t = L_{xy} + 1, \cdots, T - l - m_x + 1,$$

satisfy the conditions of Lemma 1, where $n = T - L_{xy} - l - m_x + 1$ and

$$C_1(M_x + L_x, L_y, e; l) \equiv P(\| X_{t-L_x}^{M_x+L_x} - X_{t+l-L_x}^{M_x+L_x} \| < e, \| Y_{t-L_y}^{L_y} - Y_{t+l-L_y}^{L_y} \| < e),$$
$$C_2(L_x, L_y, e; l) \equiv P(\| X_{t-L_x}^{L_x} - X_{t+l-L_x}^{L_x} \| < e, \| Y_{t-L_y}^{L_y} - Y_{t+l-L_y}^{L_y} \| < e)$$
$$C_3(M_x + L_x, e; l) \equiv P(\| X_{t-L_x}^{M_x+L_x} - X_{t+l-L_x}^{M_x+L_x} \| < e)$$
$$C_4(L_x, e; l) \equiv P(\| X_{t-L_x}^{L_x} - X_{t+l-L_x}^{L_x} \| < e).$$

So $\{Z_{1t}\}$, $\{Z_{2t}\}$, $\{Z_{3t}\}$ and $\{Z_{4t}\}$ satisfy the central limit theorem.

Further, for any real number a_1, a_2, a_3 and a_4, the sequence $\{Z_t = a_1 Z_{1t} + a_2 Z_{2t} + a_3 Z_{3t} + a_4 Z_{4t}, t = L_{xy} + 1, \cdots, T - l - L_{xy} - m_x + 1\}$ also satisfies the conditions of Lemma 1 which implying that

$$\sqrt{n}\begin{bmatrix} \hat{C}_1(M_x + L_x, L_y, e; l) - C_1(M_x + L_x, L_y, e; l) \\ \hat{C}_2(L_x, L_y, e; l) - C_2(L_x, L_y, e; l) \\ \hat{C}_3(M_x + L_x, e; l) - C_3(M_x + L_x, e; l) \\ \hat{C}_4(L_x, e; l) - C_4(L_x, e; l) \end{bmatrix} \xrightarrow{d} N(0, \mathbf{\Sigma}),$$

where

$$\hat{C}_1(M_x + L_x, L_y, e; l) \equiv \frac{1}{n}\sum_{t=L_{xy}+1}^{T-l-m_x+1} \prod_{i=1}^{n_1} I(x_{i,t-L_{x_i}}^{m_{x_i}+L_{x_i}}, x_{i,t+l-L_{x_i}}^{m_{x_i}+L_{x_i}}, e) \cdot \prod_{j=1}^{n_2} I(y_{j,t-L_{y_j}}^{L_{y_j}}, y_{j,t+l-L_{y_j}}^{L_{y_j}}, e)$$

$$\hat{C}_2(L_x, L_y, e; l) \equiv \frac{1}{n}\sum_{t=L_{xy}+1}^{T-l-m_x+1} \prod_{i=1}^{n_1} I(x_{i,t-L_{x_i}}^{L_{x_i}}, x_{i,t+l-L_{x_i}}^{L_{x_i}}, e) \cdot \prod_{j=1}^{n_2} I(y_{j,t-L_{y_j}}^{L_{y_j}}, y_{j,t+l-L_{y_j}}^{L_{y_j}}, e)$$

$$\hat{C}_3(M_x + L_x, e; l) \equiv \frac{1}{n}\sum_{t=L_{xy}+1}^{T-l-m_x+1} \prod_{i=1}^{n_1} I(x_{i,t-L_{x_i}}^{m_{x_i}+L_{x_i}}, x_{i,t+l-L_{x_i}}^{m_{x_i}+L_{x_i}}, e)$$

$$\hat{C}_4(L_x, e; l) \equiv \frac{1}{n}\sum_{t=L_{xy}+1}^{T-l-m_x+1} \prod_{i=1}^{n_1} I(x_{i,t-L_{x_i}}^{L_{x_i}}, x_{i,t+l-L_{x_i}}^{L_{x_i}}, e)$$

and $\mathbf{\Sigma}$ is a 4 × 4 symmetric matrix. Denote

$$h_1(L_x, L_y, M_x, l, k) = I(x_{L_{xy}+1+k-L_x}^{L_x+M_x}, x_{L_{xy}+1+k+l-L_x}^{L_x+M_x}, e),$$

$$h_2(L_x, L_y, l, k) = I(y_{L_{xy}+1+k-L_y}^{L_y}, y_{L_{xy}+1+k+l-L_y}^{L_y}, e),$$

We have

$$\begin{aligned}
\mathbf{\Sigma}_{11} &= E[(h_1(L_x, L_y, M_x, l, 0)h_2(L_x, L_y, l, 0) - C_1(M_x + L_x, L_y, e; l))^2] \\
&+\sum_{k=1}^{n-1} 2\left(1-\frac{k}{n}\right)E[(h_1(L_x, L_y, M_x, l, 0)h_2(L_x, L_y, l, 0) - C_1(M_x + L_x, L_y, e; l)) \\
&\quad (h_1(L_x, L_y, M_x, l, k)h_2(L_x, L_y, l, k) - C_1(M_x + L_x, L_y, e; l))],\\
\mathbf{\Sigma}_{12} &= E[(h_1(L_x, L_y, M_x, l, 0)h_2(L_x, L_y, l, 0) - C_1(M_x + L_x, L_y, e; l)) \\
&\quad (h_1(L_x, L_y, 0, l, 0)h_2(L_x, L_y, l, 0) - C_2(L_x, L_y, e; l))] \\
&+\sum_{k=1}^{n-1}\left(1-\frac{k}{n}\right)E[(h_1(L_x, L_y, M_x, l, 0)h_2(L_x, L_y, l, 0) - C_1(M_x + L_x, L_y, e; l)) \\
&\quad (h_1(L_x, L_y, 0, l, k)h_2(L_x, L_y, l, k) - C_2(L_x, L_y, e; l))] \\
&+\sum_{k=1}^{n-1}\left(1-\frac{k}{n}\right)E[(h_1(L_x, L_y, M_x, l, k)h_2(L_x, L_y, l, k) - C_1(M_x + L_x, L_y, e; l)) \\
&\quad (h_1(L_x, L_y, 0, l, 0)h_2(L_x, L_y, l, 0) - C_2(L_x, L_y, e; l))],\\
\mathbf{\Sigma}_{13} &= E[(h_1(L_x, L_y, M_x, l, 0)h_2(L_x, L_y, l, 0) - C_1(M_x + L_x, L_y, e; l)) \\
&\quad (h_1(L_x, L_y, M_x, l, 0) - C_3(M_x + L_x, e; l))] \\
&+\sum_{k=1}^{n-1}\left(1-\frac{k}{n}\right)E[(h_1(L_x, L_y, M_x, l, 0)h_2(L_x, L_y, l, 0) - C_1(M_x + L_x, L_y, e; l)) \\
&\quad (h_1(L_x, L_y, M_x, l, k) - C_3(M_x + L_x, e; l))] \\
&+\sum_{k=1}^{n-1}\left(1-\frac{k}{n}\right)E[(h_1(L_x, L_y, M_x, l, k)h_2(L_x, L_y, l, k) - C_1(M_x + L_x, L_y, e; l)) \\
&\quad (h_1(L_x, L_y, M_x, l, 0) - C_3(M_x + L_x, e; l))],\\
\mathbf{\Sigma}_{14} &= E[(h_1(L_x, L_y, M_x, l, 0)h_2(L_x, L_y, l, 0) - C_1(M_x + L_x, L_y, e; l)) \\
&\quad (h_1(L_x, L_y, 0, l, 0) - C_4(L_x, e; l))] \\
&+\sum_{k=1}^{n-1}\left(1-\frac{k}{n}\right)E[(h_1(L_x, L_y, M_x, l, 0)h_2(L_x, L_y, l, 0) - C_1(M_x + L_x, L_y, e; l)) \\
&\quad (h_1(L_x, L_y, 0, l, k) - C_4(L_x, e; l))] \\
&+\sum_{k=1}^{n-1}\left(1-\frac{k}{n}\right)E[(h_1(L_x, L_y, M_x, l, k)h_2(L_x, L_y, l, k) - C_1(M_x + L_x, L_y, e; l)) \\
&\quad (h_1(L_x, L_y, 0, l, 0) - C_4(L_x, e; l))],
\end{aligned}$$

$$\begin{aligned}
\boldsymbol{\Sigma}_{22} &= E[(h_1(L_x, L_y, 0, l, 0)h_2(L_x, L_y, l, 0) - C_2(L_x, L_y, e; l))^2] \\
&\quad + \sum_{k=1}^{n-1} 2\left(1 - \frac{k}{n}\right) E[(h_1(L_x, L_y, 0, l, 0)h_2(L_x, L_y, l, 0) - C_2(L_x, L_y, e; l)) \\
&\quad (h_1(L_x, L_y, 0, l, k)h_2(L_x, L_y, l, k) - C_2(L_x, L_y, e; l))] \;, \\
\boldsymbol{\Sigma}_{23} &= E[(h_1(L_x, L_y, 0, l, 0)h_2(L_x, L_y, l, 0) - C_2(L_x, L_y, e; l)) \\
&\quad (h_1(L_x, L_y, M_x, l, 0) - C_3(M_x + L_x, e; l))] \\
&\quad + \sum_{k=1}^{n-1} \left(1 - \frac{k}{n}\right) E[(h_1(L_x, L_y, 0, l, 0)h_2(L_x, L_y, l, 0) - C_2(L_x, L_y, e; l)) \\
&\quad (h_1(L_x, L_y, M_x, l, k) - C_3(M_x + L_x, e; l))] \\
&\quad + \sum_{k=1}^{n-1} \left(1 - \frac{k}{n}\right) E[(h_1(L_x, L_y, 0, l, k)h_2(L_x, L_y, l, k) - C_2(L_x, L_y, e; l)) \\
&\quad (h_1(L_x, L_y, M_x, l, 0) - C_3(M_x + L_x, e; l))] \;, \\
\boldsymbol{\Sigma}_{24} &= E[(h_1(L_x, L_y, 0, l, 0)h_2(L_x, L_y, l, 0) - C_2(L_x, L_y, e; l)) \\
&\quad (h_1(L_x, L_y, 0, l, 0) - C_4(L_x, e; l))] \\
&\quad + \sum_{k=1}^{n-1} \left(1 - \frac{k}{n}\right) E[(h_1(L_x, L_y, 0, l, 0)h_2(L_x, L_y, l, 0) - C_2(L_x, L_y, e; l)) \\
&\quad (h_1(L_x, L_y, 0, l, k) - C_4(L_x, e; l))] \\
&\quad + \sum_{k=1}^{n-1} \left(1 - \frac{k}{n}\right) E[(h_1(L_x, L_y, 0, l, k)h_2(L_x, L_y, l, k) - C_2(L_x, L_y, e; l)) \\
&\quad (h_1(L_x, L_y, 0, l, 0) - C_4(L_x, e; l))] \;, \\
\boldsymbol{\Sigma}_{33} &= E[(h_1(L_x, L_y, M_x, l, 0) - C_3(M_x + L_x, e; l))^2] \\
&\quad + \sum_{k=1}^{n-1} 2\left(1 - \frac{k}{n}\right) E[(h_1(L_x, L_y, M_x, l, 0) - C_3(M_x + L_x, e; l)) \\
&\quad (h_1(L_x, L_y, M_x, l, k) - C_3(M_x + L_x, e; l))] \;,
\end{aligned}$$

$$\begin{aligned}
\Sigma_{34} &= E[(h_1(L_x, L_y, M_x, l, 0) - C_3(M_x + L_x, e; l)) \\
&\quad (h_1(L_x, L_y, 0, l, 0) - C_4(L_x, e; l))] \\
&\quad + \sum_{k=1}^{n-1}\left(1 - \frac{k}{n}\right) E[(h_1(L_x, L_y, M_x, l, 0) - C_3(M_x + L_x, e; l)) \\
&\quad (h_1(L_x, L_y, 0, l, k) - C_4(L_x, e; l))] \\
&\quad + \sum_{k=1}^{n-1}\left(1 - \frac{k}{n}\right) E[(h_1(L_x, L_y, M_x, l, k) - C_3(M_x + L_x, e; l)) \\
&\quad (h_1(L_x, L_y, 0, l, 0) - C_4(L_x, e; l))],
\end{aligned}$$

$$\begin{aligned}
\Sigma_{44} &= E[(h_1(L_x, L_y, 0, l, 0) - C_4(L_x, e; l))^2] \\
&\quad + \sum_{k=1}^{n-1} 2\left(1 - \frac{k}{n}\right) E[(h_1(L_x, L_y, 0, l, 0) - C_4(L_x, e; l)) \\
&\quad (h_1(L_x, L_y, 0, l, k) - C_4(L_x, e; l))].
\end{aligned}$$

Under the null hypothesis, applying the delta method (Serfling, 1980), we have

$$\sqrt{n}\left(\frac{\hat{C}_1(M_x + L_x, L_y, e, l)}{\hat{C}_2(L_x, L_y, e, l)} - \frac{\hat{C}_3(M_x + L_x, e, l)}{\hat{C}_4(L_x, e, l)}\right) \xrightarrow{d} N(0, \sigma^2(M_x, L_x, L_y, e, l)),$$

where $\sigma^2(M_x, L_x, L_y, e, l) = \nabla' \boldsymbol{\Sigma} \nabla$, in which

$$\nabla = \left(\frac{1}{C_2(L_x, L_y, e, l)}, -\frac{C_1(M_x + L_x, L_y, e, l)}{C_2^2(L_x, L_y, e, l)}, -\frac{1}{C_4(L_x, e, l)}, \frac{C_3(M_x + L_x, e, l)}{C_4^2(L_x, e, l)}\right)'.$$

An consistent estimator $\hat{\sigma}^2(M_x, L_x, L_y, e, l)$ of the asymptotic variance can be got by replacing all the parts in the sandwich $\nabla' \boldsymbol{\Sigma} \nabla$ by their empirical estimates.

ACKNOWLEDGMENTS

Zhidong Bai's research was supported by The National Science Foundation of China (11571067 and 11471140). Yongchang Hui's research was supported by The National Science Foundation of China (11401461), Ministry of Science and Technology of China (2015DFA81780) and Fundamental Research Funds for the Central Universities (2015gjhz15). Shurong Zheng's research was supported by The National Science Foundation of China(11522105 and 11690012). Dandan Jiang's research was supported by The National Science Foundation of China (11471140).

REFERENCES

1. Granger CWJ. Investigating causal relations by econometric models and cross-spectral methods. Econometrica. 1969; 37(3): 424–438. doi: 10.2307/1912791
2. Hurlin C, Venet B. Granger causality tests in panel data models with fixed coefficients. processed Paris University Paris IX. 2001; 10: 1–30.
3. Ghysels E, Hill JB, Motegi K. Testing for Granger causality with mixed frequency data. Journal of Econometrics. 2016; 192(1): 207–230. doi: 10.1016/j.jeconom.2015.07.007
4. Dufour JM, Renault E. Short run and long run causality in time series: theory. Econometrica. 1998; 66: 1099–1125. doi: 10.2307/2999631
5. Dufour JM, Pelletier D, Renault E. Short run and long run causality in time series: inference. Journal of Econometrics. 2006; 132: 337–362. doi: 10.1016/j.jeconom.2005.02.003
6. Granger CWJ. Forecasting in business and economics. Academic Press. 1989; 32(1): 223–226.
7. Baek EG, Brok WA. A general test for nonlinear Granger causality: bivariate model. working paper, Korea Development Institute and University of Wisconsin Madison. 1992.
8. Hiemstra C, Jones JD. Testing for linear and nonlinear Granger causality in the stock price-volume relation. Journal of Finance. 1994; 49(5): 1639–1664. doi: 10.2307/2329266
9. Bai ZD, Wong WK, Zhang BZ. Multivariate linear and nonlinear causality tests. Mathematics and Computers in simulation. 2010; 81: 5–17. doi: 10.1016/j.matcom.2010.06.008
10. Lam NX. Twin deficits hypothesis and Feldstein-Horioka puzzle in Vietnam. International Research Journal of Finance and Economics. 2012; 101: 169–179.
11. Zheng XL, Chen BM. Stock market modeling and forecasting:a system adaptation approach Lecture Notes in Control and Informatioon Sciences. London: Springer-Verlag; 2013.
12. Choudhry T, Hassan S, Shabi S. Relationship between gold and stock markets during the global crisis: evidence from linear and nonlinear causality tests. International Review of Finance Analysis. 2015; 41: 247–256. doi: 10.1016/j.irfa.2015.03.011
13. Choudhry T, Papadimitriou F, Shabi S. Stock market volatility and

business cycle: evidence from linear and nonlinear causality tests. Journal of Banking and Finance. 2016; 66: 89–101. doi: 10.1016/j.jbankfin.2016.02.005

14. Diks C, Panchenko V. A note on the Hiemstra-Jones test for Granger non-causality. Studies in Nonlinear Dynamics and Econometrics. 2005; 9(2) No.4: 1–7.
15. Diks C, Panchenko V. A new statistic and practical guidelines for nonparametric Granger causality testing. Journal of Economic Dynamics and Control. 2006; 30: 1647–1669. doi: 10.1016/j.jedc.2005.08.008
16. Bai ZD, Hui YC, Lv ZH, Wong WK, Zhu ZZ. The Hiemstra-Jones test tevisited. 2016; arXiv:1701.03992.
17. Ibragimov IA. On the average of real zeroes of random polynomials.I. The coefficients with zero means. Theory of Probability and its Applications. 1971; 16(2): 228–248. doi: 10.1137/1116023
18. Diks C, Wolski M. Nonlinear granger causality: guidelines for multivariate analysis. Journal of Applied Econometrics. 2016; 31: 1333–1351. doi: 10.1002/jae.2495

8

Multivariate Time Series Similarity Searching

Jimin Wang, Yuelong Zhu, Shijin Li, Dingsheng Wan, and Pengcheng Zhang

College of Computer & Information, Hohai University, Nanjing 210098, China

ABSTRACT

Multivariate time series (MTS) datasets are very common in various financial, multimedia, and hydrological fields. In this paper, a dimension-combination method is proposed to search similar sequences for MTS. Firstly, the similarity of single-dimension series is calculated; then the overall similarity of the MTS is obtained by synthesizing each of the single-dimension similarity based on weighted BORDA voting method. The dimension-combination method could use the existing similarity searching method. Several experiments, which used the classification accuracy as a measure, were performed on six datasets from the UCI KDD Archive to validate the method. The results show the advantage of the approach compared to the traditional similarity measures, such as Euclidean distance (ED), cynamic time warping (DTW), point distribution (PD), PCA similarity factor (S_{PCA}),

Citation: Wang, J., Zhu, Y., Li, S., Wan, D., & Zhang, P. (2014). Multivariate time series similarity searching. The Scientific World Journal, 2014. (9 pages), DOI/URL: https://doi.org/10.1155/2014/851017.

Copyright: © 2014 Jimin Wang et al. This is an open access article distributed under the Creative Commons Attribution 3.0 Unported (CC BY 3.0) License.

and extended Frobenius norm (Eros), for MTS datasets in some ways. Our experiments also demonstrate that no measure can fit all datasets, and the proposed measure is a choice for similarity searches.

INTRODUCTION

With the improving requirement of industries for information and the rapid development of the information technology, there are more and more datasets obtained and stored in the form of multidimensional time series, such as hydrology, finance, medicine, and multimedia. In hydrology, water level, flow, evaporation, and precipitation are monitored for hydrological forecasting. In finance, stock price information, which generally includes opening price, average price, trading volume, and closing price, is used to forecast stock market trends. In medicine, electroencephalogram (EEG) from 64 electrodes placed on the scalp is measured to examine the correlation of genetic predisposition to alcoholism [1]. In multimedia, for speech recognition, the Australian sign language (AUSLAN) is gathered from 22 sensors on the hands (gloves) of a native Australian speaker using high-quality position trackers and instrumented gloves [2].

A time series is a series of observations, $x_i(t)$; $[i = 1, \ldots, n; t = 1, \ldots, m]$, made sequentially through time where i indexes the measurements made at each time point t [3]. It is called a univariate time series when n is equal to 1 and a multivariate time series (MTS) when n is equal to or greater than 2.

Univariate time series similarity searches have been broadly explored and the research mainly focuses on representation, indexing, and similarity measure [4]. A univariate time series is often regarded as a point in multidimensional space, so one of the goals of time series representation is to reduce the dimensions (i.e., the number of data points) because of the curse of dimensionality. Many approaches are used to extract the pattern, which contains the main characteristics of original time series, to represent the original time series. Piecewise linear representation (PLA) [5, 6], piecewise aggregate approximation (PAA) [7], adaptive piecewise constant approximation (APCA) [8], and so forth use l adjacent segments to represent the time series with length m ($m \gg l$). Furthermore, perceptually important points (PIP) [9, 10], critical point model (CMP) [11], and so on reduce the dimensions by preserving the salient points. Another common family of time series representation approaches transform time series into discrete symbols and perform string operations on time series, for example, symbolic aggregate approximation (SAX) [12], shape description alphabet

(SDA) [13], and other symbols generated method based on clustering [14, 15]. Representing time series in the transformation is another large family, such as discrete Fourier transform (DFT) [4] and discrete wavelet transform (DWT) [16] which transform the original time series into frequency domain. After transformation, only the first few or the best few coefficients are chosen to represent the original time series [3]. Many of the representation schemes are incorporated with different multidimensional spatial indexing techniques (e.g., k-d tree [17] and r-tree and its variants [18, 19]) to index sequences to improve the query efficiency during similarity searching. Given two time series *S* and *Q* and their representations *PS* and *PQ*, a similarity measure function *D* calculates the distance between the two time series, denoted by *D*(*PQ*, *PS*) to describe the similarity/dissimilarity between *Q* and *S*, such as Euclidean distance (ED) [4] and the other *Lp* norms, dynamic time warping (DTW) [20, 21], longest common subsequence (LCSS) [22], the slope distance [23], and the pattern distance [24].

The multidimensional time series similarity searches study mainly two aspects, the overall matching and match-by-dimension. The overall matching treats the MTS as a whole because of the important correlations of the variables in MTS datasets. Many overall matching similarity measures are based on principal component analysis (PCA). The original time series are represented by the eigenvectors and the eigenvalues after transformation. The distance between the eigenvectors weighted by eigenvalues is used to describe the similarity/dissimilarity, for example, Eros [3], S_{PCA} [25], and S_{PCA} [26]. Lee and Choi [27] combined PCA with the hidden Markov model (HMM) to propose two methods, PCA + HMM and PCA + HMM + SVM, to find similar MTS. With the principal components such as the input of several HMMs, the similarity is calculated by combining the likelihood of each HMM. Guan et al. [28] proposed a pattern matching method based on point distribution (PD) for multivariate time series. Local important points of a multivariate time series and their distribution are used to construct the pattern vector. The Euclidean distance between the pattern vectors is used to measure the similarity of original time series.

By contrast, match-by-dimension breaks MTS into multiple univariate time series to process separately and then aggregates them to generate the result. Li et al. [29] searched the similarity of each dimensional series and then synthesized the similarity of each series by the traditional BORDA voting method to obtain the overall similarity of the multivariate time series. Compared to the overall matching, match-by-dimension could take advantage of present univariate time series similarity analysis approaches.

In this paper, a new algorithm based on the weighted BORDA voting method for the MTS k nearest neighbor (kNN) searching is proposed. Each MTS dimension series is considered as a separate univariate time series. Firstly, similarity searching approach is used to search the similarity sequence for each dimension series; then the similar sequences of each dimensional series are synthesized on the weighted BORDA voting method to generate the multivariate similar sequences. Compared to the measure in [29], our proposed method considers the dimension importance and the similarity gap between the similar sequences and generates more accurate similar sequences.

In the next section, we briefly describe the BORDA voting method and some similarity measures widely used. Section 3 presents the proposed algorithm to search the kNN sequences. Datasets and experimental results are demonstrated in Section 4. Finally, we conclude the paper in Section 5.

RELATED WORK

In this section, we will briefly discuss BORDA voting method, the method in [29], and the DTW, on which our proposed techniques are based. Notations section contains the notations used in this paper.

BORDA Voting Method

BORDA voting, a classical voting method in group decision theory, is proposed by Jena-Charles de BORDA [30]. Supposing k is the number of winners, c is the number of candidates; e electors express their preference from high to low in the sort of candidates. To every elector›s vote, the candidate ranked first is provided e points (called voting score), the second candidate e-1 points, followed by analogy, and the last one is provided 1 point. The accumulated voting score of the candidate is BORDA score. The candidates, BORDA scores in the top k, are called BORDA winners.

Similarity Measure on Traditional BORDA Voting

Li et al. [29] proposed a multivariate similarity measure based on BORDA voting, denoted by S_{BORDA}; the measure is divided into two parts: the first one is the similarity mining of univariate time series and the second one is the integration of the results obtained in the first stage by BORDA voting. In the first stage, a certain similarity measure is used to query *k*NN sequences on univariate series of each dimension in the MTS. In the second stage, the scores of each univariate similar sequence are provided through the rule of BORDA voting. The most similar sequence scores *i* points, the second scores *i*-1, followed by a decreasing order, and the last is 1. The sequences with same time period or very close time period will be found in different univariate time series. According to the election rule, the sequences whose votes are less than the half of dimension are eliminated; then the BORDA voting of the rest of sequences is calculated. If a sequence of some certain time period appears in the results of *p* univariate sequences and its scores are $s_1, s_2, \ldots, s_p$, respectively, then the similarity score of this sequence is the sum of all the scores. In the end, the sequence with the highest score is the most similar to the query sequence.

Dynamic Time Warping Distance

Dynamic programming is the theoretical basis for dynamic time warping (DTW). DTW is a nonlinear planning technique combining time and distance measure, which was firstly introduced to time series mining areas by Berndt and Clifford [20] to measure the similarity of two univariate time series. According to the minimum cost of time warping path, the DTW distance supports time axis stretching but does not meet the requirement of triangle inequality.

THE PROPOSED METHOD

In the previous section, we have reviewed the similarity measure on traditional BORDA voting, S_{BORDA}, for multivariate time series. In this section, we propose a dimension-combination similarity measure based on weighted BORDA voting, called S_{WBORDA}, for MTS datasets *k*NN searching. The similarity measure can be applied for the whole sequence matching similarity searching and the subsequence matching similarity searching.

S_{WBORDA}: Multivariate Similarity Measure on Weighted BORDA Voting

S_{BORDA} takes just the order into consideration, without the actual similarity gap between two adjacent similar sequences that may lead to rank failure for the similar sequences. For example, assuming the four candidates r_1, r_2, r_3,

r_4 take part in race, the first round position is r_1, r_2, r_3, r_4, the second is r_2, r_1, r_4, r_3, the third is r_4, r_3, r_1, r_2, and the last is r_3, r_4, r_2, r_1. The four runners are all ranked number 1 with traditional BORDA score (10 points), because of considering only the rank order, but without the speed gap of each runner in the race. In our proposed approach, we use the complete information of candidate, including the order and the actual gap to neighbor.

The multivariate data sequences S with n dimensions are divided into n univariate time series, and each dimension is a univariate time series. Given multivariate query sequence Q, to search the multivariate kNN sequences, each univariate time series is searched separately. For the jth dimension time series, the $k' + 1$ nearest neighbor sequences are $s_0, s_1, \ldots, s_{k'}$, where k' is equal or greater than k and s_0 is the jth dimension series of Q and is considered to be the most similar to itself. The distances between $s_1, \ldots, s_{k'}$ and s_0 are $d_1, \ldots, d_{k'}$, respectively, where d_{i-1} is less than or equal to d_i and $d_i - d_{i-1}$ describes the similarity gap between s_i and s_{i-1} to s_0. Let the weighted voting score of s_0 be $k' + 1$ and let $s_{k'}$ be 1; the weighted voting score of the sequence s_i, vs_i, is defined by

$$\mathrm{vs}_i = w_j\left(1 + k'\left(1 - \frac{d_i}{d_{k'}}\right)\right) \quad \left(i = 1, \ldots, k' - 1\right), \tag{1}$$

where w is a weight vector based on the eigenvalues of the MTS dataset, $\sum_{j=1}^{n} w_j = 1$, and w_j represents the importance of the jth dimension series among the MTS. vs_i is inversely proportional to d_i; that is, s_0 is the baseline; the higher similarity gap between s_i and s_0 is, the lower weighted BORDA score s_i will get.

We accumulate the weighted voting score of each item in a candidate multivariate similar sequence and then obtain its weighted BORDA score. The candidate sequences are ranked on weighted BORDA scores, and the top k are the final similar sequences to Q. The model of similarity searching based on weighted BORDA voting is shown in Figure 1.

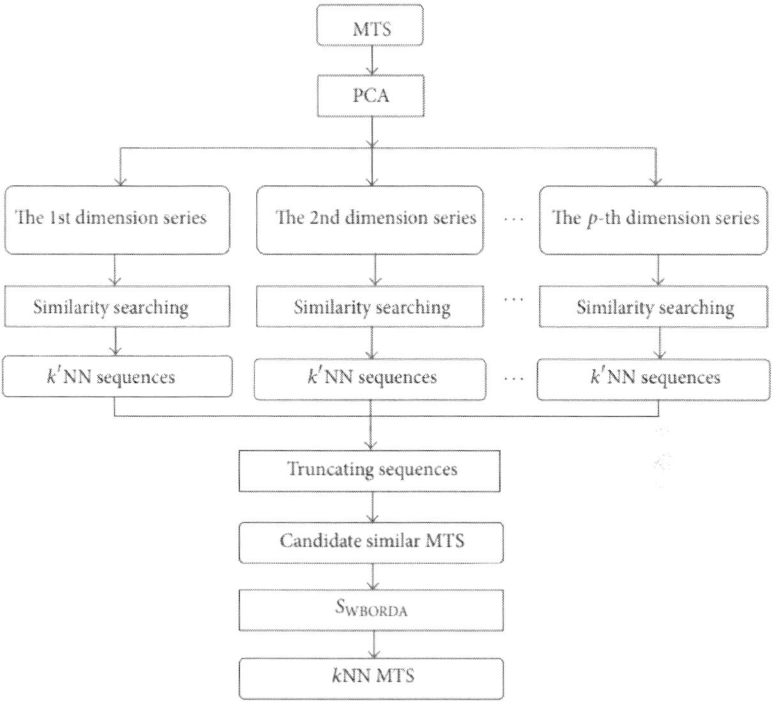

Figure 1: The model of similarity searching on weighted BORDA voting.

In the model of Figure 1, firstly, PCA is applied on original MTS and transforms it to new dataset Y whose variables are uncorrelated with each other. The first p dimensions series which contain most of characteristics of the original MTS are retained to reduce dimensions. Furthermore, univariate time series similarity searching is performed to each dimension series in Y and finds out the univariate k'NN sequences; k' should be equal or greater than the final k. Moreover, k'NN sequences are truncated to obtain the candidate multivariate similar sequences. Finally, S_{WBORDA} is performed on candidate multivariate similar sequences to obtain the kNN of query sequences. Intuitively, S_{WBORDA} measures the similarity from different aspects (dimensions) and synthesizes them. The more aspects (dimensions) from measured sequences is similar to the query sequences, the more similar the sequence is to the query sequences of the whole. The following sections describe the similarity searching in detail.

Performing PCA on Original MTS

In our proposed method, all MTS dimension series are considered independent of each other, but, in fact, correlation exists among them more or less, so PCA is applied to the MTS which can be represented as a matrix $X_{m \times n}$ and m represents the length of series, and n is the number of dimensions (variables). Each row of X can be considered as a point in n-dimensional space. Intuitively, PCA transforms dataset X by rotating the original n-dimensional axes and generating a new set of axes. The principal components are the projected coordinate of X on the new axes [3].

Performing PCA on a multivariate dataset $X_{m \times n}$ is based on the correlation matrix or covariance matrix of X and results in two matrices, the eigenvectors matrix $C_{n \times n}$ and the variances matrix $L_{n \times 1}$. Each column of $C_{n \times n}$, called eigenvector, is a unit vector, geometrically, and it presents the new axes position in the original n-dimensional space. The variances matrix element $L_{i \times 1}$, called eigenvalue, provides the variance of the ith principal component. The matrix of the new projected coordinates $D_{m \times n}$ of the original data can be calculated by $D = X \cdot C$. The first dimension univariate time series of D is the first principal component and accounts for the largest part of variances presented in the original X; the ith principal component accounts for the largest part of the remaining variances and is orthogonal to the 1st, 2nd,..., and $i - 1$th dimensions. Select the first p components $D_{m \times p}$, which retain more than, for example, 90% of the total variation presented in the original data representing X. Thus, the dimensionality reduction may be achieved, as long as $p \ll n$. Geometrically, the original X is projected on the new p-dimensional space.

In whole sequence matching similarity searching, we apply PCA to all MTS items and retain p components so that more than, for example, 90% of the total variations are retained in all MTS items at least.

Truncating Univariate Similar Sequences

In candidate similar MTS, each dimension series starts at the same time. However, the similar sequences of each dimension time series may not start at the same time. The similar sequences with close start time of each dimension could be treated as in the same candidate similar MTS and truncated. The truncation includes four steps: grouping the sequences, deleting the isolated sequences, aligning the overlapping sequences, and reordering the sequences. After truncation, the candidate multivariate similar sequences could be obtained.

The truncation for whole sequence matching similarity searching is just a special case of subsequence matching, so we introduce the truncation for subsequence matching. In Figure 2, 3NN sequences are searched for multivariate query sequences with length l, and the application of PCA on the data MTS results in the principal component series with three dimensions. 3NN searching is performed on each dimension principal component series. The 3NN sequences of first dimension are s_{11} (the subsequence from t_{11} to $t_{11} + l$), s_{12} (from t_{12} to $t_{12} + l$), and s_{13} (from t_{13} to $t_{13} + l$). The univariate similar sequences are presented according to their occurrence time, and the present order does not reflect the similarity order to the query sequence. The 3NN sequences of the second dimension are s_{21} (from t_{21} to $t_{21} + l$), s_{22} (from t_{22} to $t_{22} + l$), and s_{23} (from t_{23} to $t_{23} + l$), and these of the third dimension are s_{31} (from t_{31} to $t_{31} + l$), s_{32} (from t_{32} to $t_{32} + l$), and s_{33} (from t_{33} to $t_{33} + l$).

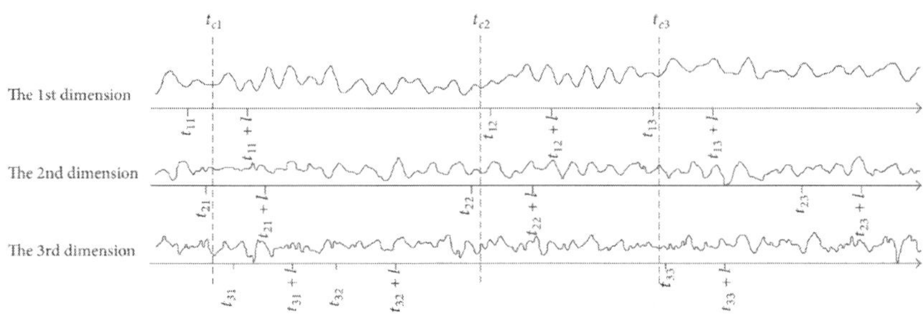

Figure 2: Truncating similar sequences for subsequence matching.

(1) *Grouping the Univariate Similar Sequences.* The univariate similar sequences of all dimensions are divided into groups, so that in each group, for any sequence s, at least one sequence w, which overlaps with s over the half length of sequence l, could be found. The univariate similar sequence, which does not overlap with any other similar sequences, will be put into a single group just including itself. In Figure 2, all the similar sequences are divided into five groups. The group g1 includes s_{11}, s_{21}, s_{31}. s_{11}, s_{21} overlaps with s_{21}, s_{31}, respectively, and the overlapping lengths are all over half of the length l, group g2 includes s_{32}, group g3 includes s_{12}, s_{22}, group g4 includes s_{13}, s_{33}, and group g5 includes s_{23}.

(2) *Deleting the Isolated Sequences.* The group, in which the number of similar sequences is less than half number of the dimensions,

is called an isolated group, and the similar sequences in isolated group are called isolated similar sequences. In Figure 2, the number of similar sequences in group g2 or g5 is less than half of the number of dimensions, that is, 3, so the similar sequences in them are deleted.

(3) *Aligning the Overlapping Sequences*. The sequences in the same group are aligned to generate the candidate multivariate similar sequences. For one group, the average start time t of all the included sequences is calculated; then the subsequence from t to $t + l$, denoted by cs, is the candidate multivariate similar sequence. Each dimension series of cs is regarded as the univariate similar sequences. The similarity distance between cs and query sequence is recalculated by the selected univariate similarity measure dimension by dimension; if the group contains the ith dimension similar sequence, then the corresponding similarity distance is set to the ith dimension series of cs to reduce computation. In Figure 2, for group g1, the average of t_{11}, t_{21}, t_{31} t_{c1} is calculated; then the subsequence s_{tc1}, from t_{c1} to $t_{c1} + l$, is the candidate multivariate similar sequence. For group g3, the similarity distance between the 2nd dimension series of s_{tc2} and the query sequence should be recalculated. The same alignment operation is performed on group g4 to obtain the candidate multivariate sequence s_{tc3}.

(4) *Reordering the Candidate Similar Sequences*. For each dimension, the candidate univariate similar sequences are reordered by the similarity distance calculated in Step (3). After reordering, S_{WBORDA} is used to synthesize the candidate similar sequences and generate the multivariate kNN sequences.

In whole matching kNN searching, the similar sequences are either whole overlapping or not overlapping each other at all, and the truncation steps are the same as those of the subsequence matching.

Computing Weights

By applying PCA to the MTS, the principal components series and the eigenvalues, which can represent the variances for principal components, are obtained. When we calculate the weighted BORDA score, we take into consideration both the similarity gap and the dimension importance for S_{WBORDA}. The heuristics proposed algorithm in [3] is used to calculate the weight vector w based on the eigenvalues. Variances are aggregated by a certain strategy, for example, min, mean, and max, on eigenvalues vectors dimension by dimension, and the vector $w\langle w_1, w_2, \ldots, w_k\rangle$ is obtained. The weight vector element is defined by

$$w_i = \frac{f(V_i)}{\sum_{j=1}^{k} f(V_j)} \qquad i = 1, \ldots, p, \tag{2}$$

where *f*() denotes the aggregating strategy. V_i means the variance vector of *i*th dimension of MTS items. Generally, for subsequence matching, V_i includes one element, and whole matching is greater than 1. Intuitively, each w_i in the weight vector represents the aggregated variance for all the *i*th principal components. The original variance vector could be normalized before aggregation.

EXPERIMENTAL EVALUATION

In order to evaluate the performance of our proposed techniques, we performed experiments on six real-world datasets. In this section, we first describe the datasets used in the experiments and the experiments methods followed by the results.

Datasets

The experiments have been conducted on four UCI datasets, [31] electroencephalogram (EEG), Australian sign language (AUSLAN), Japanese vowel (JV), and robot execution failure (REF), which are all labeled MTS datasets.

The EEG contains measurements from 64 electrodes placed on the scalp and sampled at 256 Hz to examine EEG correlates of genetic predisposition to alcoholism. Three versions, the small, the large, and the full, are included in this dataset according to the volume of the original data. We utilized the large dataset containing 600 samples and 2 classes.

The AUSLAN2 consists of samples of AUSLAN (Australian sign language) signs. 27 examples of each of 95 AUSLAN signs were captured from 22 sensors placed on the hands (gloves) of a native signer. In total, there are 2565 signs in the dataset.

The JV contains 640 time series of 12 LPC cepstrum coefficients taken from nine male speakers. The length of each time series is in the range 7–29. It describes the uttering of Japanese vowels /ae/ by a speaker successively.

The dataset contains two parts: training and test data; we utilized the training data which contains 270 time series.

The REF contains force and torque measurements on a robot after failure detection. Each failure is characterized by 6 forces/torques and 15 force/torque samples. The dataset contains five subdatasets LP1, LP2, LP3, LP4, and LP5; each of them defines a different learning problem. The LP1, LP4, and LP5 subdatasets were utilized in the experiment. LP1 which defines the failures in approach to grasp position contains 88 instances and 4 classes, LP4 contains 117 instances and 3 classes, and LP5 contains 164 instances and 5 classes. A summary is shown in Table 1.

Table 1: Summary of datasets used in the experiments

Dataset	Number of variables	Mean length	Number of instances	Number of classes
EEG	64	256	600	2
AUSLAN2	22	90	2565	95
JV	12	16	270	9
LP1	6	15	88	4
LP4	6	15	117	3
LP5	6	15	164	5

Method

In order to validate our proposed similarity measure S_{WBORDA}, 1NN classification, and 10-fold cross validation are performed [32]. That is, each dataset is divided into ten subsets, 1-fold for testing and the rest 9 for training. For each query item in the testing set, 1NN is searched in the training set and the query item is classified according to the label of the 1NNs, and the average precision is computed across all the testing items. The experiment is repeated 10 times for different testing set and training set, and 10 different error rates are obtained; then the average error across all 10 trials is computed for 10-fold cross validation. We performed 10 times 10-fold cross validation and computed the average error across all 10-fold cross validations to estimate the classification error rate. The similarity measures tested on our experiments include S_{BORDA}, PD, Eros, and S_{PCA}. DTW is selected as the univariate similarity measure for S_{BORDA} and S_{WBORDA}. They are denoted as S_{BORDA_DTW} and S_{WBORDA_DTW}, respectively. For DTW, the maximum amount of warping Q is decreased to 5% of the length. DTW has been extensively employed in various applications and time series similarity searching, because DTW can be applied to two MTS items warped in the time axis and with different lengths.

All other measures except PD and Eros require determining the number of components *p* to be retained. Classification has been conducted for consecutive values of *p* which retain more than 90% of total variation, until the error rate reaches the minimum. The number which retains less than 90% of total variation is not considered. For Eros, the experiments are conducted as proposed in [3]. The Wilcoxon signed-rank test is used to ascertain if S_{BORDA_DTW} yields an improved classification performance on multiple data sets in general. PCA is performed on the covariance matrices of MTS datasets.

Results

The classification error rates are presented in Table 2 in the form of percentages. Although experiments have been conducted for various *p*, that is, the number of principal components (for S_{BORDA_DTW}, S_{WBORDA_DTW}, and S_{PCA}), only the best classification accuracies are presented.

Table 2: Classification error rate (%)

	EEG	AUS-LAN2	JV	LP1	LP4	LP5
S_{BORDA_DTW}	$_{(14)}27.7$	$_{(4)}52.6$	$_{(5)}53.4$	$_{(3)}15.9$	$_{(3)}9.8$	$_{(4)}24.0$
S_{WBORDA_DTW}	$_{(14,max)}27.7$	$_{(4,max)}50.4$	$_{(6,max)}52.4$	$_{(4,mean)}13.0$	$_{(3,mean)}8.1$	$_{(3,mean)}21.3$
PD	38.4	68.9	45.2	14.1	12.3	41.2
Eros	$_{(max)}5.8$	$_{(mean)}11.1$	$_{(mean)}30.6$	$_{(max)}20.6$	$_{(mean)}10.3$	$_{(mean)}35.5$
S_{PCA}	$_{(14)}1.2$	$_{(4)}13.2$	$_{(12)}29.5$	$_{(3)}24.7$	$_{(4)}20.0$	$_{(3)}35.2$

(Numbers in parentheses indicate the *p*, i.e., the number of principal components retained, "max" and "mean" indicate the aggregating functions for weight *w*.)

Firstly, we will compare similarity measures with respect to each dataset, respectively. For the EEG dataset, S_{PCA} produces the best performance and performs significantly better than the others. With regard to the AUSLAN2 dataset, Eros produces the lowest classification error rate and PD gives very poor performance. For the overall matching method, for example, Eros, *S*

$_{PCA}$ performs better than the others. For the JV dataset, S_{PCA} gives the best performance and the S_{BORDA_DTW} makes the poorest performance. For the LP1 dataset, S_{WBORDA_DTW} makes the best performance. For the LP4 dataset, S_{WBORDA_DTW} makes the best performance and S_{BORDA_DTW} and S_{WBORDA_DTW} perform better than others. In the end, for the LP5 dataset, S_{WBORDA_DTW} gives the best performance.

Finally, the similarity measures are compared for all the datasets. Between S_{BORDA_DTW} and S_{WBORDA_DTW}, the Wilcoxon signed-rank test reports that P value equals 0.043 (double side) and shows that the algorithms are significantly different. With 5% significance level, S_{WBORDA_DTW} has made better performance over S_{BORDA_DTW}. Compared to Eros and S_{PCA}, the S_{WBORDA_DTW} has better performance on LP1, LP4, and LP5. But it shows poor performance on EEG, AULSAN2, and JV. Table 3 shows the number of principal components, which just retain more than 90% of total variation, in experiment datasets. For LP1, LP4, and LP5, the first few principal components retained most of the variation after PCA performing, but for EEG, AUSLAN2, and JV, to retain more than 90% of total variation, more principal components should be retained. S_{WBORDA_DTW} searches the similar sequences dimension by dimension and then synthesizes them; it is hard to generate the aligned candidate multivariate similar sequences when many principal components are contained in the principal component series. Furthermore, for the datasets, for example, EEG, AUSLAN2, and JV, the first few principal components could not retain sufficient information of original series, and S_{WBORDA_DTW} produces poor precision. S_{WBORDA_DTW} could make better performance on the datasets which aggregate the most variation in the first few principal components after PCA.

Table 3: Contribution of principal components

	EEG	AUSLAN2	JV	LP1	LP4	LP5
Number of variables	64	22	12	6	6	6
Number of retained principal components*	14	4	5	3	3	3
Retained variation (%)	90.1	92.7	93.9	92.4	94.8	91.2

(*For every dataset, if the number of principal components is less than the number in Table 3, the retained variation will be less than 90%.)

CONCLUSION AND FUTURE WORK

A match-by-dimension similarity measure for MTS datasets, S_{WBORDA}, is proposed in this paper. This measure is based on principal component analysis, weighted BORDA voting method, and univariate time series similarity measure. In order to compute the similarity between two MTS items, S_{WBORDA} performs PCA on the MTS and retains *k* dimensions principal component series, which present more than 90% of the total variances. Then a univariate similar analysis is applied to each dimension series, and the univariate similar sequences are truncated to generate candidate multivariate similar sequences. At last, the candidate sequences are ordered by weighted BORDA score, and the *k*NN sequences are obtained. Experiments demonstrate that our proposed approach is suitable for small datasets. The experimental result has also shown that the proposed method is sensitive to the number of the classes in datasets. In the further work, it will be investigated furtherly.

In the literature, at present, there are still not so many studies in similarity analysis for MTS. In the future, we will explore new integration methods.

ACKNOWLEDGMENTS

This research is supported by the Fundamental Research Funds for the Central Universities (no. 2009B22014) and the National Natural Science Foundation of China (no. 61170200, no. 61370091, and no. 61202097).

REFERENCES

1. Zhang XL, Begleiter H, Porjesz B, Wang W, Litke A. Event related potentials during object recognition tasks. *Brain Research Bulletin*. 1995;38(6):531–538.
2. Kadous MW. *Temporal Classification: Extending the Classification Paradigm to Multivariate Time Series*. Kensington, Australia: The University of New South Wales; 2002.
3. Yang K, Shahabi C. A PCA-based similarity measure for multivariate time series. Proceedings of the 2nd ACM International Workshop on Multimedia Databases (MMDB '04); November 2004; pp. 65–74.
4. Agarwal R, Faloutsos C, Swami A. Efficient similarity search in sequence databases. Proceedings of the International Conference on Foundations of Data Organization and Algorithms (FODO '93); 1993; pp. 69–84.
5. Keogh E, Pazzani M. An enhanced representation of time series which allows fast and accurate classification, clustering and relevance feedback. Proceedings of the 4th International Conference of Knowledge Discovery and Data Mining; 1998; pp. 239–243.
6. Keogh E, Smyth P. A probabilistic approach to fast pattern matching in time series databases. Proceedings of the 3rd International Conference of Knowledge Discovery and Data Mining; 1997; pp. 24–30.
7. Keogh E, Pazzani M. A Simple dimensionality reduction technique for fast similarity search in large time series databases. Proceedings of the 4th Pacific-Asia Conference on Knowledge Discovery and Data Mining; 2000; pp. 122–133.
8. Keogh E, Chakrabarti K, Mehrotra S, Pazzani M. Locally adaptive dimensionality reduction for indexing large time series databases. Proceedings of the 2001 ACM SIGMOD International Conference on Management of Data; May 2001; pp. 151–162.
9. Chung FL, Fu TC, Luk R, Ng V. Flexible time series pattern matching based on perceptually important points. Proceedings of the International Joint Conference on Artificial Intelligence Workshop on Learning from Temporal and Spatial Data; 2001; pp. 1–7.
10. Fu T. A review on time series data mining. *Engineering Applications of Artificial Intelligence*. 2011;24(1):164–181.
11. Bao D. A generalized model for financial time series representation and prediction. *Applied Intelligence*. 2008;29(1):1–11.

12. Lin J, Keogh E, Lonardi S, Chiu B. A symbolic representation of time series, with implications for streaming algorithms. Proceedings of the 8th ACM SIGMOD International Conference on Management of Data Workshop on Research Issues in Data Mining and Knowledge Discovery (DMKD '03); June 2003; pp. 2–11.
13. Jonsson HA, Badal Z. Using signature files for querying time-series data. Proceedings of the 1st European Symposium on Principles and Practice of Knowledge Discovery in Databases; 1997; pp. 211–220.
14. Hebrail G, Hugueney B. Symbolic representation of long time-series. Proceedings of the Applied Stochastic Models and Data Analysis Conference; 2001; pp. 537–542.
15. Hugueney B, Meunier BB. Time-series segmentation and symbolic representation, from process-monitoring to data-mining. Proceedings of the 7th International Conference on Computational Intelligence, Theory and Applications; 2001; pp. 118–123.
16. Chan K, Fu AW. Efficient time series matching by wavelets. Proceedings of the 1999 15th International Conference on Data Engineering (ICDE '99); March 1999; pp. 126–133.
17. Ooi BC, McDonell KJ, Sacks-Davis R. Spatial kd-tree: an indexing mechanism for spatial databases. Proceedings of IEEE International Computers, Software, and Applications Conference (COMPSAC '87); 1987; pp. 433–438.
18. Guttman A. R-trees: a dynamic index structure for spatial searching. Proceedings of the 1984 ACM SIGMOD international conference on Management of data (SIGMOD '84); 1984.
19. Beckmann N, Kriegel H, Schneider R, Seeger B. R-tree. An efficient and robust access method for points and rectangles. Proceedings of the 1990 ACM SIGMOD International Conference on Management of Data; May 1990; pp. 322–331.
20. Berndt DJ, Clifford J. Using dynamic time warping to find patterns in time series. Proceedings of the AAAI KDD Workshop; 1994; pp. 359–370.
21. Keogh E, Pazzani M. Derivative dynamic time warping. Proceedings of the 1st SIAM International Conference on Data Mining; 2001.
22. Paterson M, Dančík V. *Longest Common Subsequences*. Berlin, Germany: Springer; 1994.

23. Zhang J-Y, Pan Q, Zhang P, Liang J. Similarity measuring method in time series based on slope. *Pattern Recognition and Artificial Intelligence*. 2007;20(2):271–274.
24. Wang D, Rong G. Pattern distance of time series. *Journal of Zhejiang University*. 2004;38(7):795–798.
25. Krzanowski WJ. Between-groups comparison of principal components. *Journal of the American Statistical Association*. 1979;74(367):703–707.
26. Johannesmeyer MC. *Abnormal Situation Analysis Using Pattern Recognition Techniques and Historical Data*. Santa Barbara, Calif, USA: University of California; 1999.
27. Lee H, Choi S. PCA+HMM+SVM for EEG pattern classification. *Signal Processing and Its Applications*. 2003;1(7):541–544.
28. Guan H, Jiang Q, Wang S. Pattern matching method based on point distribution for multivariate time series. *Journal of Software*. 2009;20(1):67–79.
29. Li S-J, Zhu Y-L, Zhang X-H, Wan D. BORDA counting method based similarity analysis of multivariate hydrological time series. *ShuiliXuebao*. 2009;40(3):378–384.
30. Black D. *The Theory of Committees and Elections*. 2nd edition. London, UK: Cambridge University Press; 1963.
31. Begleiter H. UCI Machine Learning Repository. Irvine, CA: University of California, School of Information and Computer Science, 1999, http://mlr.cs.umass.edu/ml/datasets.html.
32. Fábris F, Drago I, Varejão FM. A multi-measure nearest neighbor algorithm for time series classification. Proceedings of the 11th Ibero-American Conference on AI; 2008; pp. 153–162.

9

A Method for Comparing Multivariate Time Series with Different Dimensions

Avraam Tapinos[1], **Pedro Mendes**[1,2]

[1]School of Computer Science and Manchester Institute of Biotechnology, The University of Manchester, Manchester, United Kingdom,
[2]Virginia Bioinformatics Institute, Virginia Tech, Blacksburg, Virginia, United States of America

ABSTRACT

In many situations it is desirable to compare dynamical systems based on their behavior. Similarity of behavior often implies similarity of internal mechanisms or dependency on common extrinsic factors. While there are widely used methods for comparing univariate time series, most dynamical systems are characterized by multivariate time series. Yet, comparison of multivariate time series has been limited to cases where they share a common dimensionality. A semi-metric is a distance function that has the properties of non-negativity, symmetry and reflexivity, but not sub-additivity. Here we develop a semi-metric – SMETS – that can be used for comparing groups

Citation: Tapinos, A., & Mendes, P. (2013). A method for comparing multivariate time series with different dimensions. PloS one, 8(2) (11 pages)

Copyright: © 2013 Tapinos, Mendes. This is an open-access article distributed under the terms of the Creative Commons Attribution License, which permits unrestricted use, distribution, and reproduction in any medium, provided the original author and source are credited.

of time series that may have different dimensions. To demonstrate its utility, the method is applied to dynamic models of biochemical networks and to portfolios of shares. The former is an example of a case where the dependencies between system variables are known, while in the latter the system is treated (and behaves) as a black box.

INTRODUCTION

The term 'time series' is used to describe a set of data points that vary over time. The analysis of different time series is an important activity in many areas of science and engineering. Methods like the Autoregressive Moving Average (ARMA) and Fourier analysis, [1] are widely used for forecasting future values based on the existing time series. Another important application is the comparison of different time series. The underlying aim of this kind of analysis is to uncover similarities and patterns that might exist in the data. This translates to four specific activities: 1) indexing is used to identify the most similar time series in a dataset from a query time series; 2) classification is used to categorize data into predefined groups [2]; 3) clusteringis an unsupervised categorization of data [3], [4]; 4) anomaly detection is the identification of abnormal or unique data items [5]. For most of these activities it is necessary to compare time series using an appropriate similarity measure [6]. By similarity measure we mean any method, metric or non-metric, which compares two time series objects and returns a value that encodes how similar the two objects are. Distance metrics are commonly used similarity measures to define if two time series are similar [7].

For method d to be categorized as a metric, or distance metric, it must fulfill the following conditions for all x and y [8]:

- $d(x,y) \geq 0$ Non-negativity
- $d(x,y) = d(y,x)$ Symmetry
- $d(x,x) = 0$ Reflexivity
- $d(x,y) = 0$ only if $x = y$ Identity
- $d(x,z) \leq d(x,y) + d(y,z)$ Triangle Inequality

However, the use of metrics is not always possible or desirable. Different non-metric similarity measures provide a different perspective on comparing time series. Depending on the nature of the data one might need to use a similarity method that is not metric (does not fulfill all the distance conditions). In some cases the use of different non-metric similarity methods is more desirable since i) these non-metrics may be able to process data that metrics cannot and/or ii) provide more meaningful results than the metric methods [9], [10]. In the next section we define a semi-metric that we propose to be valuable to compare multidimensional time series.

Often it is computationally expensive (in time or storage) to apply the analysis directly to the original time series. In those cases it is more desirable to carry out the data mining analysis on shorter representations of the time series. Many methods exist for creating such representations and estimating the distance between pairs of time series approximations, such as discrete Fourier transform [11], discrete wavelet transform [12], piecewise aggregate approximation [13], or symbolic aggregate approximation [14]. These methods are widely used in many fields, including econometrics, bioinformatics and signal processing.

Of particular interest are dynamical systems composed of several variables that can be measured or simulated as a function of time. For example, models of chemical reaction networks are composed of variables representing different chemical species; stock portfolios are sets of individual stocks that are nonetheless interdependent (even though these dependencies are not known explicitly); temporal gene expression data sets represent observations of levels of different genes or gene products from an organism's genome; models of the behavior of electronic circuits are composed of several variables that represent voltages at different points in the circuit. Up until now data mining in the context of these dynamical systems has been limited to comparisons of single time series: two particular chemical species of two biochemical models, the time series of two particular stocks, or the voltages of two points in two separate circuits. Multidimensional time series comparisons are also possible [15] but only if the various time series have the same dimensionality. These methods allow us to compare two dynamical models as long as they contain the same number of variables.

However, existing approaches [16]–[19] are not applicable when the two dynamical models have different numbers of component variables. In that case the only method that has been applied is to establish the (weighted) average behavior of each model (group of time series) and then compare the two average univariate time series [20]. While this approach may be satisfactory for some applications, it does not satisfy the needs of many others. One may be interested in comparing two groups of time series using all of the information contained therein, yet allowing for the two groups to have a different number of components. For example one may want to know whether a 3-variable model of calcium oscillations is more 'similar' to a model of calcium oscillations with 4 variables or another one with 10 variables. Equally we may want to know if the behavior of the group of 100 shares included in the Financial Times and (London) Stock Exchange (FTSE) is more similar to the group of 30 shares included in the New York Stock Exchange (NYSE) or the 50 shares included in the Shanghai Stock Exchange (SSE).

Figure 1 illustrates the problem addressed here: three models are presented which contain different numbers of components. Clearly (and purposely) these models have some similar features: both A and C have oscillating variables with a similar frequency and relative amplitude, while both A and B have components that are monotonic. A has similarities to both B and C, but which one is 'closer' to or 'more like' A?

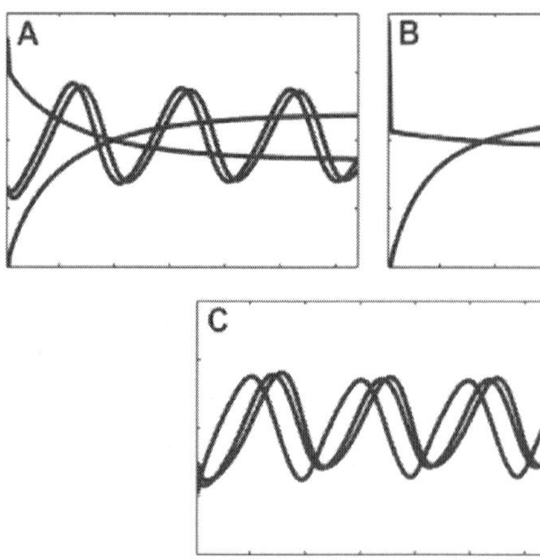

Figure 1. Three dynamic models with different dimensionality. **A** model with 4 variables, **B** model with 2 variables and **C** model with 3 variables. A has similarities with both B and C, however the distance between B and C is large. The question that SMETS addresses is which of B and C is closest to A?

MODEL

Distance between Univariate Time Series

Numerous methods have been proposed for calculating the distance between univariate time series. Some of the most used are the Euclidean distance, the

Manhattan distance (taxicab distance), Dynamic Time Warping (DTW), and the Longest Common Subsequence (LCSS).

Most applications in time series data mining require or benefit from some level of compression of the data since e.g. they may not fit in memory together or we may have grounds for first removing higher-frequency noise. Methods that create shorter representations of the original time series, like the Discrete Fourier Transform [11], the Discrete Wavelet Transform (DWT) [12], the Piecewise Aggregate Approximation (PAA) [13], or the Symbolic Aggregation Approximation (SAX) [14] are thus widely used. Lower bounding is a required property of these representations [21], i.e. the distance between two time series representations must be smaller or equal to the distance between the original time series. Here we use the Haar wavelet transformation method from the DWT family of representations. We then use the Euclidean distance in DWT space to measure distance between univariate time series.

SMETS

A new method, SMETS (**S**emi **M**etric **E**nsemble **T**ime **S**eries), is proposed to compare multivariate time series of arbitrary dimensions. The method is designed to provide numerical indices that translate the level of similarity between two multivariate time series: this is achieved by matching the most similar univariate time series component between each model. The method also takes into account the differences that arise from unmatched univariate components when one of the time series has a higher dimensionality than the other.

SMETS consists of two parts: the first identifies the similarity between the two models. This is achieved by partially matching all the univariate time series components from one model (the one with the smallest number of variables) with the most similar univariate time series components from the second model. The second part of the method adds two penalties that account for the complexity of the unmatched time series and for the difference in cardinality between models. These penalties are computed from the remaining unmatched time series of the second model and the difference between the dimensions of the two time series. Consequently, the partial matching of the two models means that, in general, SMETS does not satisfy the triangle inequality rule. Since it satisfies the rest of the metric conditions (non-negativity, symmetry, identity and reflexivity), SMETS is a semi-metric method [22], [23]. In the special case where the two time series have the same dimension, then the triangle inequality is also fulfilled and SMETS is a metric.

Part 1, partial matching.

The aim of this step is to link all the univariate time series from the model with the smallest cardinality to the most similar univariate time series from the second model. Since we are using time series representations, the distance metric used is particular to each one. The examples included here use the Haar Wavelet Transform and so the distance is simply the Euclidean distance between the DWT representations of each univariate time series. It is also possible to apply the method directly on the original time series rather than on their transformations. The partial matching proceeds according to the following algorithm:

- Calculate the distance between each of the component time series or their representations from the model with the largest cardinality and every time series from the model with the smallest cardinality. Distances between the component (univariate) time series can be measured using any of the methods discussed above. Here the Euclidean distance in Haar DWT space is used.
- Identify the two time series (one from each model) with the smallest distance and record that distance.
- Remove the two component time series that were matched from further calculations.
- Repeat steps 1 to 3 until all time series from the model with the smallest cardinality have been matched.

Two univariate time series are considered as the most similar if they share the smallest distance among all univariate time series across the two groups. Every time a pair of component time series is matched, their distance is recorded in a vector d and both time series are removed from the process. This step is important because it eliminates the possibility of multiple matchings of the univariate time series. Each component of the multivariate time series with the smallest dimension will therefore be matched to one and only one component of the multivariate time series with the largest dimension. Some of the components of the multivariate time series with the largest dimension will thus not be matched to a counterpart in the other multivariate time series.

After matching the most similar univariate time series, their overall distance is calculated using a p-norm of d [8] (Equation 1).

$$\|d\|_n = \sqrt[n]{\sum_{i=1}^{n} d_m^n} \quad 1$$

In this case p=n, the dimension of the smallest time series. In a set of multivariate time series, all of different dimensionality, the p-norm used in each comparison is different. The use of a p-norm here is beneficial because it provides a normalized distance value that depicts the similarity level of the partially matched time series.

The p-norm value calculated from Equation 1 provides an indication of the level of similarity between the matched univariate time series. However, Equation 1 does not take into consideration the influence of the unmatched component time series. Based on that, a penalty must be added to the p-norm to account for the dissimilarity that arises from the unmatched time series.

Part 2, penalization.

In the second step penalties are added to account for differences between the multivariate time series. A simple way to account for the unmatched components would be to add their distance to the closest counterparts in the other multivariate time series. However it is important to account for how much information (in the sense of Shannon) is contained in the unmatched components. Thus we weight the distance between the unmatched components to the closest counterpart in the other multivariate time series by the proportion of information contained in that component. This means that unmatched time series with high information content will contribute to making the overall distance larger. Unmatched time series with little information content (e.g. constant traces) will contribute little to the overall distance. Equation 2 measures the relative information of a univariate component time series:

$$H_j = -\sum_{i=1}^{q} p(t_{j,i}) \log_2 p(t_{j,i}) \quad 2$$

Where H_j is the entropy of the (univariate) j component time series; $t_{j,i}$ is the i-th data point of the component time series t_j; q is the length of the component time series, and $p(t_{j,i})$ is the frequency of the value $t_{j,i}$ in the time series. The relative information content RE_j of each unmatched component time series j is then:

$$RE_j = \frac{min(d_j)'(H_j)}{\sum_{j=1}^{m} H_i} \quad 3$$

Where d_j is the smallest distance between the j-th unmatched component time series and any component time series from the smallest model; m is the dimension of the larger time series. Therefore the overall entropy penalty EP that accounts for the distance of the unmatched components is:

$$EP = \sum_{j=1}^{m-n} RE_j \quad 4$$

This EP value is then added to the p-norm value obtained from Equation 1.

The EP penalty however would be zero if all unmatched univariate component time series were constant (since they would have zero information content), but this would violate the identity condition (see Figure 2 for an example). To avoid this and comply with the identity condition, another penalty is therefore added to account for the difference in dimensionality between the two time series. This is done through the ratio of the difference of dimensions to the sum of the dimensions:

$$P = \frac{m-n}{m+n} \quad 5$$

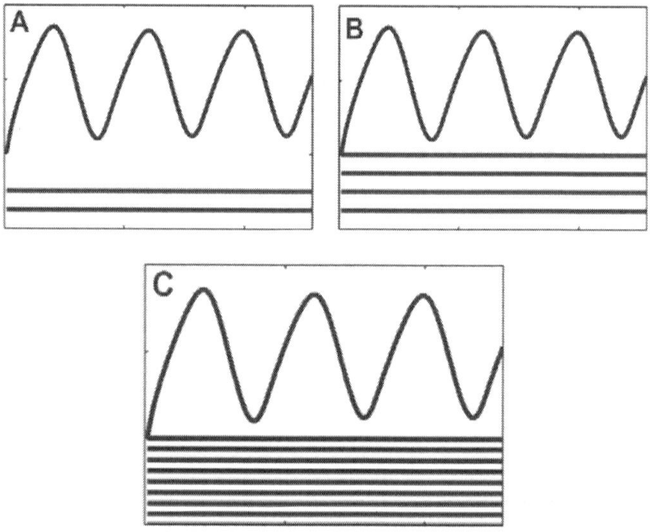

Figure 2. Three similar models. Models A, B and C are very similar; all three models contain an oscillating variable which behaves exactly the same and a

different number of variables that are constant (zero entropy). Because SMETS also takes into account the difference of dimensions it can distinguish between these models: the distance A–B is the smallest (0.25), followed by the distance B–C (0.33) and then the distance A–C (0.54).

Yet, simply adding P to EP gives too much weight to the difference of dimensions and would result in that most multivariate time series of different dimension would never be similar, despite how well their components could be matched. Thus this last penalty needs to be made weaker, which is achieved with a 2-norm. Finally SMETS is described by Equation 6:

$$SMETS = \sqrt{(\|d\|_n + EP)^2 + P^2}, \qquad 6$$

which fulfills all the conditions of a semi-metric and is therefore an appropriate means for indexing multidimensional time series of arbitrary dimensions.

The reason for the addition of the second penalty (Equation 5) is best explained using the graphical example of Figure 2. Three models are presented, each of which contains a component time series with an oscillation, plus a number of other components that are static; the only difference between the models is the number of components that are static. Thus, model A has two static components, model B four, and model C nine, while all have exactly one oscillating component. Without adding the penalty of Equation 5, the distance between any pair would be exactly zero. This is the case because the unmatched components are static and therefore have zero entropy, so that in this case Equation 4 adds no penalty. However, intuitively, model C is less similar to model A than is model B because C contains a larger number of unmatched components. Equation 5 thus deals with this by taking into account the number of unmatched components. This penalty ensures the property that only objects that are exactly the same have zero distance, a requirement for semi-metrics [22], [23].

Complexity

The time complexity of algorithms is important to ascertain whether they scale to large problems. The SMETS algorithm described here scales with the cube of the dimension of the largest time series (i.e. the one of higher dimensionality): $O(n^3)$. This makes the algorithm applicable to most practical applications, even in the presence of large data sets.

RESULTS

To demonstrate the application of SMETS we analyze four data sets from different types of activities. The first is a financial data set of stock market financial data where SMETS is used to compare five different indices. In second place we analyze a set of time series produced from dynamic models of biochemical networks. The third data set is composed of economic data representing trade of various commodities. Finally we analyze a data set composed of electrophysiological sleep data.

Financial Time Series Data

Financial data represent an area where SMETS seems to be well suited, as it consists largely of time series data analysis. We illustrate how it can be applied to the estimation of similarities between different stock indices. A number of stock market indices are used as benchmarks to evaluate the 'performance' of financial markets. Each index contains a certain number of stocks and a weighted average is usually calculated to reflect their collective performance, taken to reflect the overall performance of that market. Thus the Dow Jones Industrial Average lists 30 stocks representative of the American market, the NASDAQ-100 is an index that tracks the 100 largest non-financial companies in the National Association of Securities Dealers Automated Quotations (NASDAQ) market, the FTSE100 is an index of the 100 companies with the largest capitalization traded in the London market, the Deutscher Aktien indeX (DAX) includes 30 German companies traded in the Frankfurt market, and the SSE-50 lists the 50 major Chinese companies traded in Shanghai. Each of these can be seen as a set of connected stocks whose performance is linked (it is not important here to discuss any mechanisms of how they are linked), and therefore we consider their historic financial data to consist of multivariate time series. Given that each of these indices have different number of components, SMETS is appropriate to compare them. Up until now they have been compared only by the method of weighted averages (where the weights are often the relative capital of each stock). Since the weighted average destroys information, we think it may be useful to apply SMETS since this uses all of the information contained in all stocks.

Daily adjusted closing stock price data for each company represented in these indices for the period May 19, 2010 to April 18, 2011 was obtained from Yahoo Finance [24]. The data included consists of: a) FTSE 100 index and all stocks included in it have 234 data points, b) DAX 30 and all stocks included in it have 238 data points, c) Dow Jones 30 and NASDAQ 100 and all stocks included in both indices have 232 data points, d) SSE 50 and all stocks included in it have 229 data points. The differences in number of data points is due to different markets having different number of closing days (holidays, etc.).

Before applying the DWT the data were normalized by subtracting their mean value and dividing by the standard deviation. This operation is carried out on each univariate component. This normalization results in that time series are only different in their shape [25], since the differences in amplitude have been removed.

The DWT requires sequences of length that are powers of two [26]. For these data, we therefore added zeros to the end of each component time series such that the length was 256 and then transformed each one with the Haar DWT to a length 16 by keeping only the 16 coefficients with largest magnitude. In every component time series representation, the effect of zero padding affects the last symbol of the representation, so we truncated the representations to a length 15 by removing the last symbol of each one [27]. This is important to eliminate the bias that the zero padding would otherwise introduce in the comparisons.

SMETS was applied to the multivariate time series for each index, which were constructed by grouping the appropriate sets of companies. A distance matrix was established based on SMETS and in parallel we used the traditional weighted averages (official indices provided by each stock market) that represent each stock (and are therefore univariate time series) and constructed a Euclidean distance matrix between them. Hierarchical clustering was applied to the two distance matrices. Figure 3 depicts the dendrograms constructed based on the clustering results that used weighted averages versus clustering results that used SMETS. The corresponding distance matrices are shown as heat maps in Figure 4. The results obtained from both the weighted average method and SMETS are not too different, however with SMETS the NASDAQ and Dow Jones indices are clearly within the same cluster, while FTSE100 and DAX group in a different one. With the weighted average method the four group within a single cluster. It is also interesting that with SMETS the FTSE100 is quite distant from the NASDAQ100. Both methods identify the SSE50 as the most dissimilar of all the indices. Plausibly these facts are related to the composition of the indices (some stocks are present in both NASDAQ100 and Dow Jones) and the nature and frequency of trades within and between specific markets.

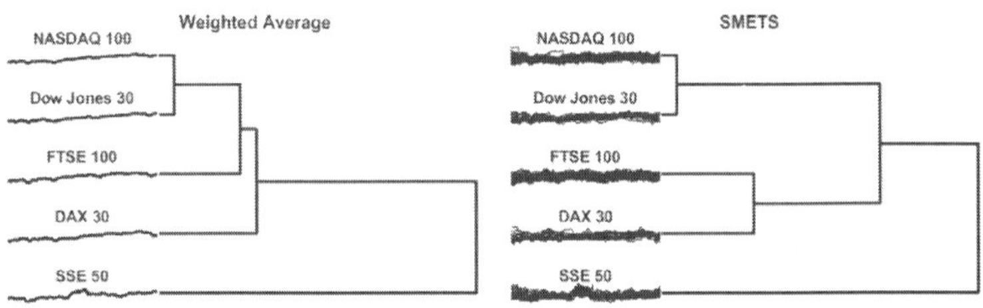

Figure 3. Hierarchical clustering of five stock indices. Indices were clustered based on the traditional weighted average method and on SMETS. The dendrogram reveals the relative distances between each entity. The time series considered by each method are represented to the left.

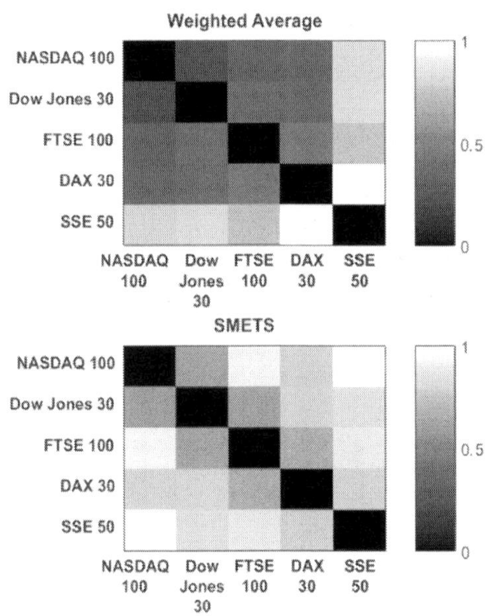

Figure 4. Distance matrices for the five stock indices. Distance values were measured using the weighted average and SMETS and are encoded in grayscale.

Biochemical Network Model Dynamics

Another area where SMETS will be useful is in modeling and simulation. Dynamical models, for example based on ordinary differential equations, represent various physical systems, such as electronic circuits or biochemical networks. Such models can be easily simulated given a certain initial condition producing time series with the behavior of the model variables. During the process of constructing and refining models it is often useful to seek other models that have similar behavior to some target. SMETS is thus well suited to this task as it allows one to find models that have some overall behavior similar to some arbitrary specification.

In systems biology there is an initiative that is collecting all published models in a database, BioModels [28], that are made available in a standard markup language (SBML) [29]. Currently this database is indexed using a number of chemical properties of the parameters and variables in the models, but not by their behavior. It would be ideal if one could ask which model in this database behaves most similar to the one a modeler is developing. This task can be easily carried out with SMETS. To illustrate this we have extracted a small subset of eight random models from the BioModels database (models 4, 21, 131, 152, 217, 331, 357 and 405). These were then loaded into the COPASI simulator [30] which produced time series for each model by integration of their differential equations. Note that each model has a different dimension, the smallest having 3 variables and the largest 64 variables. We then applied SMETS (using the same data preprocessing as above: normalization by subtracting mean and dividing by standard deviation, followed by the Haar DWT representation using the largest 16 coefficients) to these data and used the resulting distances to establish a hierarchical clustering. In parallel we applied the average method to calculate distances that were also clustered with the same algorithm. Figure 5 depicts the classifications of the models based on each approach and Figure 6 represents the distance matrices as heat maps. It is obvious that the classification based on SMETS is different from the one based on averages. We argue that the SMETS-based classification is superior. Model 357 is clearly the most similar to 405, as identified by SMETS, however the averages method pairs it with model 4. Even qualitatively it is obvious that model 4 has sustained oscillations while model 357 does not. Model 217 is also similar to 357 and 405– its variables go through large changes in the early part of the time series and relax towards a steady state in the final part, just like the other two. But the average method pairs model 217 with model 152, yet model 152's variables display large changes in the initial part as well as in the end of the time series (SMETS paired this one with model 131, which has a similar behavior)

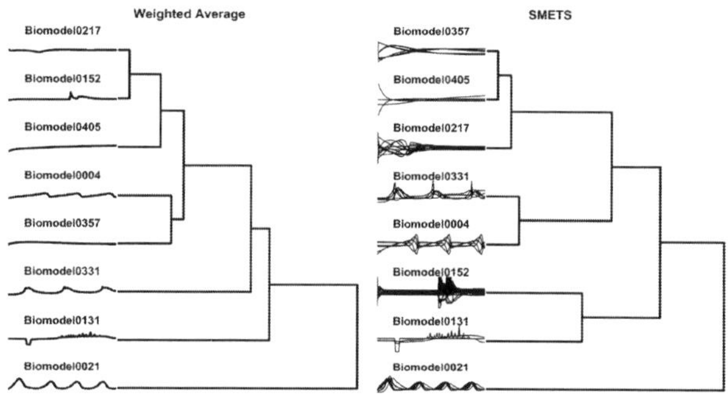

Figure 5. Hierarchical clustering of eight systems biology models. Models were obtained from the BioModels database [28] using average versus SMETS. The dendrogram reveals the relative distances between each entity. The time series considered by each method are represented to the left.

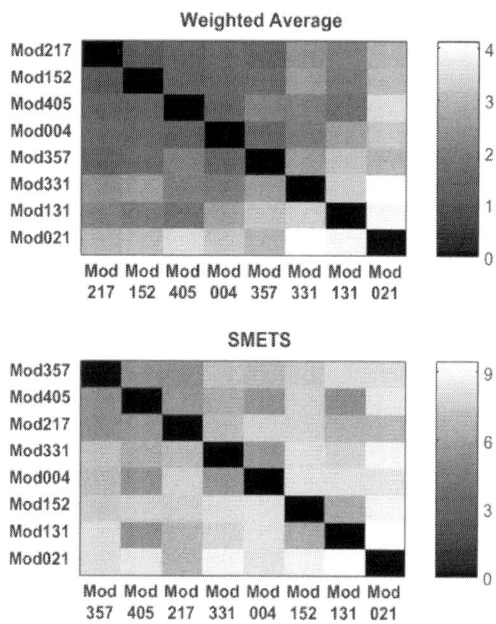

Figure 6. Distance matrices for the systems biology models. Distance values were measured using the average and SMETS distances and are encoded in grayscale.

Economic Time Series Data

One of the main types of data studied in economics is the volume of trade of various commodities. Much like the financial data discussed earlier, these data are published both as time series of single commodities (coffee, oil, etc.) as well as weighted averages of certain groupings of commodities (energy, food, etc.). Primary commodities are a set of raw materials that can be processed and transformed to manufacture goods. Fluctuations in the price of a primary commodity can influence the price of the rest of the commodities or the prices of the final goods and have a significant influence in global economics. Therefore, different sets of primary commodities time series can be treated as multivariate time series.

The International Monetary Fund (IMF) collects the prices of primary commodities and studies the economic development of different countries. The primary commodities are categorized in groups in order to investigate the status and trends of the global economy. For each group of primary commodities' time series a weighted average is also published that reflects the overall performance of the group.

We obtained commodity price time series data, and the group weighted averages, from the IMF website [31]. This consisted of monthly average prices of the primary commodities and the indices of different commodity groups for the period of January 2002 to August 2012. Each univariate time series has a length of 249 data points; a total of 10 groups of commodities are provided, each one having different number of component time series. Additionally some individual time series appear in more than one group, for example "bananas" appears in the following groupings: "food", "food and beverage", "non-fuel commodities" and "all commodities". The groupings of the primary commodities, i.e. the multivariate time series, are: a) All Commodities, b) Non-Fuel, c) Food and Beverage, d) Food, e) Beverages, f) Industrial Input, g) Agricultural Raw Material, h) Metals, i) Energy, j) Crude Oil.

Before creating the Haar wavelet representations, each component time series was normalized by subtracting the mean value and divided by the standard deviation. Time series were padded with 7 zeros at the end of each component time series to make a length of 256. In order to eliminate bias form the zero padding, the representation was truncated to a length of 15 data points.

A SMETS distance matrix was created for the different sets of commodities. In parallel, an Euclidean distance matrix was created by using the IMF indices (time series weighted averages) for comparison. Agglomerative hierarchical clustering was applied to each distance table. Figure 7 illustrates the results of hierarchical clustering in terms of dendrograms of the weighted averages and SMETS. Figure 8 depicts the distance matrices as heat maps.

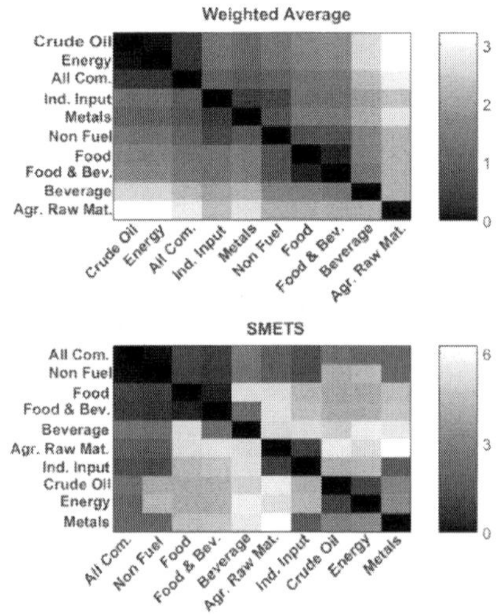

Figure 7. Hierarchical clustering of primary commodity prices. Distances were measured using the weighted average method versus SMETS. The dendrogram reveals the relative distances between each entity. The time series considered by each method are represented to the left.

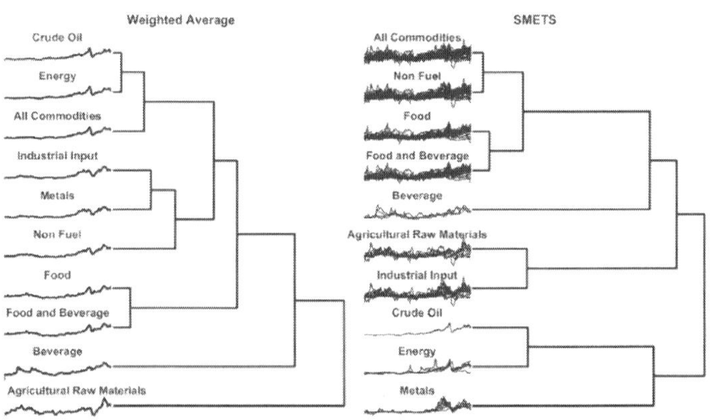

Figure 8. Distance matrices for the primary commodity prices. Distance values were measured using the average and SMETS distances and are encoded in grayscale.

The results of the two approaches are significantly different. With the classical weighted average approach the Energy, Crude Oil and All Commodities are grouped together, whereas with SMETS, All Commodities are clustered with Non-Fuel commodities. It should be noted that All Commodities includes all of the univariate time series that are also included in all other groups. Obviously there are common components between itself and any of Energy, Crude Oiland Non-fuel commodities. But there is nothing in common between Non-fuel commodities and either of Energy or Crude Oil. When SMETS encounters a component that is exactly equal in the two multivariate time series, it will be always matched. So the SMETS distance is smaller for the case when two multivariate time series will have the largest number of common components. In this case it is clearly All Commodities and Non-Fuel commodities, which share 45 common components. While Energy has only 7 in common, and Crude Oil only 3 in common. Because the weight of the Crude Oil and Energy components is very large, then the weighted average causes the effects of all other commodities to be minimized.

Electrophysiological Sleep Data

Neurophysiology studies the function of the nervous system and its underlying dynamics. Various nervous system functions are investigated by means of recording and analyzing the time-dependent electric signals.

PHYSIONET [32] is a resource that gives access to many electrophysiological data sets obtained experimentally [33]. In this example sleep data from the Sleep-EDF database [34]–[35] is used. The study of sleep has identified several stages that healthy individuals go through while asleep. These studies may also provide insight into pathologies that manifest during sleep.

We obtained data from the SLEEP-EDF dataset in PHYSIONET, which consists of 8 sleep recordings, where 4 of them were obtained from healthy volunteers with no sleep difficulties [35] and the other 4 were obtained from healthy volunteers with mild difficulties in falling asleep [34]. The recordings from the volunteers with no sleep difficulties contained the following component time series EOG, FpzCz, PzOz, EEG, submental-EMG envelope, oro-nasal airflow and rectal body temperature components [35]. The recordings from the individual with the sleeping difficulties contain measurements of EOG, FpzCz, PzOz, EEG and submental-EMG envelope [34]. Thus half of the data are 7-dimensional time series, while the other half are 5-dimensional time series. Since 5 dimensions are common among all data, one could think that removing the two extra dimensions (the oro-nasal airflow and rectal body temperature, in half of the data) would provide a better classification. This is, of course, not needed for application of SMETS since it deals well with the extra dimensions. To demonstrate this, the data were analyzed in two different ways. First we apply SMETS to the unmodified data set (half of the data 7D, the other half 5D), and then we removed the 2 extra component time series in the data from normal volunteers [35] and applied SMETS to the resulting data set entirely consisting of 5-dimensional time series.

All time series were composed of 6000 data points, which were zero-padded to a length of 8192. The Haar wavelet transform was applied and the 64 largest coefficients were retained. Then the representations were truncated to a length of 47 time points (to remove the effect of zero-padding).

The resulting distance matrices obtained by applying a) Euclidean distance between the averages of all the component time series, b) SMETS applied to the unmodified data set, c) Euclidean distance between the averages of the 5 common component time series, and d) SMETS applied to a data set that was entirely composed of the 5 common component time series. Clustering of these data resulted in dedrograms depicted in Figures 9 and 10 and heat maps in Figure 11, and 12.

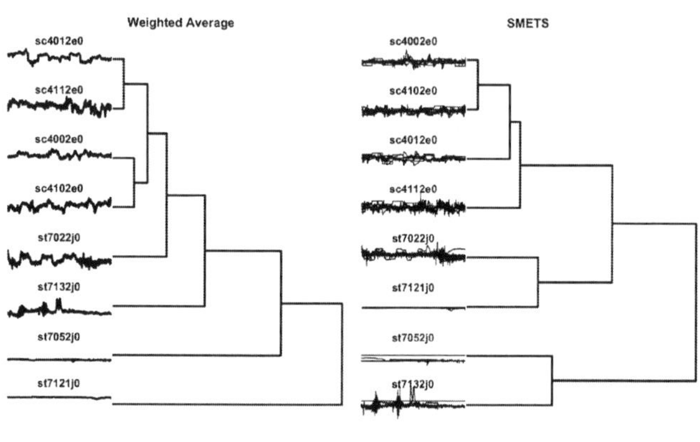

Figure 9. Hierarchical clustering of unmodified electrophysiological sleep data. Distances were measured using the weighted average method versus SMETS. The dendrogram reveals the relative distances between each entity. The time series considered by each method are represented to the left. Note that series sc4102e0, st7022j0 st7121j0 contain only 5 dimensions, while the other four contain 7 dimensions (see Results section for details).

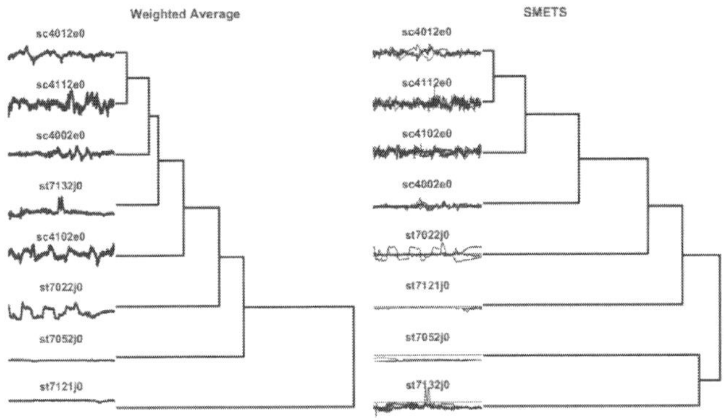

Figure 10. Hierarchical clustering of modified electrophysiological sleep data. Distances were measured using the weighted average method versus SMETS. The dendrogram reveals the relative distances between each entity. The time series considered by each method are represented to the left. All time series have only 5 dimensions, by removing the two extra dimensions from series sc4012e0, sc4112e0, sc4102e0 and sc4002e0 (see Results section for details).

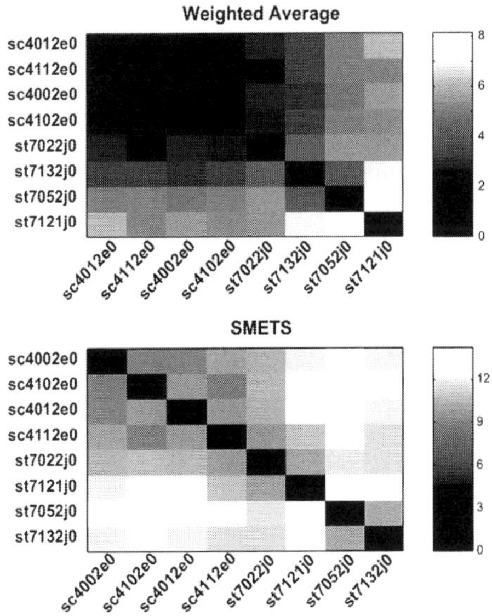

Figure 11. Distance matrices unmodified electrophysiological sleep data. Distance values were measured using the average and SMETS distances and are encoded in grayscale.

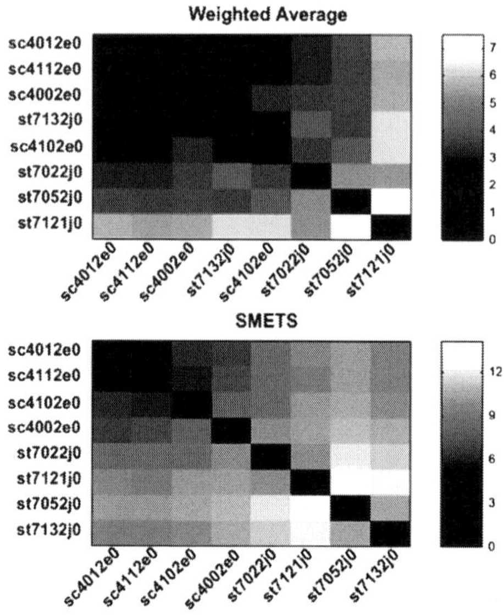

Figure 12. Distance matrices modified electrophysiological sleep data. Distance values were measured using the average and SMETS distances and are encoded in grayscale. Here all time series contain 5 dimensions (see Results section for details).

The results are not too different with any of the four methods; essentially all cluster the normal individuals together. In the complete data set (7D/5D) SMETS shows a better separation between normal individuals and those with sleep problems. However it is possible that this is the result of the bias introduced by the difference of dimensions (because all normals are 7D and all sleep disorders 5D). To remove this possible bias in the data, we eliminated the extra two components in the data of normal individuals. In this case both the averages method and SMETS show a somewhat less demarked separation. But clearly both methods still are capable of separating normals from disorders.

Discussion

We propose a method – SMETS – for comparing multivariate time series with different dimensionalities. It calculates the distance between the most similar components of two multivariate time series, and then adds penalty values to account for the difference in their dimensionalities. The penalty value is calculated using Shannon's entropy of the unmatched components. Thus, SMETS uses all of the information contained in both time series, despite their different dimensionality, which makes this method unique.

Current methods for comparing multivariate time series like the Euclidean distance, dynamic time warping [16], weighted sum singular value decomposition (WSSVD) [17], principal component analysis similarity factor (SPCA) [18] and extended Frobenius norm (EROS) [19]are limited to applications where the time series are of equal dimensionality. SMETS removes this restriction and allows distances to be calculated even when the data are of different dimensions. The examples presented demonstrate that SMETS can identify similarities without being too influenced by the difference in dimensions. A distinctive example is the case of the behavior of two biological models from the BioModels database: Model 131 contains only 3 variables while model 152 contains 64 variables, yet despite this large difference, their SMETS distance is small, allowing them to cluster together (Figures 5 and 6). This is entirely justified because both models display similar temporal behavior: variables from both models change rapidly in the initial stage and then again towards the end of the observation, while in between they have little variation. By contrast, the traditional weighted average obscures their similarity.

Both the financial and biological model examples reveal an advantage of using SMETS over the weighted averages method. Averaging all of the component time series destroys a great deal of information but SMETS avoids this and uses all of the data contained in all components. The matched components all contribute to the calculation of similarity, while the unmatched components add a penalty to the distance. Both Figures 3 and 5 show cases where the original multivariate time series are very different, but the average of their components is similar. This is especially obvious in the biological models example where even visual inspection (Figure 5) shows that the classification is more accurate with SMETS. For example the BioModels 217 and 152 have a similar average behavior but are quite distinct when considering all their component time series. This is less clear in the dendrograms of the financial data, probably because those time series are quite similar to start with (i.e. the stocks included in those indices are strongly correlated). However both distance matrices, when viewed as heat maps (Figures 4 and 6), show that SMETS reveals more structure in the data than method of averages.

The example with economic data presents an interesting case where some component time series are common between multivariate time series. This is because the classes are hierarchical and, for example the component West_

Texas_Intermdiate belongs to Crude Oil, as well as to Energy and to All Commodities. When applying SMETS these components are guaranteed to be matched. The SMETS analysis puts emphasis on the similarity of time-dependent patterns, whereas the IMF weighted average puts more emphasis on commodities that have large trades. The result is that with SMETS All Commodities is closer to Non Fuel Commodities while with the IMF weighted averaging All Commodities is closer to Crude Oil and Energy. If the objective of the comparison is to find what part of the economy has the largest weight then the weighted averages is the most suitable. On the other hand, SMETS is best to identify which multivariate time series are most similar based on their time dependent patterns.

One of the growing trends in data mining is the use of very large data sets (sometimes known as “big data”). Searching for patterns in such datasets is often hard due to their size and dimensionality. SMETS is applicable to such datasets because it can easily be combined with time series representations that compress the data by orders of magnitude. In the examples above we used a wavelet transform representation and the distance calculations were carried out in that space, allowing for the full time series to be discarded as only the representations are needed for calculations.

SMETS is, to our knowledge, the only method that allows comparing multivariate time series of different dimensionality that uses all of the information contained therein. Therefore we propose that SMETS will be a useful tool for time series data mining.

ACKNOWLEDGMENTS

We thank Douglas B. Kell and Kieran Smallbone for comments on the manuscript and for early discussions on the topic of time series data mining.

AUTHOR CONTRIBUTIONS

Conceived and designed the experiments: AT PM. Performed the experiments: AT. Analyzed the data: AT. Contributed reagents/materials/analysis tools: AT PM. Wrote the paper: AT PM.

REFERENCES

1. Shumway RH, Stoffer DS (2000) Time series analysis and its applications. New York: Springer Verlag.
2. Wei L, Keogh E (2006) Semi-supervised time series classification. In Proceedings of the 12th ACM SIGKDD International Conference on Knowledge Discovery and Data Mining (KDD '06). New York, NY:ACM. 748–753.
3. Alon J, Sclaroff S, Kollios G, Pavlovic V (2003) Discovering clusters in motion time-series data. In Proceedings of IEEE Computer Society Conference on Computer Vision and Pattern Recognition. vol.1 I375–I381.
4. Warren Liao T (2005) Clustering of time series data: a survey. Pattern Recognition 38: 1857–1874.
5. Chin SC, Ray A, Rajagopalan V (2005) Symbolic time series analysis for anomaly detection: a comparative evaluation. Signal Proc 85: 1859–1868.
6. Ye N (2003) The handbook of data mining. Mahwah, NJ: Lawrence Erlbaum.
7. Keogh E, Kasetty S (2003) On the need for time series data mining benchmarks: A survey and empirical demonstration. Data Mining and Knowledge Discovery 7: 349–371.
8. Deza MM, Deza E (2009) Encyclopedia of Distances. Berlin Heidelberg:Springer Verlag.
9. Veltkamp RC (2001) Shape matching: similarity measures and algorithms. International Conference on Shape Modeling and Applications, 188–197.
10. Vlachos M, Kollios G, Gunopulos D (2002) Discovering similar multidimensional trajectories. Proceedings of the 18th International Conference on Data Engineering. IEEE 673–684.
11. Agrawal R, Faloutsos C, Swami A (1993) Efficient similarity search in sequence databases. Lecture Notes Comp Sci 730: 69–84.
12. Chan KP, Fu AWC (1999) Efficient time series matching by wavelets. Proceedings of the 15th International Conference on Data Engineering. IEEE. 126–135.
13. Keogh E, Chakrabarti K, Pazzani M, Mehrotra S (2001) Dimensionality reduction for fast similarity search in large time series databases.

Knowledge Inf Sys 3: 263–286.

14. Lin J, Keogh E, Wei L, Lonardi S (2007) Experiencing SAX: a novel symbolic representation of time series. Data Mining and Knowledge Discovery 15: 107–144.
15. Vlachos M, Hadjieleftheriou M, Gunopulos D, Keogh E (2003) Indexing multi-dimensional time-series with support for multiple distance measures. Proceedings of the ninth ACM SIGKDD International Conference on Knowledge Discovery and Data Mining. New York, NY:ACM. 216–225.
16. Rath TM, Manmatha R (2002) Lower-Bounding of Dynamic Time Warping Distances for Multivariate Time Series. Technical Report MM-40, Amherst:University of Massachusetts.
17. Shahabi C, Yan D (2003) Real-time pattern isolation and recognition over immersive sensor data streams. In Proceedings of the 9th International Conference on Multi-Media Modeling. 93,113.
18. Krzanowski WJ (1979) Between-Groups Comparison of Principal Components. J Am Stat Assoc 74: 703–707.
19. Yang K, Shahabi C (2004) A PCA-based similarity measure for multivariate time series. Proceedings of the 2nd ACM International Workshop on Multimedia Databases. New York, NY: ACM. 65–74.
20. Sutcliffe CMS (2006) Stock index futures. Aldershot, England:Ashgate Publishing Ltd.
21. Faloutsos C, Ranganathan M, Manolopoulos Y (1994) Fast subsequence matching in time-series databases. In Snodgrass RT and Winslett M, editors. Proceedings of the 1994 ACM SIGMOD International Conference on Management of Data (SIGMOD '94), New York, NY :ACM 23: 419–429.
22. Hart KP, Nagata J-I, Vaughan JE (2004) Encyclopedia of General Topology. Amsterdam: Elsevier.
23. Wilson WA (1931) On Semi-Metric Spaces. Am J Maths 53: 361–373.
24. Yahoo! Finance UK Available: http://uk.finance.yahoo.com.Accessed 2011 Oct 26.
25. Kalpakis K, Gada D, Puttagunta V (2001) Distance measures for effective clustering of ARIMA time-series. In Proceedings IEEE International Conference on Data Mining, 273–280.
26. Jensen A, Cour-Harbo AL (2001) Ripples in Mathematics: The Discrete Wavelet Transform. Berlin Heidelberg:Springer-Verlag.

27. Chakrabarti K, Keogh E, Mehrotra S, Pazzani M (2002) Locally adaptive dimensionality reduction for indexing large time series databases. ACM Trans Database Sys 27: 188–228.

28. Le Novere N, Bornstein B, Broicher A, Courtot M, Donizelli M, et al. (2006) BioModels Database: a free, centralized database of curated, published, quantitative kinetic models of biochemical and cellular systems. Nucleic Acid Res 34: D689–D691.

29. Hucka M, Finney A, Sauro HM, Bolouri H, Doyle JC, et al. (2003) The systems biology markup language (SBML): a medium for representation and exchange of biochemical network models. Bioinformatics 19: 524–531.

30. Hoops S, Sahle S, Gauges R, Lee C, Pahle J, et al. (2006) COPASI–a COmplex PAthway SImulator. Bioinformatics 22: 3067–3074.

31. International Monetary Fund website. Available: http://www.imf.org/. Accessed 2012 Oct 3.

32. PhysioNet website. Available: http://www.physionet.org/. Accessed 2012 Oct 25.

33. Goldberger AL, Amaral LAN, Glass L, Hausdorff JM, Ivanov PCh, et al. (2000) PhysioBank, PhysioToolkit, and PhysioNet: Components of a New Research Resource for Complex Physiologic Signals. Circulation 101: e215–e220.

34. Kemp B, Zwinderman AH, Tuk B, Kamphuisen HAC, Oberyé JJL (2000) Analysis of a sleep-dependent neuronal feedback loop: the slow-wave microcontinuity of the EEG. IEEE Transact. Biomed. Eng. 47: 1185–1194.

35. Mourtazaev MS, Kemp B, Zwinderman AH, Kamphuisen HAC (1995) Age and gender affect different characteristics of slow waves in the sleep EEG. Sleep 18: 557–564.

10

Network Structure of Multivariate Time Series

Lucas Lacasa, Vincenzo Nicosia & Vito Latora

School of Mathematical Sciences, Queen Mary University of London, Mile End Road, London, E14NS, UK

ABSTRACT

Our understanding of a variety of phenomena in physics, biology and economics crucially depends on the analysis of multivariate time series. While a wide range tools and techniques for time series analysis already exist, the increasing availability of massive data structures calls for new approaches for multidimensional signal processing. We present here a non-parametric method to analyse multivariate time series, based on the mapping of a multidimensional time series into a multilayer network, which allows to extract information on a high dimensional dynamical system through the analysis of the structure of the associated multiplex network. The method is simple to implement, general, scalable, does not require ad hoc phase space partitioning and is thus suitable for the analysis of large, heterogeneous and

Citation: Lacasa, L., Nicosia, V., &Latora, V. (2015). Network structure of multivariate time series. Scientific reports, 5, 15508. (9 pages)

Copyright: © This work is licensed under a Creative Commons Attribution 4.0 International (CC BY 4.0) License.

non-stationary time series. We show that simple structural descriptors of the associated multiplex networks allow to extract and quantify nontrivial properties of coupled chaotic maps, including the transition between different dynamical phases and the onset of various types of synchronization. As a concrete example we then study financial time series, showing that a multiplex network analysis can efficiently discriminate crises from periods of financial stability, where standard methods based on time-series symbolization often fail.

INTRODUCTION

Time series analysis is a central topic in physics, as well as a powerful method to characterize data in biology, medicine and economics and to understand their underlying dynamical origin. In the last decades, the topic has received input from different disciplines such as nonlinear dynamics, statistical physics, computer science or Bayesian statistics and, as a result, new approaches like nonlinear time series analysis1 or data mining2 have emerged. More recently, the science of complex networks3[,4,5] has fostered the growth of a novel approach to time series analysis based on the transformation of a time series into a network according to some specified mapping algorithm and on the subsequent extraction of information about the time series through the analysis of the derived network. Within this approach, a classical possibility is to interpret the interdependencies between time series (encapsulated for instance in cross-correlation matrices) as the weighted edges of a graph whose nodes label each time series, yielding so called functional networks, that have been used fruitfully and extensively in different fields such as neuroscience6 or finance7[,8,9]. A more recent perspective deals with mapping the particular structure of univariate time series into abstract graphs10[,11,12,13,14,15,16], with the aims of describing not the correlation between different series, but the overall structure of isolated time series, in purely graph-theoretical terms. Among these latter approaches, the so called visibility algorithms15[,16] have been shown to be simple, computationally efficient and analytically tractable methods17[,18], able to extract nontrivial information about the original signal19, classify different dynamical origins20 and provide a clean description of low dimensional dynamics21[,22,23,24]. As a consequence, this particular methodology has been used in different domains including earth and planetary sciences25[,26,27,28], finance29 or biomedical fields30 (see31 for a review). Despite their success, the range of applicability of visibility methods has been so far limited to univariate time series (see however24[,28]), whereas the most challenging problems in several areas of nonlinear science concern systems governed by a large number of degrees of freedom, whose evolution is indeed described by multivariate time series.

In order to fill this gap, in this work we introduce a visibility approach to analyze multivariate time series based on the mapping of a multidimensional signal into an appropriately defined multi-layer network32[,33,34,35,36,37], which we call multiplex visibility graph. Taking advantage of the recent developments in the theory of multilayer networks32[,34,35,36,37,38,39], new information can be extracted from the original multivariate time series, with the aims of describing signals in graph-theoretical terms or to construct novel feature vectors to feed automatic classifiers in a simple, accurate and computationally efficient way. We will show that, among other possibilities, a particular projection of this multilayer network produces a (single-layer) network similar in spirit to functional networks, while being more accurate than standard methodologies to construct those ones. We validate our method by investigating the rich high-dimensional dynamics displayed by canonical models of spatio-temporal chaos and then apply our framework to describe and quantify periods of financial instability from a set of empirical multivariate financial time series.

RESULTS

Let us start by recalling that visibility algorithms are a family of geometric criteria which define different ways of mapping an ordered series, for instance a temporal series of N real-valued data $\{x(t)\}_{t=1}^{N}$, into a graph of N nodes. The standard linking criteria are the natural visibility15 (a convexity criterion) and the horizontal visibility16 (an ordering criterion). In the latter version, two nodes i and j are linked by an edge if the associated data x(i) and x(j) have horizontal visibility, i.e. if every intermediate datum x(k) satisfies the ordering relation $x(k)<\inf\{x(i), x(j)\}$, $\forall$ k: i<k<j. The resulting Horizontal Visibility Graph (HVG) is an outerplanar graph (indeed a subgraph of the original visibility graph), whose topological properties have been shown to be analytically tractable for a large class of different dynamical processes16[,17]. Both the natural and horizontal graphs are undirected, however directed graphs can be easily constructed by distinguishing ingoing from outgoing links with respect to the arrow of time, something which has proven useful to assess time asymmetries and to quantify time series irreversibility40[,41,42].

Consider a M-dimensional real valued time series $\{\mathbf{x}(t)\}_{t=1}^{N}$, with $\mathbf{x}(t) = (x^{[1]}(t), x^{[2]}(t), \ldots, x^{[M]}(t)) \in \mathbb{R}^M$ for any value of t, measured empirically or extracted from a M-dimensional, either deterministic or stochastic, dynamical system. An M-layer multiplex network, that we call the multiplex visibility graph $\mathcal{M}$ is then constructed, where layer α corresponds to the HVG associated to the time series of state variable $\{x^{[\alpha]}(t)\}_{t=1}^{N}$. We illustrate this procedure for M=3 in Fig. 1. Note that in this work we focus on the undirected, horizontal visibility, but other visibility linking criteria can be analogously used to define different multiplex graphs. $\mathcal{M}$ is represented by the vector of adjacency matrices of its layers $\mathcal{A} = \{A^{[1]}, A^{[2]}, \ldots, A^{[M]}\}$, where $A^{[\alpha]} = \{a_{ij}^{[\alpha]}\}$ and $a_{ij}^{[\alpha]}=1$ if and only if node i and node j are connected by a link at layer α33[,37]. Such a mapping builds a bridge between multivariate series analysis and the recent developments in the theory of multilayer networks32[,34,35,36,37,38,39], making it possible to employ the structural descriptors introduced to study multiplex networks as a toolbox for the characterisation of multivariate signals. In the following we will focus on two measures which capture, respectively, the abundance of single edges across layers and the presence of inter-layer correlations of node degrees39. In such a way we can characterize information shared across variables (layers) of the underlying high dimensional system, this aspect being indeed of capital importance in fields such as neuroscience or economics and finance.

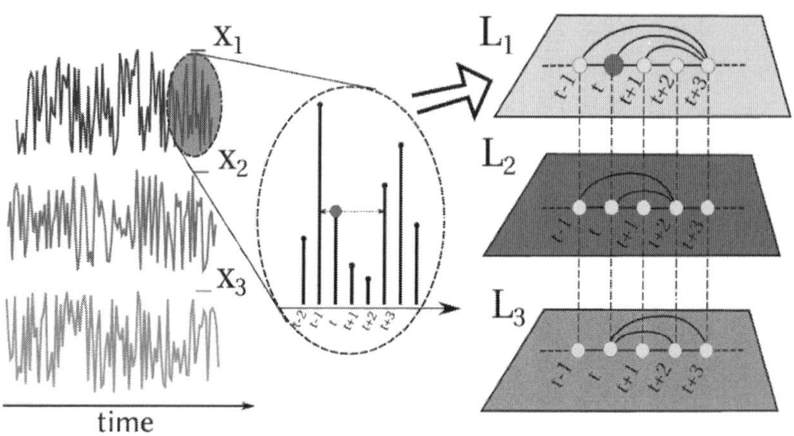

Figure 1. The Horizontal Visibility Graph (HVG) algorithm maps a M-dimensional time series $\{x^{[\alpha]}(t)\}_{t=1}^{N}, \alpha = 1, \ldots, M,$, into a multiplex visibility graph $\mathcal{M}$, i.e. a multi-layer network where each layer α is the HVG of the α-th component of the time series.

The first measure we introduce is the average edge overlap

$$\omega = \frac{\sum_i \sum_{j>i} \sum_\alpha a_{ij}^{[\alpha]}}{M \sum_i \sum_{j>i} \left(1 - \delta_{0, \sum_\alpha a_{ij}^{[\alpha]}}\right)} \quad 1$$

which computes the expected number of layers of the multiplex on which an edge is present. Note that ω takes values in [1/M, 1] and in particular ω = 1/M if each edge (i, j) exists in exactly one layer, i.e. if there exist a layer α such that $a_{ij}^{[\alpha]} = 1$ and $a_{ij}^{[\beta]} = 0 \ \forall \beta \neq \alpha$, while ω = 1 only if all the M layers are identical. As a consequence, the average edge overlap of a multiplex visibility graph can be used as a proxy of the overall coherence of the original multivariate time series, with higher values of ω indicating high correlation in the microscopic structure of the signal.

The second measure we use quantifies the presence of interlayer correlations between the degrees of the same node at two different layers. Given a pair of layers α and β of $\mathcal{M}$, respectively characterized by the degree distributions $P(k^{[\alpha]})$ and $P(k^{[\beta]})$, we can define an interlayer mutual information $I_{\alpha,\beta}$ as:

$$I_{\alpha,\beta} = \sum_{k^{[\alpha]}} \sum_{k^{[\beta]}} P(k^{[\alpha]}, k^{[\beta]}) \log \frac{P(k^{[\alpha]}, k^{[\beta]})}{P(k^{[\alpha]}) P(k^{[\beta]})} \quad 2$$

where $P(k^{[\alpha]}, k^{[\beta]})$ is the joint probability to find a node having degree $k^{[\alpha]}$ at layer α and degree $k^{[\beta]}$ at layer β (see Methods for details). In general, the higher $I_{\alpha,\beta}$ the more correlated the degree distributions of the two layers and, therefore, the structure of the associated time series. If we then average the quantity $I_{\alpha,\beta}$ over every possible pair of layers of $\mathcal{M}$, we obtain a scalar variable $I = \langle I_{\alpha,\beta} \rangle_{\alpha,\beta}$ which captures the typical amount of information flow in the system.

It is worth noticing that the values $\{I_{\alpha,\beta}\}$ can be considered as the weights of the edges of a graph of layers $\mathcal{G}$, this being a projection of the original multiplex visibility graph $\mathcal{M}$ into a (single-layer) weighted graph of M nodes, where each node represents one layer. The weights of the edges of such graph denote the magnitude of mutual information, but our approach can be easily generalized so that edges can represent different types of interdependence, such as correlation6, causality43, etc.44 between layers. By using this particular projection, where each layer is represented by a node and relationships between layers are indicated by weighted edges, we are actually analyzing the visibility analog of classical functional networks. However, we shall indeed see that the projection obtained through the multiplex visibility algorithm is genuinely different and often works better than other ways to construct functional networks. This is mainly because the latter ones often require the symbolization of each time series and this pre-processing is usually afflicted by several limitations and ambiguities1. As we show in the online Supplementary Information (SI), the technique proposed here also appears to be free from these well known ambiguities.

Information flow and phase diagram in Coupled Map Lattices

As a case study we first consider diffusively coupled map lattices (CMLs), high-dimensional dynamical systems with discrete time and continuous state variables, widely used to model complex spatio-temporal dynamics45 in as disparate contexts as turbulence46,[47], financial markets48, biological systems49 or quantum field theories50. Namely we consider a ring of M sites and we assume that the dynamical evolution of the state $x^{[\alpha]}$ of each site α is determined by a compromise between an internal chaotic evolution and an external diffusive coupling among nearest-neighbor sites:

$$x^{[\alpha]}(t+1) = (1-\varepsilon)f\left[x^{[\alpha]}(t)\right] + \frac{\varepsilon}{2}\left(f\left[x^{[\alpha-1]}(t)\right] + f\left[x^{[\alpha+1]}(t)\right]\right), \quad 3$$

$\forall\alpha = 1,\ldots, M$, where $\varepsilon \in [0, 1]$ is the coupling strength and f(x) is typically a chaotic map. In the right hand of the equation one can easily recognize the discrete version of the Laplacian (that drives diffusion). The control parameter ε is a quantity that tunes the relative strength of the diffusive term in the evolution of the system, acting as an effective viscosity constant and separating different dynamical phases where the system gets trapped into different attractors as ε varies. It is well known45 that for different values of ε and M, CMLs display a very rich phase diagram, which includes different degrees of synchronization and dynamical phases such as Fully Developed Turbulence (FDT, a phase with incoherent spatiotemporal chaos and high dimensional attractor), Pattern Selection (PS, a sharp suppression of chaos in favor of a randomly selected periodic attractor), or different forms of spatio-temporal intermittency (STI, a chaotic phase with low dimensional attractor, which is indeed a pseudo-phase that interpolates between FDT and PS). The origin of such a rich and intertwined structure comes from the interplay between the local tendency towards inhomogeneity, induced by the chaotic dynamics of each variable and the global tendency towards homogeneity in space, induced by the diffusive coupling.

Figure 2 summarizes the analysis and results obtained for a CML of M=5 diffusively coupled, fully chaotic logistic maps $f(x)=4x(1-x)$, which exhibits several transitions from high dimensional chaos, to pattern selection, to several forms of partially synchronized chaotic states when ε is increased. Sample multivariate time series of the system in three different states (Fully Developed Turbulence (FDT) at ε=0.05, Periodic Pattern Selection (PS) at ε=0.17 and Spatio-temporal Intermittency (STI) at ε=0.37) are shown in the top of panel (a). In Fig. 2(c) we plot, as a function of ε, the average mutual information I of the multiplex visibility graph associated to a multivariate time series of 2^{14} data extracted from eq. (2) (results are averaged over 100 realizations, see Methods for details and SI equivalent analysis using average edge overlap). The different dynamical phases that emerge as a function of ε are also depicted as different background colors. We find that I is able to distinguish between different dynamical phases. In particular, I is a monotonically increasing function of ε in the FDT phases and therefore quantifies the amount of information flow among units, which increases in the FDT as the contribution of the diffusion increases. Notably, it also detects qualitative changes in the underlying dynamics (such as the chaos suppression in favor of a randomly selected periodic pattern, or the onset of a multi-band chaotic attractor during intermittency) and therefore constitutes a scalar order parameter of the system.

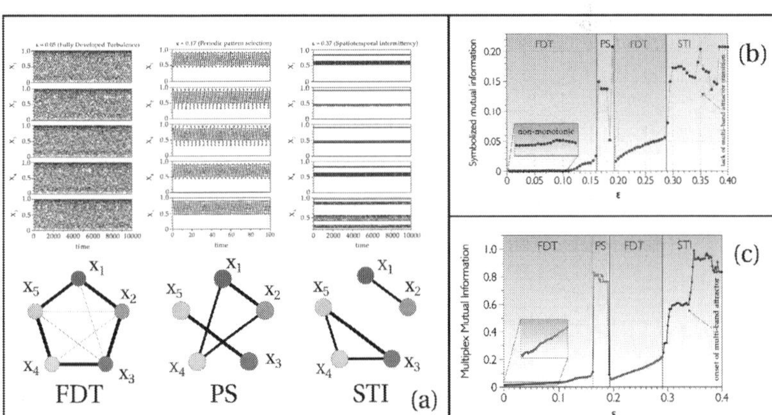

Figure 2. (**a**) Sample time series generated by five diffusively coupled fully chaotic logistic maps at different values of the coupling strength ε, showing respectively Fully Developed Turbulence (FDT) at ε=0.05, Pattern Selection (PS) at ε=0.17 and Spatio-temporal Intermittency (STI) at ε=0.37. Information flow among units is well captured by projecting the multiplex visibility graph

$\mathcal{M}$ into a layer graph $\mathcal{G}$ (bottom), whose nodes represent layers and the edges weights (their thickness in the figure) the mutual information among them. (**b**) The average pair wise mutual information I^{SYMB} computed from pre-symbolized time series, or (**c**) the corresponding version I computed from the associated multiplex visibility graph can both be used as order parameters to distinguish different dynamical phases (see SI for additional analysis with average edge overlap). However, only the multiplex measure is capable of detecting fine-grain structures such as the onset of multi-band chaotic attractors.

For comparison, in Fig. 2(b) we also plot the corresponding quantity I^{SYMB}, derived from a standard functional network analysis. Namely, I^{SYMB} is the average mutual information computed directly on the multivariate time series, after performing the necessary time series symbolization (see methods for details). Although there are qualitative similarities with I, subtle aspects such as the monotonic increase of synchronization with ε in FDT, or the onset of multiband attractors in STI are not captured by I^{SYMB} (additional details comparing our method to standard functional network approaches can be found in the SI). In the bottom of panel (a) of the same figure we also report the projections of the three multiplex networks $\mathcal{M}$ into the corresponding graphs of layers $\mathcal{G}$, whose edge widths are proportional to the values $I_{\alpha,\beta}$ of mutual information between layers α and β. A simple visual inspection of such graphs reveals the different type of information flow among units, depending on the dynamical phases of the system. In particular, notice that the diffusive nature of the coupling emerges in the ring-like structure of graph $\mathcal{G}$ corresponding to weakly interacting maps (FDT) (the analysis is extended in SI to globally coupled maps, these being a mean-field version of CML where complete synchronisation is possible, showing that our method correctly detects the onset of this new regime).

Scaling Up the System

The previous study suggests that the quantities $\{I_{\alpha,\beta}\}$ (see Fig. 2(a)) accurately capture relevant information of the underlying dynamics. To further explore this aspect and to assess scalability, we considered a chain of M=200 diffusively coupled logistic maps, each governed by Eq. (3). New short dynamical phases, such as the so called Brownian motion of Defects (BD) –a transient phase between FDT and PS–, emerge when the dimension of the system is increased, whereas others such as intermittency are now just mere interfaces advancing the onset of periodicity (PS). These phases have been extensively analysed in ref. 45. As the description gets more cumbersome, projections and coarse-grained variables are needed. Since the graph of layers $\mathscr{G}$ is by construction a complete graph (just as any functional network), for visual reasons in Fig. 3(a)we report the structural properties of $\mathscr{G}'$, the backbone of $\mathscr{G}$ obtained starting from an empty graph of M nodes and adding edges sequentially in decreasing order of $I_{\alpha,\beta}$, until the resulting graph consists of a single connected component. The structure of $\mathscr{G}'$ is unique for each phase and qualitatively different across phases, thus providing a simple qualitative way to portrait different dynamics in high-dimensional systems.

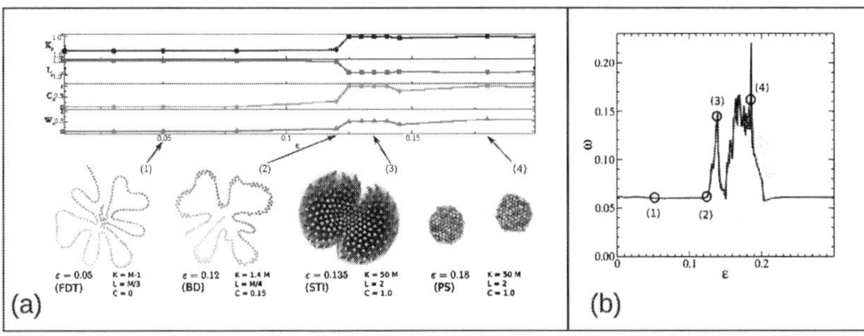

Figure 3. (**a**) Backbone graphs $\mathscr{G}'$ of the graph of layers $\mathscr{G}$ obtained from the multiplex visibility graph of M=200 diffusively coupled chaotic maps for different values of ε. For each dynamical phase, $\mathscr{G}'$ has a different structure. In FDT, $\mathscr{G}'$ is a chain, revealing the diffusive nature of the coupling in this weakly interacting situation. In PS, $\mathscr{G}'$ is formed by non-interacting communities, whereas in STI -where the ghost of the periodic attractor is perturbed by chaotic excursions-, $\mathscr{G}'$ is formed of slightly overlapping dense communities. We also report several topological properties of $\mathscr{G}'$ (number of edges K, average shortest path length L, clustering coefficient C and total weight of the edges) that have different qualitative values in each phase. Nodes colours determine the map label, so similar colours denote maps at close distance in the CML. (**b**) The four different phases are also visible in the plot of the average edge overlap of the multiplex as a function of the coupling strength ε (numerical simulations are averaged over 50 realizations). In particular, for $\varepsilon<0.12$ we have ω at its minimum, as expected in FDT. Then, for $0.12<\varepsilon<0.15$ we observe a consistent increase in ω, which is in agreement with the development of BD first and STI afterwards. For $0.16<\varepsilon<0.2$ the system is in the periodic pattern selection phase, characterised by relatively larger values of edge overlap, while for $\varepsilon>0.2$ the systems is again in FDT. The values of ε corresponding to the four sample graphs shown in panel (**a**) are indicated by empty circles.

We might complement this analysis by addressing now the average edge overlap ω, which is a somehow more basic structural indicator. This measure effectively plays the role of an order parameter able to distinguish these different dynamical phases quite accurately, as shown by the plot reported in Fig. 3(b). As expected, the FDT phases, observed for $\varepsilon<0.12$ and for $\varepsilon>0.2$ (see ref. 45) are associated to the minimum values of ω ($\omega_{\min} \simeq 0.06$) . Conversely, the periodic PS phase ($0.16<\varepsilon<0.20$) is characterised by the largest values of edge overlap. In this phase, each edge is present, on average, in up to 45 of the 200 layers of the multiplex, indicating a higher overall level of coherence. It is interesting to stress that the average edge overlap of a multiplex network can be computed quite efficiently from the adjacency matrices of the layers. In fact, it is easy to show that the time complexity of computing ω is $O(\mathcal{K})$, where $\mathcal{K}$ is the total number of edges in the multiplex. As horizontal visibility graphs tend to be sparse for a large variety of dynamical systems16, the effective time complexity is $O(N \times M)$, i.e. linear in the length and in the dimension of the original time series. If we take into account the fact that the time complexity of the horizontal visibility graph algorithm is linear for noisy series in the length of the series, then we can conclude that the proposed approach to multivariate time series analysis can be readily employed also for the analysis large data sets.

Multiplex analysis of financial instabilities

As an example of the possible applications of the multiplex visibility graph approach to the analysis of real-world multivariate time series, we report a study of the prices of financial assets. Namely, we considered the time evolution of stock prices of M=35 of the largest US companies by market capitalization from NYSE and NASDAQ (see SI for details) over the period 1998–2013. The M time series have a very high resolution (one data per minute), yielding $O(2 \cdot 10^6)$ data per company. We divided each multivariate time series into non-overlapping periods of three months (quarters) and we constructed a temporal multiplex visibility graph consisting of 64 multilayer snapshots, each formed by the 35-layer multiplex visibility graph corresponding to one of the 3-months periods. We then investigated the time evolution of the multiplex mutual information among layers and how this correlates with the presence of periods of financial instability.

In Fig. 4(a) we plot the value of the average multiplex mutual information I as a function of time, while in Fig. 4(b) we show the values of average edge overlap ω. For comparison we also report in Fig. 4(c) the values of I^{SYMB} computed directly on the original series, after an appropriate symbolization (see Methods and SI for details). We find that the multiplex visibility graph approach captures the onset of the major periods of financial instability (1998–1999, corresponding to the .com bubble and 2007–2012, corresponding to the great recession that took place as a consequence of the mortgage subprime crisis), which are characterised by a relatively increased synchronisation of stock prices, clearly distinguishing them from the seemingly unsynchronised interval 2001–2007, which in turn corresponds to a more stable period of the economy. In direct analogy with the language used for CMLs, we could say that in periods of financial stability, the system is close to equilibrium, degrees of freedom are evolving in a quasi-independent way, reaching a fully developed turbulent state of low mutual information (hence unpredictable and efficient from a financial viewpoint). On the other hand, during periods of financial instability (bubbles and crisis) the system is externally perturbed, hence driven away from equilibrium and the degrees of freedom share larger mutual information (the system is therefore less unpredictable and inefficient from a financial viewpoint). Notice that this picture is confirmed by the plot of the average edge overlap ω, reported in Fig. 4(b), where peaks correspond quite closely to the major periods of financial crises. As shown in Fig. 4(c), instead, an analogous analysis based on the symbolization of the time series fails to capture all such details (see SI for additional analysis). Finally, as also seen in the case of the multiplex visibility graphs associated to CMLs, the differences in the values of average mutual information corresponding to different phases are indeed related to a different underlying structure of the network of layers. In Fig. 4(a) we show the Maximal Spanning Trees (MST) of the networks of layers associated to six typical time windows7[,51]. The three networks at the bottom of the Figure represent periods of financial stability, while those at the top of the Figure correspond to the three local maxima of mutual information. Interestingly, the MSTs in periods of financial instability all have a massive hub which is directly linked to as much as 50% of all the other nodes. Conversely, the degree is more evenly distributed in the MSTs associated to periods of economic stability.

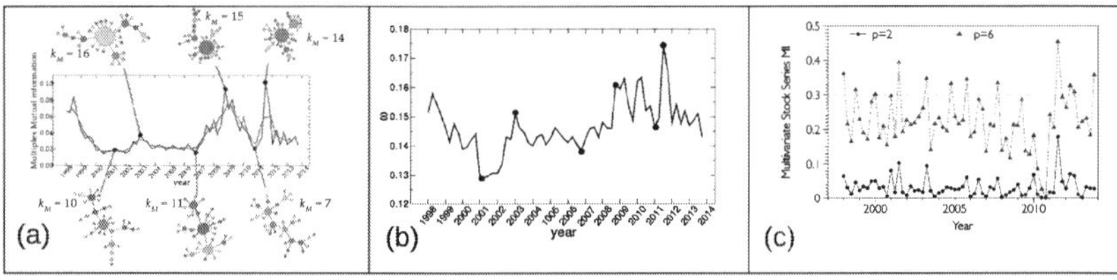

Figure 4. (**a**) The multiplex mutual information is a suitable quantity to detect global changes of behavior in multivariate financial time series. In the plot we report the value of the average information (the red line is the corresponding running average) computed from the multiplex networks constructed from price time series of 35 major assets in NYSE and NASDAQ in each 3-month period between January 1998 and December 2013 (see Methods for details). Notice that the most pronounced peaks of mutual information correspond to periods of major financial instability (the .com bubble in 1998–1999, the mortgage subprime crisis in 2007–2012). The Maximum Spanning Trees of the corresponding networks of layers (six typical examples are shown), always include a large hub during crises (the three top networks), whose degree k_M is larger than the maximum degree observed in periods of stability (the three bottom networks). Each asset is assigned the same color in all the networks, while the size of a node is proportional to its degree. (**b**) The difference between periods of stability and crises is also detected by the average edge overlap ω. For visual reference, we report in the plot the six example points examined in panel (**a**). It is evident that peaks of ω correspond to crises while dips are usually associated to normal market activity. (**c**) The mutual information among the same set of time series performed after a standard symbolization with p symbols is not able to single out crises. The resulting signal is much more herratic and not as informative as the multiplex mutual information shown in panel (**a**).

DISCUSSION

The approach based on multiplex visibility graphs introduced in this work provides an alternative and powerful method to analyze multivariate time series. We have first validated our method focusing on signals whose underlying dynamics is well known and showing that measures describing the structure of the corresponding multiplex networks (which are not affected by the usual problems of standard symbolization procedures) are able to capture and quantify the onset of dynamical phases in high-dimensional coupled chaotic maps, as well as the increase or decrease of mutual information among layers (maps) within each phase. We then have studied an application to the analysis of multivariate financial series, showing that multiplex measures, differently from other standard methods, can easily distinguish periods of financial stability from crises and can thus be used effectively as a support tool in decision making.

The proposed method is extremely flexible and can be used in all situations where the dynamics is poorly understood or unknown, with potential applications ranging from fluid dynamics to neuroscience or social sciences. In particular, the accuracy of some basic multiplex metrics (e.g., the average edge overlap and the normalised mutual information of the degree sequences) in distinguishing different dynamical phases makes it possible to efficiently analyse large multivariate timeseries. In fact, the construction of the horizontal visibility multiplex network associated to a multivariate time-series is linear in the number of nodes and in the number of variables, while the estimation of average edge overlap and mutual information are respectively linear and quadratic in the number of layers. In this article we have focused only on a particular aspect, which is the characterization of the information flow among the different variables of the system and we have consequently based our analysis on the study of the resulting networks of layers. However, our approach is quite general and the mapping of multivariate time series into multiplex visibility graphs paves the way to the study of the relationship between specific structural descriptors recently introduced in the context of multiplex networks and the properties of real-world dynamical systems. We are confident that our method is only the first step towards the construction of feature-based automatic tools to classify dynamical systems of any kind.

METHODS

Inter-layer degree correlations in multiplex networks

In a M-layer multiplex network $\mathscr{M}$ on N nodes, defined by the vector of adjacency matrices $\mathscr{A} = \{A^{[1]}, A^{[2]}, \ldots, A^{[M]}\}$, each node i is associated to a multiplex degree vector $k_i = \{k_i^{[1]}, k_i^{[2]}, \ldots, k^{[M]}\}$, where $k_i^{[\alpha]} = \sum_j a_{ij}^{[\alpha]}$ is the total number of edges incident on node i at layer α. With the usual notation, $P(k^{[\alpha]})$ is the degree distribution of layer α. Several measures have been recently proposed in the literature to quantify the sign and intensity of inter-layer degree correlations. Here we choose to characterize the degree correlation between two layers α and β by using the mutual information between the corresponding degree sequences $\{k_i^{[\alpha]}\}$ and $\{k_i^{[\beta]}\}$, i = 1,..., N. This choice was motivated by the direct dynamical interpretation of the mutual information for visibility graphs. In particular, we computed the quantities:

$$P(k^{[\alpha]}, k^{[\beta]}) = \frac{N_{k^{[\alpha]},k^{[\beta]}}}{N} \quad 4$$

where $N_{k^{[\alpha]},k^{[\beta]}}$ is the number of nodes having degree equal to $k^{[\alpha]}$ and $k^{[\beta]}$ respectively on layer α and on layer β. By definition we have:

$$\sum_{k^{[\alpha]}}\sum_{k^{[\beta]}} N_{k^{[\alpha]},k^{[\beta]}} = N \quad 5$$

Using this notation, the mutual information $I_{\alpha,\beta}$ between the degree distributions at layers α and β reads:

$$I_{\alpha,\beta} = \sum_{k^{[\alpha]}}\sum_{k^{[\beta]}} P(k^{[\alpha]}, k^{[\beta]}) \log \frac{P(k^{[\alpha]}, k^{[\beta]})}{P(k^{[\alpha]})P(k^{[\beta]})} \quad 6$$

Comparing with Measures on Symbolized Series

Symbolization is one of the standard requisites for the analysis of (multivariate) time series. The procedure consists in transforming the original (continuous state) signal $\{x(t)\} \in \mathbb{R}^d$ into a (discrete state) sequence {S(t)} whose values belong to a finite set of p symbols $\{s_1, s_2, \ldots, s_p\}$. There are several ways to construct such symbolised representations of multivariate signals, but one of the main problems which characterise symbolisation is the choice of the mapping function and the determination of the appropriate number of symbols to be used (see SI for details). The standard approach in CMLs (for which $\{x(t)\} \in [0, 1]^d$) is to make a partition of each dimension into p non-overlapping subintervals, assigning a different symbol to those values belonging to different subintervals (see SI for additional details). The resulting symbolized series is a symbol sequence. One can then estimate the mutual information between two symbol sequences x(t) and y(t) as

$$I_{x,y}^{symb} = \sum_{x}\sum_{y} P(x, y) \log \frac{P(x, y)}{P(x)P(y)} \quad 7$$

so that I^{symb} is just the pairwise average of $I_{x,y}^{symb}$ over a set of series. In Fig. 2(b) we reported the value of the averaged mutual information $I^{SYMB}(\varepsilon)$ for a system of 5 diffusively coupled chaotic map lattices. In particular, for each value of ε, we generated 100 realizations of a multivariate series $\{x(t)\}_{t=1}^{t=N}$ of the dynamics described in Eq. (3), with $N=2^{14}$ points and $x \in [0, 1]^5$. The time series corresponding to each realization l was symbolized by partitioning the phase space of each variable into p=2 non-overlapping subintervals of the same size $[0,1/2) \cup [1/2,1]$. Accordingly, an original multivariate time series is coarse-grained into a multivariate symbolic sequence. Then, we averaged the mutual information $I_\ell^{\text{symb}}(\varepsilon)$ for each realization, as explained in the previous subsection and finally $I^{SYMB} = \langle I_\ell^{\text{SYMB}} \rangle_\ell$. A similar computation is made in Fig. 4(b)for financial time series, where we have analyzed the cases with p=2 and 6 symbols.

For the construction of Fig. 2(c) we used, for each value of ε, the same 100 realizations of Eq. (3). For each realization l, we constructed the corresponding multiplex visibility graph, by considering as layers the Horizontal Visibility Graphs of each of the five components. Then, we computed the pairwise mutual information between any pair of layers $I_{\alpha,\beta\ell}$ and the corresponding average $I_\ell = \langle I_{\alpha,\beta\ell} \rangle_{\alpha,\beta}$. The value of I reported in the figure is the average over the 100 realizations, i.e. $I = \langle I_\ell \rangle_\ell$.

A similar computation is made in Fig. 4(a) in the case of financial time series.

REFERENCES

1. Kantz, H. & Schreiber, T. Nonlinear Time Series Analysis (Cambridge University Press, Cambridge, 2006).
2. Hastie, T., Tibshirani, R. & Friedman, J. Elements of Statistical Learning (Springer-Verlag, 2009).
3. Albert, R. & Barabasi, A.-L. Statistical mechanics of complex networks. Rev. Mod. Phys. 74, 47 (2002).
4. Boccaletti, S., Latora, V., Moreno, Y., Chavez, M. & Hwang, D. U. Complex networks: structure and dynamics. Phys. Rep. 424, 175 (2006).
5. Newman, M. E. J. Networks: An Introduction. (Oxford University Press, Oxford, 2010).
6. Bullmore, E. & Sporns, O. Complex brain networks: graph theoretical analysis of structural and functional systems. Nat. Rev. Neurosci 10(3), 186–198 (2009).
7. Tumminello, M., Lillo, F. & Mantegna, R. N. Correlation, hierarchies, and networks in financial markets. Journal of Economic Behavior & Organization. 75(1), 40–58 (2010).
8. Cai, S.-M., Zhou, Y.-B., Zhou, T. & Zhou, P.-L. Hierarchical organization and disassortative mixing of correlation-based weighted financial networks. Int. J. Mod. Phys. C 21(3), 433–441 (2010).
9. Gao, Y.-C., Zeng, Y. & Cai, S.-M. Influence network in the Chinese stock market. J. Stat. Mech. P03017 (2015).
10. Zhang, J. & Small, M. Complex network from pseudoperiodic time series: topology versus dynamics. Phys. Rev. Lett. 96, 238701 (2006).
11. Kyriakopoulos, F. & Thurner, S. Directed network representations of discrete dynamical maps. In Lecture Notes in Computer Science 4488, 625–632 (2007).
12. Xu, X., Zhang, J. & Small, M. Superfamily phenomena and motifs of networks induced from time series. Proc. Natl. Acad. Sci. USA 105, 19601–19605 (2008).
13. Donner, R. V., Zou, Y., Donges, J. F., Marwan, N. & Kurths, J. Recurrence networks: a novel paradigm for nonlinear time series analysis. New J. Phys. 12, 033025 (2010).
14. Donner, R. V. et al. The Geometry of Chaotic Dynamics - A Complex Network Perspective. Eur. Phys. J. B 84, 653–672 (2011).

15. Lacasa, L., Luque, B., Ballesteros, F., Luque, J. & Nuno, J. C. From time series to complex networks: The visibility graph. Proc. Natl. Acad. Sci. USA 105, 13 (2008).
16. Luque, B., Lacasa, L., Luque, J. & Ballesteros, F. J. Horizontal visibility graphs: Exact results for random time series. Phys. Rev. E 80, 046103 (2009).
17. Lacasa, L. On the degree distribution of horizontal visibility graphs associated to Markov processes and dynamical systems: diagrammatic and variational approaches. Nonlinearity 27, 2063–2093 (2014).
18. Gutin, G., Mansour, M. & Severini, S. A characterization of horizontal visibility graphs and combinatorics on words. Physica A 390, 12 (2001).
19. Lacasa, L., Luque, B., Luque, J. & Nuño, J. C. The Visibility Graph: a new method for estimating the Hurst exponent of fractional Brownian motion. Europhys. Lett. 86, 30001 (2009).
20. Lacasa, L. & Toral, R. Description of stochastic and chaotic series using visibility graphs. Phys. Rev. E 82, 036120 (2010).
21. Luque, B., Lacasa, L., Ballesteros, F. J. & Robledo, A. Analytical properties of horizontal visibility graphs in the Feigenbaum scenario. Chaos 22, 013109 (2012).
22. Luque, B., Ballesteros, F. J., Nunez, A. M. & Robledo, A. Quasiperiodic graphs: structural design, scaling and entropic properties. J. Nonlin. Sci. 23 335–342 (2013).
23. Nunez, A., Luque, B., Lacasa, L., Gomez, J. P. & Robledo, A. Horizontal Visibility graphs generated by type-I intermittency. Phys. Rev. E 87, 052801 (2013).
24. Ahmadlou, M. & Adeli, H. Visibility graph similarity: A new measure of generalized synchronization in coupled dynamic systems. Physica D 241(4), 326–332 (2012).
25. Aguilar-San Juan, B. & Guzman-Vargas, L. Earthquake magnitude time series: scaling behavior of visibility networks. Eur. Phys. J. B. 86, 454 (2013).
26. Donner, R. V. & Donges, J. F. Visibility graph analysis of geophysical time series: Potentials and possible pitfalls. Acta Geophysica 60, 589–623 (2012).
27. Zou, Y., Small, M., Liu, Z. & Kurths, J. Complex network approach to characterize the statistical features of the sunspot series. New J. Phys. 16, 013051 (2014).

28. Zou, Y., Donner, R. V., Marwan, N., Small, M. & Kurths, J. Long-term changes in the north-south asymmetry of solar activity: a nonlinear dynamics characterization using visibility graphs. Nonlin. Processes Geophys. 21, 1113–1126 (2014).
29. Qian, M. C., Jiang, Z. Q. & Zhou, W. X. Universal and nonuniversal allometric scaling behaviors in the visibility graphs of world stock market indices. J. Phys. A 43, 335002 (2010).
30. Ahmadlou, M., Ahmadi, K., Rezazade, M. & Azad-Marzabadi, E. Global organization of functional brain connectivity in methamphetamine abusers. Clinical Neurophysiology 124, 6, 1122 (2013).
31. Nuñez, A., Lacasa, L. & Luque, B. Visibility algorithms: a short review in Graph Theory (Intech) (2012).
32. Bianconi, G. Statistical mechanics of multiplex networks: Entropy and overlap. Phys. Rev. E. 87, 062806 (2013).
33. Nicosia, V., Bianconi, G., Latora, V. & Barthelemy, M. Growing multiplex networks. Phys. Rev. Lett. 111, 058701 (2013).
34. De Domenico, M. et al. Mathematical formulation of multilayer networks. Phys. Rev. X 3, 041022 (2013).
35. Kivelä, M. et al. Multilayer networks. J. Complex Networks, 2(3), 203–271 (2014).
36. Boccaletti, S. et al. The structure and dynamics of multilayer networks. Phys. Rep. 544(1), 1–122 (2014).
37. Battiston, F., Nicosia, V. & Latora, V. Structural measures for multiplex networks. Phys. Rev. E 89, 032804 (2014).
38. Nicosia, V., Bianconi, G., Latora, V. & Barthelemy, M. Non-linear growth and condensation in multiplex networks. Phys. Rev. E 90, 042807 (2014).
39. Nicosia, V. & Latora, V. Measuring and modelling correlations in multiplex networks. arxiv:1403.1546 (2014).
40. Lacasa, L., Nuñez, A., Roldan, E., Parrondo, J. M. R. & Luque, B. Time series irreversibility: a visibility graph approach. Eur. Phys. J. B 85, 217 (2012).
41. Donges, J. F., Donner, R. V. & Kurths, J. Testing time series irreversibility using complex network methods. Europhys. Lett. 102, 10004 (2013).
42. L. Lacasa & R. Flanagan Time reversibility from visibility graphs of

non-stationary time series. Phys. Rev. E 92, 022817 (2015).
43. Granger, C. W. J. Investigating causal relations by econometric models and cross-spectral methods. Econometrica 37(3), 424–438 (1969).
44. Sporns, O. Structure and function of complex brain networks. Dialogues Clin. Neurosci. 15(3), 247–262 (2013).
45. Kaneko, K. Pattern dynamics in spatiotemporal chaos: Pattern selection, diffusion of defect and pattern competition intermettency. Physica D 34, 1–41 (1989).
46. Kaneko, K. Theory and applications of coupled map lattices (Vol. 12). John Wiley & Son Ltd. (1993).
47. Hilgers, A. & Beck, C., Hierarchical coupled map lattices as cascade models for hydrodynamic turbulence. Europhys. Lett. 45, 552 (1999).
48. Hilgers, A. & Beck, C. Turbulent behavior of stock market indices and foreign currency exchange rates. Int. J. Bif. Chaos 7, 1855 (1997).
49. Bevers, M. & Flather, C. H. Numerically exploiting habitat fragmentation effects on populations using cell-based coupled map lattices, Theor. Pop. Biol. 55, 61 (1999).
50. Beck, C. Chaotic scalar fields as models for dark energy. Phys. Rev. D 69(12), 123515 (2004).
51. Mantegna, R. Hierarchical structure in financial markets. Europhys J B. 11, 193–197 (1999).

11

Networks: On the Relation of bi- and Multivariate Measures

Wolfgang Mader[1,2], Malenka Mader[1,2,3], Jens Timmer[1,2,4], Marco Thiel[5] & Björn Schelter[1,5]

[1]Institute of Physics, University of Freiburg, Germany
[2]Freiburg Center of Data Analysis and Modeling, University of Freiburg, Germany
[3]University Medical Center Freiburg, Germany
[4]BIOSS, Center for Biological Signalling Studies, University of Freiburg, Germany
[5]University of Aberdeen, UK

ABSTRACT

A reliable inference of networks from observations of the nodes' dynamics is a major challenge in physics. Interdependence measures such as a the correlation coefficient or more advanced methods based on, e.g., analytic phases of signals are employed. For several of these interdependence measures, multivariate counterparts exist that promise to enable distinguishing direct and indirect connections. Here, we demonstrate analytically how bivariate measures relate to the respective multivariate ones; this knowledge will in turn be used to demonstrate the implications of thresholded bivariate measures for network inference. Particularly, we

Citation: MMader, W., Mader, M., Timmer, J., Thiel, M., &Schelter, B. (2015). Networks: On the relation of bi-and multivariate measures. Scientific reports, 5, 10805. (7 pages)

Copyright: © This work is licensed under a Creative Commons Attribution 4.0 International (CC BY 4.0) License.

show, that random networks are falsely identified as small-world networks if observations thereof are treated by bivariate methods. We will employ the correlation coefficient as an example for such an interdependence measure. The results can be readily transferred to all interdependence measures partializing for information of thirds in their multivariate counterparts.

INTRODUCTION

Complex systems are ubiquitous. Typically they are represented by networks and also visualized as such. Modern societies rely on the understanding of the dynamics of these networks, for instance, to be able to interfere with them. Examples of complex systems modeled as networks include power grids, traffic systems, climate, or neuronal oscillators.

A network G(V,L) is defined as a set of nodes $V=\{n_1,n_2,\ldots,n_N\}$ with $|V|=N$, and, in general, an ordered set of links L, $L\subset\{(n_i,n_j)\in V\times V\}$, with $|L|=D$, where $|\cdot|$ denotes the number of elements in a set. The number of links d_i attached to n_i is called the degree of the node. The links of an undirected network can be represented in a symmetric adjacency matrix A with entries $a_{ij}=1$ if $l_{ij}\in L$ and 0 otherwise. By collecting the nodes in rows and the links in columns, the $N\times D$incidence matrix is constructed. It is $b_{ij}=1$ if the j-th link is attached to i-th node and zero otherwise.

Here, we are concerned with the typical scenario, that the network itself needs to be inferred from measurements; no prior knowledge of the network topology is available. The nodes are identified with the processes of interest; often but not necessarily the locations of their measurements. The links between them are determined using interdependence measures; the correlation coefficient is among the most frequently applied ones1[,2].

Classifying interdependence measures into two classes, (i) bivariate and (ii) (truly) multivariate approaches, provides the basis for our investigations. Here, we use the notion "multivariate" for the simultaneous analysis of interdependences between three or more processes. We demonstrate that bivariate network reconstruction techniques, if applied naively, fall short in reconstructing the network's topology as well as its classification, for example as a small world network. We note that, e.g., in directed acyclic graphs an apt combination of bivariate measures can also reveal the correct network topology.

Prior related work can be found in e.g.1[,3,4,5] The findings presented in these publications are based on data obtained by simulating different models.

In this manuscript, we suggest a mathematical framework that independent of any specific nodal dynamics enables a deeper understanding of network inference. Our analytical results are illustrated and supported by simulations.

THEORY ON DATA-BASED NETWORK INFERENCE

In this section, we outline the key differences of bi- and multivariate interdependence measures and the implications for network reconstruction. We also show how the topology resulting from a bivariate measure can be derived from the topology of the respective multivariate one, using the (partial) correction matrix as an example.

Bivariate and Multivariate Measures

The key challenge in network inference from measurements is to determine the direct links, i.e. those links that represent a true physical interaction between two nodes. A naive application of a bivariate approach would analyze all pairwise combinations and investigate whether or not there is an interaction. This is suboptimal as indirect links are falsely classified as links in a network. Multivariate interdependence measures, in contrast, exploit all (linear) information of the N observations simultaneously. For various bivariate techniques, multivariate counterparts have been suggested, enabling a direct comparison between the two. The results presented here hold for all multivariate measures whose estimation can be reduced to inverting the matrix of the corresponding bivariate measure6 such as the multivariate versions of correlation, cross-spectral analysis7, mean phase coherence8 and recurrence analysis9.

Taking the network in Fig. 1 (a) as an example, a multivariate measure detects only direct links (black lines). Applying the corresponding bivariate measure, the network in Fig. 1 (b) would result, where the dashed lines indicate indirect links, which are suggested from bivariate analysis.

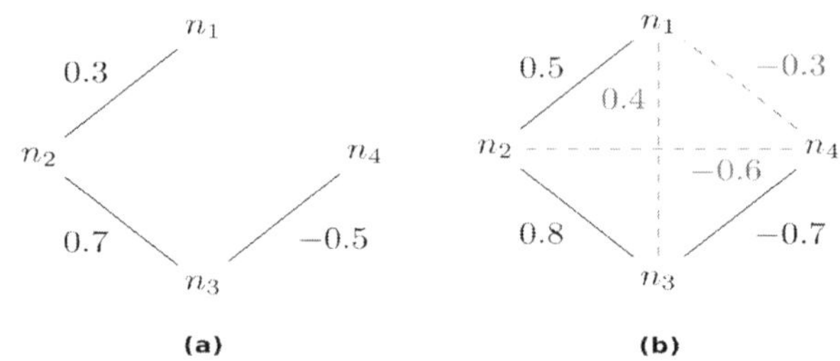

Figure 1. (**a**) Network with only direct links and arbitrarily chosen weights compatible with partial correlation coefficients.(**b**) Fully connected network including indirect links and weights of the corresponding rounded bivariate correlation coefficients.

Weights in Fig. 1 (a) are chosen arbitrarily but compatible with a multivariate correlation analysis. Therefore, the weights can be understood as partial correlation coefficients; those given in Fig. 1 (b) are the corresponding bivariate correlation coefficients. In the following we characterize the relation between the two.

Bivariate and Partial Correlation

Let Z be an (s+t)-, X an s- and Y a t-vector valued stationary process with covariance matrix

$$\Sigma_{ZZ} = \begin{pmatrix} \Sigma_{XX} & \Sigma_{XY} \\ \Sigma_{YX} & \Sigma_{YY} \end{pmatrix} \quad 1$$

such that X and Y are a partition of Z.

The partial covariance matrix6[,10]

$$\varepsilon_{YY} = \Sigma_{YY} - \Sigma_{YX}\Sigma_{XX}{}^{-1}\Sigma_{XY} \quad 2$$

measures the covariance within Y after removing the linear effects of X. By normalizing ε_{YY}, the partial correlation matrix of Y6.

$$\pi_{YY} = h \cdot \varepsilon_{YY} \cdot h \quad 3$$

is obtained. The matrix h consists purely of the diagonal entries $\varepsilon_{YY_{ii}}{}^{-1/2}$, i=1,..., t of the t-dimensional process Y. Partial correlation is a linear multivariate symmetric measure, $\pi_{YY_{ij}} = \pi_{YY_{ji}}$, , quantifying the strength of the direct connection between Y_i and Y_j conditioned on the linear information of X. Often Y is chosen to be two dimensional, investigating whether or not there is a link between two processes.

Correlation is normalized covariance, the following line of argument applies to both quantities. It is convenient to start with the covariance matrix ϱ. Given ϱ for the full s+t-dimensional system, the partial correlation matrix can be obtained by the following steps: (i) matrix inversion6

$$g = \varrho^{-1} \quad 4$$

and (ii) a subsequent normalization , $\tilde{\pi} = \tilde{h} \cdot g \cdot \tilde{h}$, with $\tilde{h}$ a diagonal matrix and $\tilde{h}_{ii} = g_{ii}^{-1/2}$, i=1,…, s+t. To arrive at the partial correlation matrix π, (iii) the off-diagonal elements of π have to be multiplied by −16[,11],

$$\pi_{ij} = -\tilde{\pi}_{ij}, \text{ for } i \neq j . \quad 5$$

The algorithm given in Eqs. (4)–(5), relates correlation and partial correlation coefficients. It is interesting to note that given this procedure all partial correlation coefficients are obtained in one step; further calculations that would result when applying Eq. (2) are not needed.

THEORETICAL RESULTS

Reversing the argument given above, we can derive the correlation matrix from the partial correlation matrix. This is exploited in the following to investigate how indirect links emerge when applying bivariate techniques naively.

Any partial correlation matrix can be separated into

$$\pi = 1 + R , \quad 6$$

where the off-diagonal entries of R are populated with the partial correlation coefficients while its diagonal is zero. Multiplying the off-diagonals by − 1 [cf. Eq.(5)] leads to

$$\breve{\pi} = 1 - R . \quad 7$$

Expanding the matrix inversion

$$\breve{\pi}^{-1} = \breve{\varrho} = f(R) = (1 - R)^{-1} \quad 8$$

in a Taylor series about R=0 results in

$$\breve{\varrho} = f(R) = 1 + \sum_{m=1}^{\infty} R^m . \qquad 9$$

The matrix $\breve{\varrho}$ is up to normalization the bivariate correlation matrix. In the first order approximation

$$\varrho^{1\text{st}} = 1 + R , \qquad 10$$

the correlation matrix equals the partial correlation matrix [Eq. (6)]. This shows that all indirect links are encoded in second and higher order terms of the matrix expansion [Eq. (9)]. This fact has key consequences for the nature of indirect links:

- Typically, powers of R lead to non-zero values for all entries of ϱ leading to a fully connected network.
- The bivariate correlation coefficients typically differ from the multivariate ones also for the direct interactions.
- If R is a nilpotent matrix of degree z, the sum in Eq. (9) has only $m=z-1$ terms. Network topologies with a nilpotent matrix R of degree 2 yield the same result for the correlation and partial correlation matrix.
- For a directed chain $n_1 \rightarrow n_2 \rightarrow \ldots \rightarrow n_N$ the link between bivariate and multivariate measure can be understood intuitively. For N=4, for instance,

$$R = \begin{pmatrix} 0 & 0 & 0 & 0 \\ c & 0 & 0 & 0 \\ 0 & c & 0 & 0 \\ 0 & 0 & c & 0 \end{pmatrix}, R^2 = \begin{pmatrix} 0 & 0 & 0 & 0 \\ 0 & 0 & 0 & 0 \\ c^2 & 0 & 0 & 0 \\ 0 & c^2 & 0 & 0 \end{pmatrix}, R^3 = \begin{pmatrix} 0 & 0 & 0 & 0 \\ 0 & 0 & 0 & 0 \\ 0 & 0 & 0 & 0 \\ c^3 & 0 & 0 & 0 \end{pmatrix} \qquad 11$$

- it becomes obvious, that the indirect links resulting from R^2 are passing one node and $R^3$are passing two nodes. Note, that the strengths of the indirect links ($|c|<1$) scale as c^i where i is the order of the link; intuitively it is the shortest path using direct links that manifests this indirect link.
- For an undirected chain, R is no longer nilpotent. Still, the lengths of paths inducing indirect links scale with the powers of R. Of special interest is the situation of a three node undirected chain

$$R = \begin{pmatrix} 0 & c & 0 \\ c & 0 & d \\ 0 & d & 0 \end{pmatrix}, \; R^2 = \begin{pmatrix} c^2 & 0 & cd \\ 0 & c^2 + d^2 & 0 \\ cd & 0 & d^2 \end{pmatrix}, \qquad 12$$

since such open triangles are closed when treated by a bivariate measure. In a local second order approximation, the weight of the indirect link emerging from an open triangle is proportional to the product of the two direct links involved.

In case c and d are strong direct connections, they have large partial correlation coefficients compared to the average partial correlation coefficient appearing in the network. The corresponding indirect connection is then likely to have a large correlation coefficient, compared to the average correlation coefficient. If a threshold is used to select relevant links, a strong but indirect link might be selected, while a weak direct link which would represent a true physical connection in the network is missed. The reason for the so called transitivity property of bivariate measures1[,4] is therefore founded in the fact, that indirect links constituted by only two direct ones are introduced already in the second order Taylor approximation; making their magnitude comparable to correlation coefficients of direct links.

In the following paragraphs, the effects of these properties are illustrated in a simulation study. The focus is on the effect of naive bivariate analysis onto the network topology and network classification. We emphasis again that our simulations are free of any assumptions about nodal dynamics, which demonstrates the universality of our framework.

SIMULATION SETUP AND RESULTS

Using the framework introduced in the last section, a simulation study is conducted. We compare the outcome of correlation coefficient and partial correlation coefficient when used to infer links in different network topologies. To this end, a regular network and two versions of random networks are investigated. In the first version, all partial correlation coefficient have the same value, in the second version, those values vary. The section is completed by a result showing that random networks are falsely classified as small world networks in the frame of bivariate interdependence measures.

Estimating Links in Regular Networks

The first system investigated is a regular network with N=80 nodes. In this network, each node is connected to its nearest neighbors only, resulting in a ring with D=80 direct links. To demonstrate that indirect links of increasing length correspond to increasing orders of the Taylor series [Eq. (9)], all nodes must be identical. Hence, all links have to share the same strength. A correlation matrix must be positive semidefinite. Derived from Sylvester's criterion, such a matrix can be obtained by ensuring, that the matrix is Hermitian, has real and non-negative diagonal elements and is diagonal dominant. The first two criteria are naturally meet by correlation matrices, while the last one must be cared for. Since in the regular network considered here, each node has degree 2, the value of the partial correlation coefficient was chosen to 0.49, which is near the largest possible value ($2 \cdot 0.49 < 1$).

The corresponding bivariate correlation coefficients, indicating direct and indirect links, are obtained from the partial correlation matrix using the framework introduced above[Eqs. (6) –(9)]. The expected clustering of the correlation coefficients with respect to the length of their constituting path is visible from their histogram [Fig. 2 , left]. The 80 "truly" direct links are shown in dark gray; these are the strongest links as expected from our framework. In light gray, the correlation coefficients of indirect links are presented. The right most (light gray) cluster refers to first order indirect links n_i—n_{i+2} and so on. In this scenario, thresholding the correlation coefficients would provide the correct network topology if the threshold were chosen "correctly" between 0.67 and 0.8.

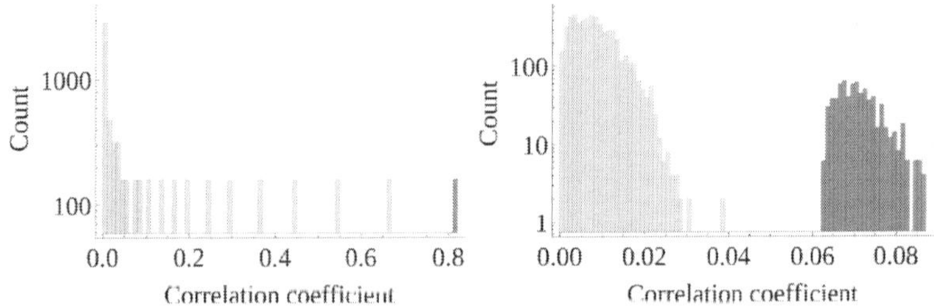

Figure 2. To all links in a regular (left) and a random (right) network of 80 nodes each, the same weight representing a partial correlation coefficient was attached.

The corresponding correlation coefficients are calculated and summarized in these histogram. (left) Due to the regularity of the network, the correlation coefficients cluster, where the 160 largest represent the direct (dark gray) the other the indirect links (light gray). (right) Because the network topology is random, nodes differ in the number of links attached to them. This influences the values of the correlation coefficients and destroys the strict clustering with respect to path length. Direct links (dark gray) exhibit larger correlation coefficients than indirect ones (light gray). In both networks a threshold allows to separate direct and indirect links.

Estimating Links in Random Networks

The second system investigated assumes an Erdös-Rényi network with N=80 nodes and D= =351 links. Using about Nlog(N) links guarantees no unconnected node in the network12[,13]. As in the last simulation, all links share the same partial correlation value in the partial correlation matrix describing the network. For the matrix to be diagonal dominant and hence positive definite, this value must be chosen smaller than $1/d_{Max}$, where d_{Max} is the maximum node degree in the network. As predicted by our framework, the orders of the Taylor expansion are no longer preserved in the histogram of correlation coefficients [Fig. 2 , right]; but still all correlation coefficients representing indirect links (light gray) are smaller than the ones representing direct ones (dark gray). Again a threshold can be found which separates direct and indirect links; this allows for the correct inference of the true network topology.

Connections of Varying Strength

When dealing with random networks and varying strengths of links, a more complicated scenario emerges. To obtain the partial correlation matrix, we start with the incidence matrix of an Erdös-Rényi network of N=80 nodes and D=351 links. The strengths of links are drawn from the uniform distribution over the interval [0,1]. The partial correlation matrix is obtained as the normalized matrix square of this weighted incidence matrix. According to Eq. (12), a weak direct link can have a smaller correlation coefficient than an indirect link resulting from two strong direct links. The situation is visualized in Fig. 3 on the left. Again, correlation coefficients of direct links are shown in dark gray, correlation coefficients of indirect ones in light gray. Since their distributions overlap, they are not separable by a threshold.

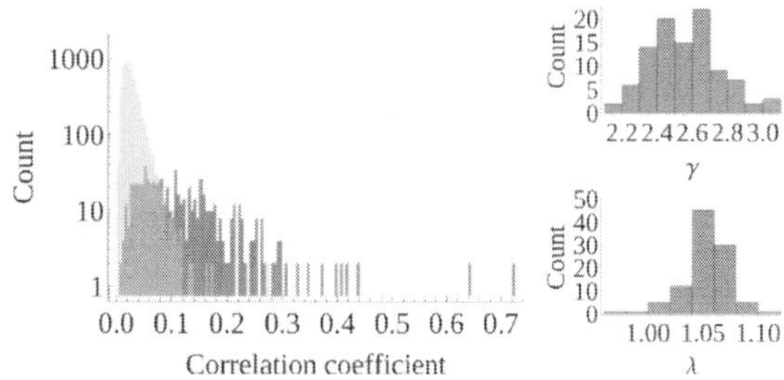

Figure 3. (left) In an 80 nodes random network, different weights representing partial correlation coefficients are attached to its links.

The histogram summarizes the corresponding correlation coefficients. In dark gray, the correlation coefficients of direct links, in light gray the ones of indirect links are presented. Because their values overlap, applying a threshold to correlation coefficients will report a mixture of direct and indirect links and hence a wrong network topology; wrong conclusion about e.g. the small world property of a network are drawn. (right) This is demonstrated by the histograms of the small world indicators γ and λ which are calculated from 100 realizations of a random network in which the largest 351 correlation coefficients were used to establish its links. Since $\gamma > 1$ and $\lambda \sim 1$ in all cases, every random network observed through correlation coefficients was falsely classified as small world.

When estimating links from measurements, the underlying network is unknown. Typically, the network exhibits non-regular topology and varying link strengths. As demonstrated, thresholding bivariate interdependence measures fails in these typical cases to reveal the correct network topology.

To demonstrate that this has broad implications for network inference, the consequences of using bivariate interdependence measures for inferring the small world property is investigated as an example.

Small-world network

The local clustering coefficient

$$c_i = \begin{cases} \frac{1}{d_i(d_i-1)} \sum_{kl} a_{ik} a_{kl} a_{li} & : d_i \geq 2 \\ 0 & : otherwise \end{cases} \quad 13$$

and the average shortest path length

$$L = \frac{2}{N(N+1)} \sum_{\substack{i=1 \\ j \leq i}}^{N} l_{ij}, \tag{14}$$

characterize the network topology on the local and global scale. In a network exhibiting the small world property, the mean clustering coefficient $C = \frac{1}{N}\sum c_i$ is high while the average shortest path length L scales with the number of nodes N as L(N)≤log N12[,14,15]. According to Eq. (14), the average shortest path lengths would be infinity in networks possessing at least one unconnected node. Similar to15 the path lengths of pairs of nodes involving an isolated node are excluded when evaluating Eq. (14).

Since in applications the scaling of L(N) is often not accessible, two proxy indicators $\gamma = C/\overline{C}_r$ and $\lambda = L/\overline{L}_r$ are employed. The mean values $\overline{C}_r$ and $\overline{L}_r$ are obtained from an ensemble of random networks. A network is said to be small world if λ≈1 and γ>1 holds1[,15].

From the partial correlation matrix of a random network (N=80), generated in the same way as described in the last section, the corresponding correlation matrix is calculated. A threshold is applied to the correlation coefficients to obtain the D=$\lceil N \log(N) \rceil$=351 strongest links. This network is tested for small worldness. In order to get a statistically meaningful result, a Monte Carlo simulation of 100 different realizations of the partial correlation matrix is carried out.

The histograms of the resulting λ and γ are presented in Fig. 3 on the right. For all realizations, γ>1 [Fig. 3, top] while λ≈1 [Fig. 3, bottom]. Therefore, all random networks are falsely classified to be small world. The classification is based on the bivariate networks alone. The explanation of the result is the transitive property for which the mathematical foundation was presented in our framework in Eqs. (9) and (12) . Hence, this result applies for all networks in which links are estimated by a naive application of a bivariate interdependence measure irrespective of the dynamics observed, the measurement technique used, or the specific bivariate measure employed.

CONCLUSION

In this manuscript, we inferred the nature of indirect links introduced by bivariate interdependence measures. We developed a mathematical framework that we motivated for correlation analysis. To this end, the inverse partial correlation matrix was expanded into a Taylor series. We demonstrated that, in general, bivariate measures are not able to reveal the true network topology from measurements; indirect links which have no underlying physical connection in the observed system are incorrectly inferred, altering the network topology.

The clustering coefficient is strongly affected by investigating networks based on the naive application of bivariate approaches. It influences high-level characterizations; a random network is e.g. falsely classified as small world in this context. The reason for this is the tendency of bivariate measures to close open triangles, i.e., the transitivity property.

Multivariate measures promise to overcome this limitation. Only links which are direct given the information of all measurements give rise to values of the interdependence measure which are not compatible with zero. Accordingly, a statistically meaningful critical value for a given level of significance can be obtained, rendering the choice of an arbitrary threshold unnecessary. Surrogate or bootstrapping approaches present alternatives to analytically derived critical values. Therefore, links reported by a multivariate measure assuming observation of all relevant processes are supposed to represent true connections, constituting the correct network topology. Since such reconstructed networks do not suffer from an increased clustering coefficient network classification can be done successfully.

REFERENCES

1. Hlinka, J., Hartman, D. & Palus, M. Small-world topology of functional connectivity in randomly connected dynamical systems. Chaos 22, 033107 (2012).
2. Tewarie, P. et al. Functional brain networks: Linking thalamic atrophy to clinical disability in multiple sclerosis, a multimodal fMRI and MEG Study. Hum. Brain Mapp. 36, 603–618 (2015).
3. Bialonski, S., Wendler, M. & Lehnertz, K. Unraveling Spurious Properties of Interaction Networks with Tailored Random Networks. PLoS One 6, e22826 (2011).
4. Zalesky, A., Fornito, A. & Bullmore, E. On the use of correlation as a measure of network connectivity. NeuroImage 60, 2096–2106 (2012).
5. Zerenner, T., Friederichs, P., Lehnertz, K. & Hense, A. A Gaussian graphical model approach to climate networks. Chaos 24, 023103 (2014).
6. Lauritzen, S. Graphical Models (Oxford University Press, 1996).
7. Dahlhaus, R. Graphical interaction models for multivariate time series. Metrika 51, 157–172 (2000).
8. Schelter, B., Winterhalder, M., Dahlhaus, R., Kurths, J. & Timmer, J. Partial Phase Synchronization for Multivariate Synchronizing Systems. Phys. Rev. Lett. 96, 208103 (2006).
9. J. Nawrath et al. Distinguishing Direct from Indirect Interactions in Oscillatory Networks with Multiple Time Scales. Phys. Rev. Lett. 104, 038701 (2010).
10. Brillinger, D. Time Series: Data Analysis and Theory. Classics in Applied Mathematics (Society for Industrial and Applied Mathematics, 2001).
11. Whittaker, J. Graphical Models in Applied Multivariate Statistics (Wiley, 1995).
12. Boccaletti, S., Latora, V., Moreno, Y., Chavez, M. & Hwang, D. Complex networks: Structure and dynamics. Phys. Rep. 424, 175–308 (2006).
13. Erdös, P. & Rényi, A. On random graphs. Publ. Math. Debrecen 6, 290–297 (1959).
14. Watts, D. & Strogatz, S. Collective dynamics of 'small-world' networks. Nature 393, 440–2 (1998).

15. Bialonski, S., Horstmann, M. & Lehnertz, K. From brain to earth and climate systems: Small-world interaction networks or not? Chaos 20, 013134 (2010).

12

Cubic Trigonometric Nonuniform Spline Curves and Surfaces

Lanlan Yan[1,2]

[1]School of Mathematics and Statistics, Central South University, Changsha 410083, China
[2]School of Science, East China University of Technology, Nanchang 330013, China

ABSTRACT

A class of cubic trigonometric nonuniform spline basis functions with a local shape parameter is constructed. Their totally positive property is proved. The associated spline curves inherit most properties of usual polynomial B-spline curves and enjoy some other advantageous properties for engineering design. They have C^2 continuity at single knots. For equidistant knots, they have C^3 continuity and C^5 continuity for particular choice of shape parameter. They can express freeform curves as well as ellipses. The associated spline surfaces can exactly represent the surfaces of revolution. Thus the curve and surface representation scheme unifies the representation of freeform shape and some analytical shapes, which is popular in engineering.

Citation: Yan, L. (2016). Cubic trigonometric nonuniform spline curves and surfaces. Mathematical Problems in Engineering, 2016. (10 pages), DOI/URL:https://doi.org/10.1155/2016/7067408

Copyright: © 2016 Lanlan Yan. This is an open access article distributed under the Creative Commons Attribution 4.0 International (CC BY 4.0) License.

INTRODUCTION

As a unified mathematical model with many desirable properties, B-splines are widely applied to the modeling of freeform curves and surfaces. However, there are several limitations of the B-spline model, which restrict its applications. For example, once the knot vectors are specified, the positions of B-spline curves are relatively fixed to their control points. On the other hand, B-spline curves fail to represent conic curves except the parabolas, as well as some transcendental curves such as the helix and the catenary, which are often used in engineering.

Although the nonuniform rational B-spline (NURBS) can overcome the first shortcoming of B-splines to a certain extent, it fails to model transcendental curves. The NURBS model has several other potential limitations due to the relative complexity of rational basis functions. For example, rational form may be unstable, and derivatives and integrals are hard to compute. Consequently, in order to overcome the drawbacks of B-splines, it is necessary to explore new models.

To enhance the flexibility of B-spline models, some researchers have suggested many types of curves with shape parameters incorporated into the basis functions. For instance, Xu and Wang [1] proposed three kinds of extensions of cubic uniform B-spline. The advantage of the extensions is that they have shape parameters, which can be used to adjust the shape of the curves without shifting the control points. Costantini et al. [2] presented a method for the construction of cubic like B-splines with multiple knots. The proposed B-splines are equipped with tension parameters, associated to the knots, which permit a modification of their shape. Han [3] constructed a kind of piecewise quartic polynomial curves with a local shape parameter. Han [4] defined a class of piecewise quartic spline curves with three local shape parameters. Hu et al. [5] constructed a kind of B-spline curves with two local shape parameters. To expand the scope of shape representation of B-spline models, some researchers have suggested many types of curves defined on nonpolynomial space. For instance, Han presented the quadratic [6] and cubic [7] trigonometric polynomial curves with a shape parameter. Han and Zhu [8] defined trigonometric B-spline curves with three local shape parameters and a global shape parameter. Dube and Sharma [9] constructed the cubic trigonometric polynomial B-spline curves with a shape parameter. Wang et al. [10] presented nonuniform algebraic-trigonometric B-splines. The curves in [6–10] not only enjoy adjustable shape, but also can exactly represent ellipses. Yan and Liang [11] constructed a class of algebraic-trigonometric blending spline curves with two shape parameters. Except for shape adjustability, the curves admit exact representations for several remarkable curves. Lü et al. [12] defined uniform algebraic-hyperbolic blending B-spline curves, which possess adjustable shape and can represent exactly the hyperbola and the catenary.

Totally positive property is one of the important properties of basis functions. Curves defined by totally positive basis must have variation diminishing and convexity preserving property. Goodman and Said [13] proved that the generalized Ball basis given in [14] is normalized totally positive and hence it possesses the same kind of shape preserving properties as the Bernstein basis [15]. Han and Zhu [16] proved that the cubic trigonometric Bézier basis given in [17] forms an optimal normalized totally positive basis. Zhu et al. [18] constructed four Bernstein-like basis functions, which form an optimal normalized totally positive basis. Based on the Bernstein-like basis, a class of totally positive B-spline-like basis functions is constructed. The associated B-spline curves have C^2 continuity at single knots and can be $C^2 \cap FC^{k+3}$ $(k \in Z^+)$ continuous for particular choice of shape parameters.

The curves given in [1–10, 12, 18] have adjustable shape. In addition, the curves given in [6–10, 12] can represent exactly some conic curves and transcendental curves. However, whether the blending functions in [1, 3–10, 12] have total positivity is unknown, so whether the associated curves have variation diminishing is unknown. The curves in [2, 18] have variation diminishing, but they cannot represent conic curve or transcendental curve. The purpose of this paper is to define a kind of B-spline-like curves and surfaces, which have adjustable shape and can represent some elementary analytic curves and surfaces, and the curves have variation diminishing thus having a good shape control.

The research topics of this paper and some existing documents, such as [19, 20], are similar. There are also some differences, though. By extending the global parameter to local parameter, Wu and Chen [19] presented cubic nonuniform trigonometric polynomial curves with multiple shape parameters. The curves are C^2 continuity for a nonuniform knot vector and C^2 continuity for a uniform knot vector, respectively. Han [20] presented quadratic trigonometric polynomial curves with local basis. The curves have C^2 continuity with a nonuniform knot vector and any value of the bias. Compared to [19, 20], novelty of this paper is listed as follows. First, it discusses the total positivity of the basis functions. This property makes the corresponding curves have variation diminishing property, which is one of the important properties of dominant Bézier curves and B-spline curves. Second, it provides the representation method of surface of revolution. In surface modeling, the construction of the rotational surface is a common problem. Third, it provides a class of higher order continuous curves, which can meet most of the needs in engineering.

The rest of the paper is organized as follows. Section 2 gives the definition and properties of the basis functions. Section 3 defines the associated curves and gives the representation of the ellipses and parabolas. Section 4 defines the associated surfaces and gives the representation of the rotating surfaces. Section 5 concludes the paper.

BASIS FUNCTIONS

The Construction of the Basis Functions

In [7], cubic trigonometric splines are presented for a nonuniform knot vector. These splines are used to define trigonometric spline curves. As special cases, the author also introduces a class of cubic trigonometric polynomial basis functions used to construct trigonometric Bézier curves. The original expression of the basis functions is as follows:

$$B_{i-3}(u) = (1 - \sin t)^2 (1 - \lambda \sin t),$$

$$B_{i-2}(u) = \sin t (1 - \sin t)^2 [2 + \lambda (1 - \sin t)],$$

$$B_{i-1}(u) = \cos t (1 - \cos t)^2 [2 + \lambda (1 - \cos t)],$$

$$B_i(u) = (1 - \cos t)^2 (1 - \lambda \cos t). \quad (1)$$

This set of basis functions contains only one shape parameter. In [17], it was further extended to possess two shape parameters.

By changing the λ in the last two functions in (1) to μ, a class of cubic trigonometric Bezier basis functions with two ′ shape parameters is defined in [17] as follows. Let $\lambda, \mu \in [-2, 1]$, for $t \in [0, \pi/2]$; the following four functions are defined to be the cubic trigonometric Bezier basis functions ′ (T-Bezier basis for short), with two shape parameters ′ λ and μ:

$$T_0(t;\lambda) = (1-\sin t)^2(1-\lambda\sin t),$$

$$T_1(t;\lambda) = \sin t(1-\sin t)(2+\lambda-\lambda\sin t),$$

$$T_2(t;\mu) = \cos t(1-\cos t)(2+\mu-\mu\cos t),$$

$$T_3(t;\mu) = (1-\cos t)^2(1-\mu\cos t). \tag{2}$$

In [16], the T-Bezier basis was proved to be the optimal normalized totally positive basis of the space $T_{\lambda,\mu}$:= Span{1,$\sin^2 t$, $(1-\sin t)^2(1-\lambda\sin t)$, $(1-\cos t)^2(1-\mu\cos t)$}for $\lambda, \mu \in (-2, 1]$. Hence the corresponding cubic trigonometric Bezier curves are suited for conformal curve design. However, ′ the Bezier curve is a single curve segment. When using B ′ ezier ′ curve to describe complex shapes, the problem of joining curve segments smoothly needs to be solved. The Bezier form ′ is the special case of B-spline. B-spline curves consist of many polynomial pieces, offering much more versatility than Bezier curves. Considering the ′ B-spline is more suitable for expressing complex curve and surface, this paper will discuss more generally the B-spline form.

Next we will construct a kind of cubic trigonometric spline basis function based on the T-Bézier basis.

Given knots $u_0<u_1<\cdots<u_{n+4}$, we refer to $U = (u_0, u_1, \ldots, _{n+4})$ as a knot vector. Let $h_j = u_{j+1} - u_j$, and $t(u) = (\pi/2)((u-u_j)/h_j)$, $j = 0, 1, \ldots, n+3$; we want to construct the associated spline basis functions as follows:

$$B_i(u) = \begin{cases} B_{i,0}(t_i), & u \in [u_i, u_{i+1}), \\ B_{i,1}(t_{i+1}), & u \in [u_{i+1}, u_{i+2}), \\ B_{i,2}(t_{i+2}), & u \in [u_{i+2}, u_{i+3}), \\ B_{i,3}(t_{i+3}), & u \in [u_{i+3}, u_{i+4}), \\ 0, & u \notin [u_i, u_{i+4}) \end{cases} \tag{3}$$

for$i = 0, 1, \ldots, n$. Here

$$B_{i,0}(t) = d_i T_3(t; \lambda_{i+1}),$$

$$B_{i,1}(t) = c_{i+1,0} T_0(t; \lambda_{i+1}) + c_{i+1,1} T_1(t; \lambda_{i+1})$$

$$+ c_{i+1,2} T_2(t; \lambda_{i+2}) + c_{i+1,3} T_3(t; \lambda_{i+2}),$$

$$B_{i,2}(t) = b_{i+2,0} T_0(t; \lambda_{i+2}) + b_{i+2,1} T_1(t; \lambda_{i+2})$$

$$+ b_{i+2,2} T_2(t; \lambda_{i+3}) + b_{i+2,3} T_3(t; \lambda_{i+3}),$$

$$B_{i,3}(t) = a_{i+3} T_0(t; \lambda_{i+3}), \tag{4}$$

in which $T_j(t; \lambda)$ for $j = 0, 1, 2, 3$ are the T-Bezier basis ′ given in (2) and a_{i+3}, b_{i+2},j, c_{i+1},j, d_i for $j = 0, 1, 2, 3$ are undetermined coefficients.

To determine the coefficient values, we impose two conditions on $B_i(u)$, which have C^2 continuity at each knot and form a partition of unity on the interval [$u3$, un+1]. Then, we can compute the coefficients as follows:

$$a_i = \frac{\beta_{i-1} h_i^2}{\alpha_i \gamma_i},$$

$$b_{i,0} = \frac{\beta_i (\gamma_{i+1} \varphi_{i-1} h_i + \gamma_i \phi_{i+1} h_{i-1})}{\alpha_i \gamma_i \gamma_{i+1}},$$

$$b_{i,1} = \frac{\beta_i \phi_{i+1}}{\gamma_{i+1}},$$

$$b_{i,2} = \frac{\beta_i h_{i+1}}{\gamma_{i+1}},$$

$$b_{i,3} = a_{i+1},$$

$$d_i = \frac{\beta_{i+2} h_i^2}{\alpha_{i+1} \gamma_{i+2}},$$

$$c_{i,3} = \frac{\beta_{i+1} (\gamma_{i+2} \varphi_i h_{i+1} + \gamma_{i+1} \phi_{i+2} h_i)}{\alpha_{i+1} \gamma_{i+1} \gamma_{i+2}},$$

$$c_{i,2} = \frac{\beta_{i+1} \varphi_i}{\gamma_{i+1}},$$

$$c_{i,1} = \frac{\beta_{i+1} h_{i-1}}{\gamma_{i+1}},$$

$$c_{i,0} = d_{i-1},$$

(5)

where $\alpha_i = h_{i-1} + h_i$, $\beta_i = 1 + 2\lambda_i$, $\varphi_i = h_{i-1} + \beta_i h_i$, and $\phi_i = \beta_i h_{i-1} + h_i$, $\gamma_i = \beta_i h_i - 2 + \beta_{i-1}\beta_i h_{i-1} + \beta_{i-1} h_i$.

For $u \in [u_i, u_{i+1})$, $i = 3, 4, \ldots, n$, we have $B_j(u) = 0$ for $j \neq i-3, i-2, i-1, i$, and

$$B_{i-3}(u) = B_{i-3,3}(t_i),$$

$$B_{i-2}(u) = B_{i-2,2}(t_i),$$

$$B_{i-1}(u) = B_{i-1,1}(t_i),$$

$$B_i(u) = B_{i,0}(t_i).$$

(6)

We can easily check that

$$(B_{i-3}(u), B_{i-2}(u), B_{i-1}(u), B_i(u))$$

$$= (T_0(t_i, \lambda_i), T_1(t_i, \lambda_i), T_2(t_i, \lambda_{i+1}), T_3(t_i, \lambda_{i+1})) H_i, \quad (7)$$

where

$$H_i = \begin{pmatrix} a_i & b_{i,0} & c_{i,0} & 0 \\ 0 & b_{i,1} & c_{i,1} & 0 \\ 0 & b_{i,2} & c_{i,2} & 0 \\ 0 & b_{i,3} & c_{i,3} & d_i \end{pmatrix}.$$

(8)

Theorem 1. When $\lambda_{i-1}, \lambda_i, \lambda_{i+1}, \lambda_{i+2} \in (-1/2, 1]$, H_i is a nonsingular stochastic and totally positive matrix.

Proof. When $\lambda_{i-1}, \lambda_i, \lambda_{i+1}, \lambda_{i+2} \in (-1/2, 1]$, it is obvious that a_i, $b_{i,}$, $c_{i,j}$, di are all positive for $j = 0, 1, 2, 3$. In addition, from (5) we can verify that $a_i + b_{i,0} + c_{i,0} = b_{i,1} + c_{i,1} = b_{i,2} + c_{i,2} = b_{i,3} + c_{i,3} + d_i = 1$. This means that H_i is stochastic. With direct computation we have

$$\begin{vmatrix} b_{i,0} & c_{i,0} \\ b_{i,1} & c_{i,1} \end{vmatrix} = \frac{\eta_i \varphi_{i-1} h_{i-1}}{\alpha_i \gamma_i \gamma_{i+1}} > 0,$$

$$\begin{vmatrix} b_{i,0} & c_{i,0} \\ b_{i,2} & c_{i,2} \end{vmatrix} = \frac{\eta_i \left(\gamma_{i+1} \varphi_{i-1} \varphi_i + \gamma_i \delta_i h_{i-1}\right)}{\alpha_i \gamma_i \gamma_{i+1}^2} > 0,$$

$$\begin{vmatrix} b_{i,0} & c_{i,0} \\ b_{i,3} & c_{i,3} \end{vmatrix} = \frac{\eta_i \left(\gamma_{i+1} \gamma_{i+2} \varphi_{i-1} \varphi_i h_{i+1} + \gamma_{i+1}^2 \varphi_{i-1} \phi_{i+2} h_i + \gamma_i \gamma_{i+1} \phi_{i+1} \phi_{i+2} h_{i-1} + \gamma_i \gamma_{i+2} \delta_i h_{i-1} h_{i+1}\right)}{\alpha_i \alpha_{i+1} \gamma_i \gamma_{i+1}^2 \gamma_{i+2}} > 0,$$

$$\begin{vmatrix} b_{i,1} & c_{i,1} \\ b_{i,2} & c_{i,2} \end{vmatrix} = \frac{\eta_i \delta_i}{\gamma_{i+1}^2} > 0,$$

$$\begin{vmatrix} b_{i,1} & c_{i,1} \\ b_{i,3} & c_{i,3} \end{vmatrix} = \frac{\eta_i \left(\gamma_{i+1} \phi_{i+1} \phi_{i+2} + \gamma_{i+2} \delta_i h_{i+1}\right)}{\alpha_{i+1} \gamma_{i+1}^2 \gamma_{i+2}} > 0,$$

$$\begin{vmatrix} b_{i,2} & c_{i,2} \\ b_{i,3} & c_{i,3} \end{vmatrix} = \frac{\eta_i \phi_{i+2} h_{i+1}}{\alpha_{i+1} \gamma_{i+1} \gamma_{i+2}} > 0, \tag{9}$$

where $\eta_i = \beta_i \beta_{i+1} h_i$, $\delta_i = h_{i-1} + h_i + h_{i+1} + 2\lambda_i \alpha_{i+1} + 2\lambda_{i+1} \alpha_i + 4\lambda_i \lambda_{i+1} h_i$. From these, we can easily deduce that H_i is nonsingular and all its remaining minors are nonnegative. Therefore, H_i is a nonsingular stochastic and totally positive matrix.

Definition 2. Given a knot vector $U = (u_0, u_1, \ldots, u_{n+4})$ and an array $(\lambda_1, \lambda_2, \ldots, \lambda_{n+3}) \triangleq \lambda$, where all $\lambda_i \in (-1/2, 1]$, $i = 1, 2, \ldots, n + 3$, by using the T-Bezier basis and the ′ coefficients given in (5), functions (3) are defined to be the associated cubic trigonometric B-spline basis functions (T- B-spline basis for short).

It is easy to know the shape of $B(u)$ is related to λ_{i+1}, λ_{i+2}, λ_{i+3}. We refer to the array $\lambda_i = (\lambda_{i+1}, \lambda_{i+2}, \lambda_{i+3})$ as the shape parameter. For equidistant knots, we refer to the $B(u)$ as a uniform T-B-spline basis and refer to the knot vector U as a uniform knot vector. For nonequidistant knots, $B(u)$ and U are called a nonuniform$T$-B-spline basis and a nonuniform knot vector, respectively.

Figure 1 shows some graphs of uniform T-B-spline basis functions with different shape parameters. As can be seen from the figure, with the increase of the parameter λ_{i+1}, the maximum of $B(u)$ decreases and the highest point of the curve moves toward the right. With the increase of λ_{i+2}, the maximum of $B(u)$ increases correspondingly. With the increase of λ_{i+2}, the maximum of $B(u)$ decreases and the highest point of the curve moves toward the left. When $\lambda_{i+1} = \lambda_{i+3}$, the graph of $B(u)$ is symmetry.

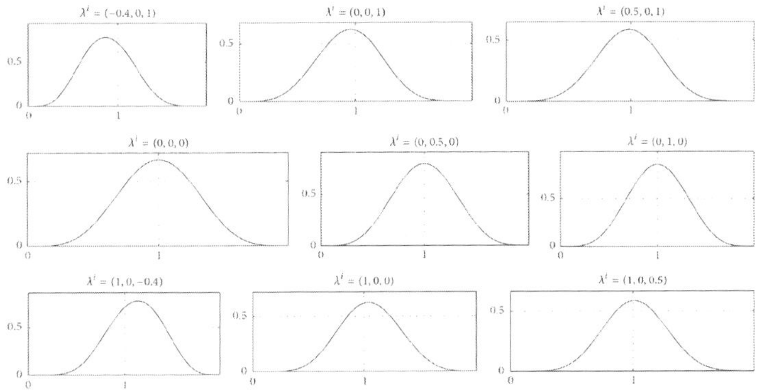

Figure 1: B-spline basis functions with different shape parameters.

THE PROPERTY OF THE BASIS FUNCTIONS

Theorem 3. *The T-B-spline basis has the following properties:*

(a) *Partition of unity.* $\sum_{i=0}^{n} B_i(u) = 1$ *for* $u \in [u_3, u_{n+1}]$.

(b) *Nonnegativity.* *If* $\lambda_{i+1}, \lambda_{i+2}, \lambda_{i+3} \in (-1/2, 1]$, *then* $B_i(u) > 0$ *for* $u \in (u_i, u_{i+4})$.

(c) *Linear independence.* *For* $\lambda_{i-1}, \lambda_i, \lambda_{i+1} \in (-1/2, 1]$, $i = 3, 4, \ldots, n+1$, *the set* $\{B_0(u), B_1(u), \ldots, B_n(u)\}$ *is linearly independent on* $[u_3, u_{n+1}]$.

(d) *Normalized total positivity.* *For* $u \in [u_i, u_{i+1}]$, $\lambda_{i-1}, \lambda_i, \lambda_{i+1}, \lambda_{i+2} \in (-1/2, 1]$, $i = 3, 4, \ldots, n$, *the system* $(B_{i-3}(u), B_{i-2}(u), B_{i-1}(u), B_i(u))$ *is a normalized totally positive basis of the space* $T_{\lambda_i, \lambda_{i+1}}$.

(e) *Continuity. With a nonuniform knot vector, the T-B-spline basis* $B(u)$ *has* C^2 *continuity for* $\lambda_{i+1}, \lambda_{i+2}, \lambda_{i+3} \in (-1/2, 1]$ *at each of the knots. With a uniform knot vector,* $B(u)$ *has C3 continuity for* $\lambda_{i+1}, \lambda_{i+2}, \lambda_{i+3} \in (-1/2, 1]$ *and* C^5 *continuity for* $\lambda_{i+1} = \lambda_{i+2} = \lambda_{i+3} = 1$ *at each of the knots.*

Proof. (a) Notice that $a_i + b_{i,0} + c_{i,0} = b_{i,1} + c_{i,1} = b_{i,2} + c_{i,2} = b_{i,3} + c_{i,3} + d_i = 1$. For $u \in [u_i, u_{i+1})$, $i = 3, 4, \ldots, n$, we have $B_j(u) = 0$ for $j \neq i-3, i-2, i-1, i$; then

$$\sum_{j=0}^{n} B_j(u) = \sum_{j=i-3}^{i} B_j(u) = \sum_{j=0}^{3} B_{i-j,j}(t_i) = 1. \tag{10}$$

(b) When $\lambda_{i+1}, \lambda_{i+2}, \lambda_{i+3} \in (-1/2, 1]$, we can see from (5) that $a_{i+3}, b_{i+2,j}, c_{i+1,j}, d_i > 0 \ (j = 0, 1, 2, 3)$. Thus, from the positivity of the T-Bézier basis, we can conclude that $B(u) > 0$ for $u \in (u_i, u_{i+4})$..

(c) For any $\xi_j \in \mathbb{R}$ $(j = 0, 1, \ldots, n)$, $u \in [u_3, u_{n+1}]$, we define

$$S(u) = \sum_{j=0}^{n} \xi_j B_j(u) = 0. \tag{11}$$

For $\lambda_{i-1}, \lambda_i, \lambda_{i+1} \in (-1/2, 1]$, $i = 3, 4, \ldots, n+1$, direct computation gives that

$$S(u_i) = a_i \xi_{i-3} + b_{i,0} \xi_{i-2} + c_{i,0} \xi_{i-1} = 0,$$

$$S'(u_i) = \frac{\pi(2+\lambda_i)}{2h_i} \left[a_i (\xi_{i-2} - \xi_{i-3}) + \frac{h_i}{h_{i-1}} d_{i-1} (\xi_{i-1} - \xi_{i-2}) \right] = 0,$$

$$S''(u_i) = \frac{\pi^2(1+2\lambda_i)}{2h_i^2} \left[a_i (\xi_{i-3} - \xi_{i-2}) + \frac{h_i^2}{h_{i-1}^2} d_{i-1} (\xi_{i-1} - \xi_{i-2}) \right] = 0. \tag{12}$$

Thus, we can obtain the following linear systems of equations with respect to $\xi_{i-3}, \xi_{i-2}, \xi_{i-1}$:

$$a_i \xi_{i-3} + b_{i,0} \xi_{i-2} + c_{i,0} \xi_{i-1} = 0,$$

$$a_i (\xi_{i-2} - \xi_{i-3}) + \frac{h_i}{h_{i-1}} d_{i-1} (\xi_{i-1} - \xi_{i-2}) = 0,$$

$$a_i (\xi_{i-3} - \xi_{i-2}) + \frac{h_i^2}{h_{i-1}^2} d_{i-1} (\xi_{i-1} - \xi_{i-2}) = 0. \tag{13}$$

Since $a_i + b_{i,0} + c_{i,0} = 1$, we have the determinant of the coefficient matrix D_i given by the above linear systems of equations as follows:

$$|D_i| = \begin{vmatrix} a_i & 1 & c_{i,0} \\ -a_i & 0 & \frac{h_i}{h_{i-1}} d_{i-1} \\ a_i & 0 & \frac{h_i^2}{h_{i-1}^2} d_{i-1} \end{vmatrix} = \frac{h_i(h_{i-1}+h_i)}{h_{i-1}^2} a_i d_{i-1} > 0. \tag{14}$$

Therefore, we can conclude that $\xi_{i-3} = \xi_{i-2} = \xi_{i-1} = 0$ for $i = 3, 4, \ldots, n + 1$.

(d) Notice that the system $(T_0(t_i, \lambda_i), T_1(t_i, \lambda_i), T_2(t_i, \lambda_{i+1}), T_3(t_i, \lambda_{i+1}))$ is the optimal normalized totally positive basis of the space $T_{\lambda_i, \lambda_{i+1}}$. Thus, by relation (7) and Theorem 1, we can see that the system $(B_{i-3}(u), B_{i-2}(u), B_{i-1}(u), B_i(u))$ is a totally positive basis.

(e) For any $\lambda_{i+1}, \lambda_{i+2}, \lambda_{i+3} \in (-1/2, 1]$ direct computation we have

$$\begin{aligned}
&B_i^{(k)}(u_i^+) = B_i^{(k)}(u_{i+4}^-) = 0, \quad k = 0, 1, 2, 3, 5. \\
&B_i(u_{i+1}^\pm) = d_i, \\
&B_i(u_{i+2}^\pm) = 1 - a_{i+2} - d_{i+1}, \\
&B_i(u_{i+3}^\pm) = a_{i+3}, \\
&B_i'(u_{i+1}^\pm) = \frac{\pi d_i (2 + \lambda_{i+1})}{2h_i}, \\
&B_i'(u_{i+2}^\pm) = \frac{\pi (2 + \lambda_{i+2})}{2} \left(\frac{a_{i+2}}{h_{i+2}} - \frac{d_{i+1}}{h_{i+1}} \right), \\
&B_i'(u_{i+3}^\pm) = -\frac{\pi a_{i+3} (2 + \lambda_{i+3})}{2h_{i+3}}, \\
&B_i''(u_{i+1}^\pm) = \frac{\pi^2 d_i (1 + 2\lambda_{i+1})}{2h_i^2}, \\
&B_i''(u_{i+2}^\pm) = -\frac{\pi^2 (1 + 2\lambda_{i+2})}{2} \left(\frac{d_{i+1}}{h_{i+1}^2} + \frac{a_{i+2}}{h_{i+2}^2} \right), \\
&B_i''(u_{i+3}^\pm) = \frac{\pi^2 a_{i+3} (1 + 2\lambda_{i+3})}{2h_{i+3}^2}.
\end{aligned} \tag{15}$$

When $h_i = h_{i+1} = h_{i+2} = h_{i+3} \triangleq h$, for any $\lambda_{i+1}, \lambda_{i+2}, \lambda_{i+3} \in (-1/2, 1]$, we have

$$B_i'''(u_{i+1}^{\pm}) = \frac{\pi^3 d_i (5\lambda_{i+1} - 2)}{8h^3},$$

$$B_i'''(u_{i+2}^{\pm}) = \frac{\pi^3 (5\lambda_{i+2} - 2)(a_{i+2} - d_{i+1})}{8h^3},$$

$$B_i'''(u_{i+3}^{\pm}) = \frac{\pi^3 a_{i+3} (2 - 5\lambda_{i+3})}{8h^3}.$$

(16)

Further, when $h_i = h_{i+1} = h_{i+2} = h_{i+3} \triangleq h$ and $\lambda_{i+1} = \lambda_{i+2} = \lambda_{i+3} = 1$, we have

$$B_i^{(4)}(u_i^{+}) = B_i^{(4)}(u_{i+4}^{-}) = 0,$$

$$B_i^{(4)}(u_{i+1}^{\pm}) = -\frac{3\pi^4 d_i}{2h^4},$$

$$B_i^{(4)}(u_{i+2}^{\pm}) = \frac{3\pi^4 (a_{i+2} + d_{i+1})}{2h^4},$$

$$B_i^{(4)}(u_{i+3}^{\pm}) = -\frac{3\pi^4 a_{i+3}}{2h^4},$$

$$B_i^{(5)}(u_{i+1}^{\pm}) = -\frac{57\pi^5 d_i}{32h^5},$$

$$B_i^{(5)}(u_{i+2}^{\pm}) = \frac{57\pi^5 (d_{i+1} - a_{i+2})}{32h^5},$$

$$B_i^{(5)}(u_{i+3}^{\pm}) = \frac{57\pi^5 a_{i+3}}{32h^5}.$$

(17)

This means that the conclusion is correct.

So far in the discussion of the *T*-*B*-spline basis, we have assumed that each knot is single. On the other hand, the *T*- *B*-spline basis also makes sense when knots are considered with multiplicity $k \leq 4$. For multiple knots, we shrink the corresponding intervals to zero and drop the corresponding pieces of the basis function. For example, if $u_i = u_{i+1}$ is a double knot, then we define

$$B_i(u) = \begin{cases} B_{i,1}(t_{i+1}), & u \in [u_{i+1}, u_{i+2}), \\ B_{i,2}(t_{i+2}), & u \in [u_{i+2}, u_{i+3}), \\ B_{i,3}(t_{i+3}), & u \in [u_{i+3}, u_{i+4}), \\ 0, & u \notin [u_i, u_{i+4}). \end{cases} \quad (18)$$

As a direct application of functions (3) and (e) in Theorem 3, we have the following corollary which shows the geometric meaning of multiple knots.

Corollary 4. *Suppose that a T-B-spline basis B(u) has a knot with multiplicity k (k = 2, 3, 4) at a parameter value u; then at this point the continuity of the T-B-spline basis is reduced from* C^2 *to* C^{3-k} *(*C^{-1} *means discontinuous) for any* $\lambda_{i+1}, \lambda_{i+2}, \lambda_{i+3} \in (-1/2, 1]$*. Moreover, the support interval of the basis is reduced from 4 segments to 5−k segments.*

SPLINE CURVES

The Construction of the Curves

Definition 5. Given control points $P_i \in \mathbb{R}^d$ $(d = 2, 3; i = 0, 1, \ldots, n)$ and a knot vector $U = (u_0, u_1, \ldots, u_{n+4})$ and an array $\lambda = (\lambda_1, \lambda_2, \ldots, \lambda_{n+3})$, where $n \geq 3$ and all $\lambda_i \in (-1/2, 1]$ $(i = 1, 2, \ldots, n + 3)$, the curve

$$P(u) = \sum_{j=0}^{n} B_j(u) P_j, \quad u \in [u_3, u_{n+1}] \quad (19)$$

is called a *T-B*-spline curve.

Obviously, for $u \in [u_i, u_{i+1}]$, $i = 3, 4, \ldots, n$, the curve $P(u)$ can be represented by curve segment

$$P(u) = \sum_{j=i-3}^{i} B_j(u) P_j. \quad (20)$$

Analogous to the cubic *B*-spline curves, since curve (u) is generated on the interval $[u_3, u_{n+1}]$, the choice of the first and last three knots is free, and these knots can be adjusted to give the desired boundary behavior of the curve; see the following descriptions.

For open curves, we choose the knot vector

$$U = (u_0 = u_1 = u_2 = u_3, u_4, \ldots, u_n, u_{n+1} = u_{n+2} = u_{n+3}$$
$$= u_{n+4}). \quad (21)$$

This ensures that the pointsP_0 and P_n are the end points of the curves. An example is given in Figure 2 (left).

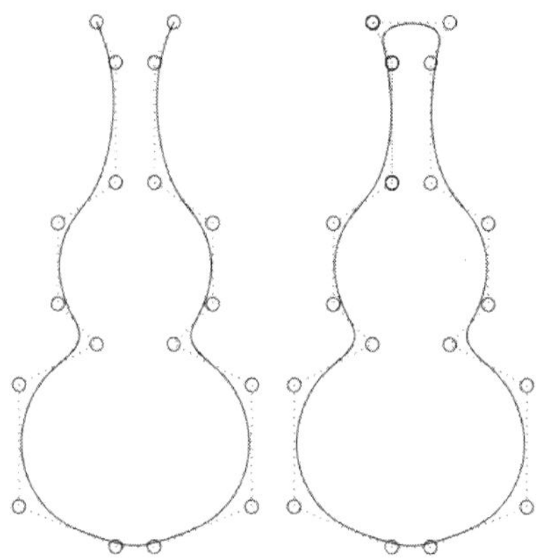

Figure 2: *T*-*B*-spline curves with all $\lambda_i = 0$.

To construct closed curves, we can extend the given points $P_0, P_{1,\ldots,n}$ by setting $P_{n+1} = P_0$, $P_{n+2} = P_1$, and $P_{n+3} = P_2$ and letting $h_j = h_{n+j+1}$ ($j = 1, 2, 3, 4$). Thus the parametric formula for a closed curve is

$$P(u) = \sum_{j=0}^{n+3} B_j(u) P_j, \quad u \in [u_3, u_{n+4}], \quad (22)$$

where$B_j(u)$ ($j = n + 1, n + 2, n + 3$) are given by expanding (3). An example is given in Figure 2 (right).

Based on the properties of the *T*-*B*-spline basis, we can know that the curve (u) has the following properties:

(a) Curve (u) lies in the convex hull of the points $P_{i-3}, P_{i-2}, P_{i-1}, P_i$ for $u \in [u_i, u_{i+1}]$.

(b) Curve (u) has variation diminishing property.

(c) If a knot $u_i$has multiplicity k ($k = 1, 2, 3, 4$), then the curve (u) has C^{3-k} continuity at this point. With a uniform knot vector, the curve (u) has C^3 continuity for any $\lambda_i \in (-1/2, 1]$ and C^5 continuity if all $\lambda_i = 1$.

The Representation of an Ellipse

Given a uniform knot vector, and given control points $P_0 = ((x_1 + x_2)/2, y_1)$, $P_1 = (x_1, (y_1 + y_2)/2)$, $P_2 = ((x_1 + x_2)/2, y_2)$, and $P_3 = (x_2, (y_1 + y_2)/2)$, where $x_1 \neq x_2, y_1 \neq y_2$. The corresponding T-B-spline curve with all $\lambda_i = 0$ is a segment of an elliptic arc, whose equation is

$$P(u) = \begin{pmatrix} x(u) \\ y(u) \end{pmatrix} = \begin{pmatrix} \frac{x_1 + x_2}{2} + \frac{x_1 - x_2}{3} \cos(t) \\ \frac{y_1 + y_2}{2} + \frac{y_2 - y_1}{3} \sin(t) \end{pmatrix},$$

$$t \in \left[0, \frac{\pi}{2}\right]. \tag{23}$$

According to the method of constructing closed curves given in Section 3.1, adding three control points $P_4 = P_0$, $P_5 = P_1$, and $P_6 = P_2$, we can obtain the entire ellipse. When $|x_1 - x_2| = |y_2 - y_1|$, we obtain an entire circle. An example is given in Figure 3.

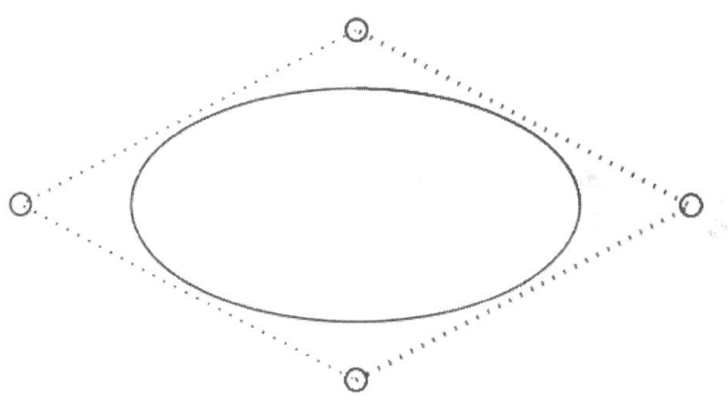

Figure 3: An ellipse represented by T-B-spline curve.

The Representation of a Parabola

Given a uniform knot vector and given control points $P_0 = ((x_1 + x_2)/2, y_1)$, $P_1 = (x_1, y_2)$, $P_2 = P_0$, and $P_3 = (x_2, y_2)$, where $x_1 \neq x_2, y_1 \neq y_2$, the corresponding T-B-spline curve with all $\lambda_i = 0$ is a segment of a parabola, whose equation is

$$P(u) = \begin{pmatrix} x(u) \\ y(u) \end{pmatrix} = \begin{pmatrix} \frac{x_1 + x_2}{2} + \frac{x_1 - x_2}{3}\cos(t) \\ \frac{2y_1 + y_2}{3} + \frac{y_2 - y_1}{3}\cos^2(t) \end{pmatrix},$$

$$t \in \left[0, \frac{\pi}{2}\right]. \quad (24)$$

Adding a control point $P_4 = P_0$, we obtain a symmetric parabolic segment composed of two curve segments. An example is given in Figure 4.

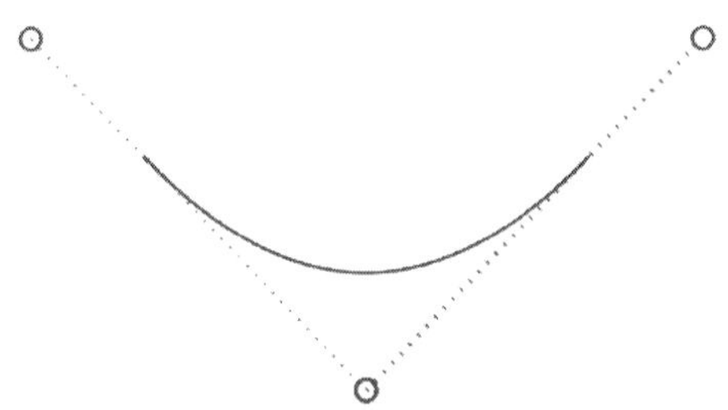

Figure 4: A parabola represented by T-B-spline curve.

SPLINE SURFACES

The Construction of the Surfaces

Exactly as in the construction of *B*-spline tensor product surfaces from *B*-spline curves, we can construct *T*-*B*-spline surface from *T*- *B*-spline curve. For *T*-*B*-spline surface, the shape parameters in u-direction and V-direction can be different.

Definition 6. Given control points $P_{ij} \in \mathbb{R}^3$ $(i = 0, 1, \ldots, m;$ $j = 0, 1, \ldots, n)$, two knot vectors $U = (u_0, u_1, \ldots, u_{m+4}), V = (v_0, v_1, \ldots, v_{n+4})$, and two arrays $\lambda^u = (\lambda_1^u, \lambda_2^u, \ldots, \lambda_{m+3}^u), \lambda^v = (\lambda_1^v, \lambda_2^v, \ldots, \lambda_{n+3}^v)$, where $m, n \geq 3$ and all $\lambda_i^u, \lambda_j^v \in (-1/2, 1]$ $(i = 1, 2, \ldots, m + 3;\ j = 1, 2, \ldots, n + 3)$, then we can define a *T*-*B*-spline surface as follows:

$$P(u,v) = \sum_{k=0}^{m}\sum_{l=0}^{n} B_k(u;\lambda^u) B_l(v;\lambda^v) P_{kl},$$

$$u \in [u_3, u_{m+1}], \quad v \in [v_3, v_{n+1}]. \tag{25}$$

For $u \in [u_i, u_{i+1}]$ and $v \in [u_j, u_{j+1}]$, $i = 3, 4, \ldots, m$, $j = 3, 4, \ldots, n$, the surface (u, V) can be represented by surface patch

$$P(u,v) = \sum_{k=i-3}^{i}\sum_{l=j-3}^{j} B_k(u;\lambda^u) B_l(v;\lambda^v) P_{kl}. \tag{26}$$

Most properties of the *T*-*B*-spline surfaces follow in a straightforward way from those of the *T*-*B*-spline curves. For example, if *U* and *V* are single knot vectors, the *T*-*B*- spline surfaces have C^2 continuity in both directions. If *U* and *V* are uniform knot vectors, the *T*-*B*-spline surfaces have C^3 continuity in general and C^5 continuity when all shape parameters are equal to 1.

The Representation of a Surface of Revolution

A surface of revolution is given by

$$P(u,v) = (r(v)\cos u, r(v)\sin u, z(v))^{\mathrm{T}}. \tag{27}$$

For fixed V, an isoparametric line V = const traces out a circle of radius (V), called a meridian. Since a circle may be exactly represented by *T*--spline curve, we may find an exact representation of a surface of revolution provided we can represent (V), *z*(V) in *T*-*B*-spline curve form.

The most convenient way to define a surface of revolution is to prescribe the (planar) generating curve, or generatrix, given by (V) = (*r*(V), 0, *z*(V)) ^{T} (take the curve in *xoz* plane as an example) and by the axis of revolution, in the same plane as *g*.

Suppose the axis of revolution is *z*-axis. Assume that *g* is given by a planar *T*-*B*-spline curve with control points $G_i = (x_i, z_i)$ ($0 \le i \le m$), knot vector *U*, and shape parameter array λ . Denote $X = [x_0, x_1, \ldots,_m]$, $Z = [z_0, z_1, \ldots, z_n]$. Here *X* and *Z* are row vectors formed by the abscissa and ordinate of the control points G_i ($0 \le i \le m$).

By the method given in Section 3.2, the *T*-*B*-spline curve with control pointsC_0 = (0, 1.5),$_1$ = (1.5, 0),C_2 = (0, −1.5), C_3 = (−1.5, 0), $C_4 = C_0$, $C_5 = C_1$, and $C_6 = C_2$, uniformknot vector *V*, and shape parameter array λ^{V} of all elements are zero which is a unit circle whose center is (0, 0). Denote C_x = [−1.5, 0, 1.5, 0, −1.5, 0, 1], C_y = [0, 1.5, 0, −1.5, 0, 1.5, 0], and *I* = [1, 1, 1, 1,

1, 1, 1]. Here C_x and C_y are row vectors formed by the abscissa and ordinate of the control points C_i ($0 \le i \le 6$). I is a row vector with all elements being equal to 1, and it has the same dimension with C_x and C_y.

Let P_x, P_y, and P_z be $(m+1)\times 7$ matrixes defined by $P_x(i, :) = X(i)C_x$, $P_y(i, :) = X(i)C_y$, and $P_z(i, :) = Z(i)I$, where $1 \le i \le m+1$, $P_*(i, :)$ denotes the ith row of the matrix P_* and $*(i)$ denotes the ith element of the vector $*$. Suppose P_x, P_y, and P_z are the coordinate component matrices of the control points P_{ij} ($0 \le i \le m$, $0 \le j \le 6$). Then the *T-B*-spline surface with control points P_{ij}, knot vectors U and V, and shape parameter arrays λ^u and λ^V is a surface of revolution. Its generatrix is (V), and its axis of revolution is z-axis.

In this way, we represent exactly "classical" surfaces such as cylinders, spheres, or tori. For example, taking $X = [1, 1, 1, 1]$ and $Z = [0, 1, 2, 3]$, we obtain a cylinder. Taking $X = [-2, -1, 1, 2]$ and $Z=X$, we obtain a circular cone. Taking $X = [-1, 0, 1, 0, -1]$ and $Z = [0, 1, 0, -1, 0]$, we obtain a sphere. Taking $X = 2[-1, 0, 1, 0, -1]$ and $Z = [0, 1, 0, -1, 0]$, we obtain an ellipsoid of revolution. Taking $X = [0, -1, 0, 1]$ and $Z = [0, 1, 0, 1]$, we obtain a paraboloid of revolution. Taking $X=C_x -4I$ and $Z=C_y$, we obtain a tori. In each case we take uniform knot vector U and shape parameter array λ^u with all elements which are equal to 0. Figure 5 shows the results.

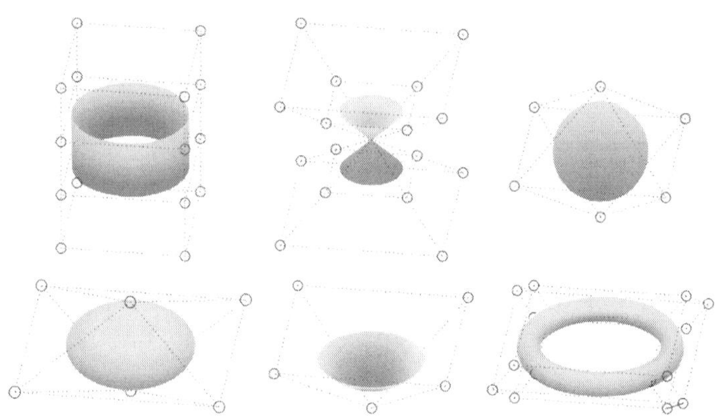

Figure 5: Surfaces of revolution represented by T-B-spline surfaces.

CONCLUSION

With total positivity, the *T-B*-spline basis is suitable for conformal design. By using the *T-B*-spline curves, we can represent ellipses and parabolas exactly. The *T-B*-spline curves can be C^3 or C^5 continuous by taking equidistant knots.By using the T-B-spline surfaces, we can represent rotating surfaces exactly. The T-B-spline surfaces can also reach C^3 or C^5 continuity. One of our future works is to apply the T-B-spline basis to generate shape preserving interpolation curves.

ACKNOWLEDGMENT

This work was supported by the National Natural Science Foundation of China (no. 11261003).

REFERENCES

1. G. Xu and G.-Z. Wang, "Extended cubic uniform B-Spline and -B-Spline," *ActaAutomaticaSinica*, vol. 34, no. 8, pp. 980–984, 2008.
2. P. Costantini, P. D. Kaklis, and C. Manni, "Polynomial cubic splines with tension properties," *Computer Aided Geometric Design*, vol. 27, no. 8, pp. 592–610, 2010.
3. X. Han, "Piecewise quartic polynomial curves with a local shape parameter," *Journal of Computational and Applied Mathematics*, vol. 195, no. 1-2, pp. 34–45, 2006.
4. X. Han, "A class of general quartic spline curves with shape parameters," *Computer Aided Geometric Design*, vol. 28, no. 3, pp. 151–163, 2011.
5. G. Hu, X. Qin, X. Ji et al., "The construction of λμ-B-spline curves and its application to rotational surfaces," *Applied Mathematics and Computation*, vol. 266, pp. 194–211, 2015.
6. X. Han, "Quadratic trigonometric polynomial curves with a shape parameter," *Computer Aided Geometric Design*, vol. 19, no. 7, pp. 503–512, 2002.
7. X. Han, "Cubic trigonometric polynomial curves with a shape parameter," *Computer Aided Geometric Design*, vol. 21, no. 6, pp. 535–548, 2004.
8. X. Han and Y. Zhu, "Curve construction based on five trigonometric blending functions," *BIT Numerical Mathematics*, vol. 52, no. 4, pp. 953–979, 2012.
9. M. Dube and R. Sharma, "Cubic TP B-spline curves with a shape parameter," *International Journal of Engineering Research in Africa*, vol. 11, pp. 59–72, 2013.
10. G. Wang, Q. Chen, and M. Zhou, "NUAT B-spline curves," *Computer Aided Geometric Design*, vol. 21, no. 2, pp. 193–205, 2004.
11. L. Yan and J. Liang, "A class of algebraic-trigonometric blended splines," *Journal of Computational and Applied Mathematics*, vol. 235, no. 6, pp. 1713–1729, 2011.
12. Y. Lü, G. Wang, and X. Yang, "Uniform hyperbolic polynomial B-spline curves," *Computer Aided Geometric Design*, vol. 19, no. 6, pp. 379–393, 2002.

13. T. N. T. Goodman and H. B. Said, "Properties of generalized Ball curves and surfaces," *Computer-Aided Design*, vol. 23, no. 8, pp. 554–560, 1991.

14. H. B. Said, "A generalized Ball curve and its recursive algorithm," *ACM Transactions on Graphics*, vol. 8, no. 4, pp. 360–371, 1989.

15. T. N. T. Goodman and H. B. Said, "Shape preserving properties of the generalised Ball basis," *Computer Aided Geometric Design*, vol. 8, no. 2, pp. 115–121, 1991.

16. X. Han and Y. Zhu, "Total positivity of the cubic trigonometric Bézier basis," *Journal of Applied Mathematics*, vol. 2014, Article ID 198745, 5 pages, 2014.

17. X.-A. Han, Y. Ma, and X. Huang, "The cubic trigonometric Bézier curve with two shape parameters," *Applied Mathematics Letters*, vol. 22, no. 2, pp. 226–231, 2009.

18. Y. Zhu, X. Han, and S. Liu, "Curve construction based on four $\alpha\beta$-Bernstein-like basis functions," *Journal of Computational and Applied Mathematics*, vol. 273, pp. 160–181, 2015.

19. H. Wu and X. Chen, "Cubic non-uniform trigonometric polynomial curves with multiple shape parameters," *Journal of Computer-Aided Design & Computer Graphics*, vol. 18, no. 10, pp. 1599–1606, 2006.

20. X. Han, "C^2 quadratic trigonometric polynomial curves with local bias," *Journal of Computational and Applied Mathematics*, vol. 180, no. 1, pp. 161–172, 2005.

13

Fitting Quadrics with a Bayesian Prior

Daniel Beale, Yong-Liang Yang, Neill Campbell, Darren Cosker& Peter Hall

University of Bath, Claverton Down, Bath, BA2 7AY, UK

ABSTRACT

Quadrics are a compact mathematical formulation for a range of primitive surfaces. A problem arises when there are not enough data points to compute the model but knowledge of the shape is available. This paper presents a method for fitting a quadric with a Bayesian prior. We use a matrix normal prior in order to favour ellipsoids when fitting to ambiguous data. The results show the algorithm copes well when there are few points in the point cloud, competing with contemporary techniques in the area.

Keywords: geometry, statistics, graphics, computer vision

Citation: Beale, D., Yang, Y. L., Campbell, N., Cosker, D., & Hall, P. (2016). Fitting quadrics with a Bayesian prior. Computational Visual Media, 2(2), 107-117. (11 pages), DOI: https://doi.org/10.1007/s41095-016-0041-9.

Copyright: © Open Access. This article is distributed under the terms of the Creative Commons Attribution 4.0 International (CC BY 4.0) License.

INTRODUCTION

Surface fitting is one of the most important research areas in computer graphics and geometric modelling. It studies how to approximate unorganized geometric data using regular surfaces with explicit algebraic forms, which benefits many important graphics applications, including shape approximation, surface reconstruction, local geometric feature analysis, etc. In the fitting process, the choice of the target surface form usually depends on the application, which might require algebraic surfaces of specific orders, or freeform surfaces, such as B-Splines.

Among all types of surfaces, quadrics and hypersurfaces represented by a quadratic polynomial in the embedding space may be the most popular form in surface fitting for several reasons. Firstly, they can represent a variety of common primitive surfaces, such as planes, spheres, cylinders, etc. Secondly, many real world shapes, ranging from industrial products to architectural designs, can be represented by a union of quadrics. Thirdly, it is the form with the least order from which it is possible to estimate second order differential properties, such as curvatures. Fourthly, the simple quadratic form allows for efficient numerical computations which are usually in closed-form.

In this work, we present a novel probabilistic quadric fitting method. Alternative to previous algorithms, we assume an additive Gaussian noise model on the algebraic distance. A Bayesian prior is placed on the parameters, allowing certain shapes to be favoured when few data are available. We have tested our quadric fitting method on various datasets, both synthetic and from the real world. The experiments and comparisons show that our method is not only efficient, but also robust to contaminated data.

RELATED WORK

There are a large number of approaches to fitting both implicit and parametric surfaces to unordered point clouds. We focus the review on existing methods in quadric fitting and then go on to show how they can be used in a variety of applications in computer graphics.

Quadric Fitting

The popularity of quadrics has led to many research papers on fitting them to unordered data points. Methods in this area can be broken into two main areas, either minimising the algebraic or geometric distance of each point to the surface. The former is simpler, since it is the value computed by evaluating the implicit equation itself (defined in Eq. (1)).

Geometric distances are formulated in terms of Euclidean distances between the points and the surface; methods in this area produce a fit which favours each point equally, although methods tend to be iterative due to the complex relationship between the implicit equation and the solution set.

One of the most popular methods to use a geometric distance is from Taubin [1], who approximates the true distance in terms of the normalised Euclidean distance to the quadric surface. Knowledge of the explicit parametric equations of the surface can be useful for more compact least-squares parameter estimates [2–8]. Fast estimation of the closest point on the surface to a general point in space is described in Ref. [9], which also makes use of RANSAC to add robustness to outliers.

Finding a least-squares estimate based on the algebraic distance yields fast and numerically stable solutions which can be solved using matrix algebra. In this case, literature in the area seeks to constrain the surface to lie on a particular surface shape, such as an ellipsoid, e.g., Refs. [10–13]. The work of Li and Griffiths [12] shows how to fit ellipsoids and Dai et al. [14] present a similar method for fitting paraboloids.

Bayesian probabilistic methods exist for fitting general quadrics: for example, Subrahmonia et al. [15] show how to fit by casting a probability distribution over the geometric Taubin-type errors. Their use of priors is to avoid overfitting or parameters which produce a minimal error but undesirable results.

Quadric Fitting Applications

Quadric fitting has been applied to many important graphics problems. In the context of shape approximation, Cohen-Steiner et al. [16] present a variational shape approximation (VSA) algorithm, where a set of planar proxies are iteratively fitted to an input mesh based on Lloyd's clustering on mesh faces, resulting in a simplified mesh representation.

To better approximate curved parts and sharp features, Wu and Kobbelt [17] extend the previous approach by using proxies other than planes, where various types of quadrics, such as spheres or cylinders, are allowed. Another extension of the VSA algorithm was presented in Ref. [18], using general quadric proxies. For surface reconstruction,quadric fitting has been used to better represent the underlying geometry and improve reconstruction quality.

Schnabel et al. [19] present a RANSACbased approach to fit different types of quadrics to noisy data. Li et al. [20] further improve the reconstruction quality by optimizing global regularities, such as symmetries, of the fitted surface arrangements. In addition to joining the geometric primitives, it is possible to create an implicit surface using a multi-level partition [21]. This kind of approach can help to smooth the resulting surface in regions which have not been accurately modelled.

In shape analysis, local geometric features (e.g., curvatures) which are invariant under certain transformations benefit greatly from quadric fitting. A common approach is to fit quadrics to a local neighbourhood so that differential properties can be estimated with the help of the fitted surface [22].

The use of computer vision in robotics applications needs the use of geometric fitting algorithms for estimating object properties for physical interaction, such as the curvature of an object [23]. Segmenting objects into primitive surfaces is also possible [24], allowing a robot to discriminate between surfaces before planning.

Our Work

Our method assumes a probability distribution over the algebraic errors, and results in a compact matrix computation for the maximum a priori estimate of the parameters, given the data points. The use of a prior allows us to choose specific primitives from the collection of quadric surfaces.

BAYESIAN QUADRICS

We begin by defining what a quadric is, and then extend it with the use of a noise model. We define a conditional distribution which provides the likelihood of the data given the parameters, from which we infer the probability of the parameters with the use of a suitable prior.

Quadrics are polynomial surfaces of degree two. Each is a variety given by the implicit equation:

$$\boldsymbol{z}^{\mathrm{T}}\boldsymbol{A}\boldsymbol{z} + \boldsymbol{b}^{\mathrm{T}}\boldsymbol{z} + c = 0 \tag{1}$$

with A is an M × M matrix, b is a vector, c is a scalar, and $z \in \mathbb{R}^{M}$.

Before defining more notation we note that A can be assumed to be symmetric, without loss ofgenerality, as explained in Appendix A. Positive definiteness is determined from the eigenvalues. Letting ψ denote each element of A, b, and c. Further, let θ_A denote the values in ψ that correspond to the values in A, i.e.,

$$\boldsymbol{\theta}_{\boldsymbol{A}} = [A_{11}, \ldots, A_{MM}, A_{12}, \ldots, A_{(M-1)M}]^{\mathrm{T}} \tag{2}$$

so

$$\psi = \begin{bmatrix} \boldsymbol{\theta}_{\boldsymbol{A}} \\ \boldsymbol{b} \\ c \end{bmatrix} \tag{3}$$

The data from the input point cloud are collected into a data matrix $\boldsymbol{X} = [\boldsymbol{x}_0, \ldots, \boldsymbol{x}_\ell, \ldots, \boldsymbol{x}_N]^{\mathrm{T}}$, so that

$$\boldsymbol{x} = (z_1^2, \ldots, z_M^2, 2z_1z_2, \ldots, z_1, \ldots, z_M, 1)^{\mathrm{T}} \tag{4}$$

where x is an arbitrary column of X, and

$$\boldsymbol{x} \cdot \psi = \boldsymbol{z}^{\mathrm{T}} \boldsymbol{A} \boldsymbol{z} + \boldsymbol{b}^{\mathrm{T}} \boldsymbol{z} + c \tag{5}$$

In order to use Bayesian statistics, an additive noise model is assumed on each of the data points, i.e., we assume that

$$\boldsymbol{X}\psi = \varepsilon \tag{6}$$

where $\varepsilon \sim N(0, I)$ is a single draw from a unit multivariate Gaussian distribution.

In order to solve the matrix equation $X\psi = 0$, without a prior on ψ one can find the eigenvector corresponding to the smallest eigenvalue of X^TX. If we consider the following eigenvalue problem:

$$\boldsymbol{X}^{\mathrm{T}} \boldsymbol{X} \boldsymbol{v} = s\boldsymbol{v} \tag{7}$$

it is sufficient to find an eigenvector v associated with eigenvalue s = 0. Existence of such a solution is guaranteed if the matrix X^TX is singular. In practice, however, we take the smallest eigenvalue.

The solution to Eq. (7) is a least-squares solution of the equation $||X\psi||^2 = 0$ since the value $||X\psi||^2$ is minimal when all of the partial derivatives with respect to ψ are zero. Consider

$$\frac{\partial}{\partial \psi}(||\boldsymbol{X}\psi||^2) = \frac{\partial}{\partial \psi}(\psi^{\mathrm{T}} \boldsymbol{X}^{\mathrm{T}} \boldsymbol{X}\psi) \tag{8}$$

$$= 2\boldsymbol{X}^{\mathrm{T}} \boldsymbol{X}\psi \tag{9}$$

Setting Eq. (9) equal to zero and dividing out the constant, one can see that a minimal solution using eigenvalue decomposition is a least-squares estimate.

The least-squares estimate introduced above places no constraint on the matrix A, and therefore can produce a matrix with negative eigenvalues. If one wishes to ensure that the matrix is an ellipsoid, for example, A is required to have a full set of

$$p(\psi|\boldsymbol{X}) \propto p(\boldsymbol{X}|\psi)p(\psi) \tag{10}$$

The distribution p(X) is constant with respect to the parameters ψ and so we replace it with the proportionality operator ($\propto$). This is useful because traditionally, one would compute p(X) = $\int_{\psi} p(\boldsymbol{X}|\psi)p(\psi)$ which can be intractable. Certain properties are maintained across proportionality, for example, the maximum of $p(\psi|X)$ is the same as the maximum of $\kappa p(\psi|X)$ for any κ. Since we only seek a maximum a priori estimate of the parameters, we only use equivalence of distributions up to proportionality.

With a careful choice of prior (discussed in Section 4), one can find the maximum a priori estimate of the parameters, given the data, by finding zeros of the log-posterior derivatives. In the context of fitting a general quadric and following from Eq. (6), we assume additive Gaussian noise on Eq. (1) in which case the likelihood is taken to be

$$p(\boldsymbol{X}|\psi) = \mathcal{N}(\boldsymbol{X}\psi|\boldsymbol{0}, \boldsymbol{I}) \tag{11}$$

In other words, the errors in computing Eq. (1) are jointly Gaussian, with a spherical unit covariance I. The assumption of a spherical covariance is equivalent to saying that the variables are independent, given the parameters ψ. This is true, since the error from the original surface of one point tells us nothing about the error of a neighbouring point. The covariances are assumed to be identical as a matter of simplicity. The identity is chosen rather than some scalar multiple, since the parameter becomes superfluous when the prior is introduced. The hyper-parameter σ (introduced later in the text) provides all of the information necessary to model the surface.

It would be possible at this point to use a prior for the vector ψ. Rather than doing this we introduce a bijection between A and ψ through the vector θ defined in Section 3. This allows the hyperparameters to be set in terms of the matrix rather than the somewhat unintuitive values of ψ. In fact, using a matrix normal on A with scalar parameters is equivalent to assuming a multivariate Gaussian on the first elements of ψ.

We only place a prior distribution over the matrix A since it contains information about the shape of the surface. This is discussed in detail in Section 7, with the choice of hyper-parameters discussed more in Section 4.

The matrix normal distribution is defined as

$$\mathcal{MN}(\boldsymbol{A}|\boldsymbol{W},\boldsymbol{U},\boldsymbol{V}) = \kappa e^{[-\frac{1}{2}\mathrm{tr}(\boldsymbol{V}^{-1}(\boldsymbol{A}-\boldsymbol{W})^{\mathrm{T}}\boldsymbol{U}^{-1}(\boldsymbol{A}-\boldsymbol{W}))]} \quad (12)$$

where $\boldsymbol{W},\boldsymbol{U},\boldsymbol{V} \in \mathbb{R}^{M\times M}$ and

$$\kappa = ((2\pi)^{np/2}|\boldsymbol{V}|^{n/2}|\boldsymbol{U}|^{n/2})^{-1} \quad (13)$$

The posterior is then found by multiplying Eq. (11) by Eq. (12) to give

$$p(\psi|\boldsymbol{X}) \propto \mathcal{N}(\boldsymbol{X}\psi|\boldsymbol{0},\boldsymbol{I})\mathcal{MN}(\boldsymbol{A}|\boldsymbol{W},\boldsymbol{U},\boldsymbol{V}) \quad (14)$$

$$\propto e^{-\frac{1}{2}||\boldsymbol{X}\psi||^2} e^{-\frac{1}{2}\mathrm{tr}(\boldsymbol{V}^{-1}(\boldsymbol{A}-\boldsymbol{W})^{\mathrm{T}}\boldsymbol{U}^{-1}(\boldsymbol{A}-\boldsymbol{W}))} \quad (15)$$

If we then choose U = V = σI, and ensure that W is diagonal and symmetric, the posterior can be simplified to

$$p(\psi|\boldsymbol{X}) \propto e^{-\frac{1}{2}||\boldsymbol{X}\psi||^2} e^{-\frac{1}{2\sigma^2}\mathrm{tr}(\boldsymbol{A}^{\mathrm{T}}\boldsymbol{A}-2\boldsymbol{W}\boldsymbol{A}+\boldsymbol{W}^2)} \quad (16)$$

$$\propto e^{-\frac{1}{2}||\boldsymbol{X}\psi||^2} e^{-\frac{1}{2\sigma^2}(\mathrm{tr}(\boldsymbol{A}^{\mathrm{T}}\boldsymbol{A})-\mathrm{tr}(2\boldsymbol{A}\boldsymbol{W})+\mathrm{tr}(\boldsymbol{W}^2))} \quad (17)$$

Taking the natural log and then the first partial derivative yields:

$$\frac{\partial \ln p(\psi|\boldsymbol{X})}{\partial \psi} = -\frac{1}{2}\frac{\partial}{\partial \psi}||\boldsymbol{X}\psi||^2 - \frac{1}{2\sigma^2}\frac{\partial}{\partial \psi}\mathrm{tr}(\boldsymbol{A}^{\mathrm{T}}\boldsymbol{A}) - \frac{1}{\sigma^2}\frac{\partial}{\partial \psi}\mathrm{tr}(\boldsymbol{W}^{\mathrm{T}}\boldsymbol{A}) \quad (18)$$

Before computing the vector derivatives, we first note the matrix derivatives for a symmetric A and W can be written as (see Ref. [25] for more details):

$$\frac{\mathrm{d}(\mathrm{tr}(\boldsymbol{W}^{\mathrm{T}}\boldsymbol{A}))}{\mathrm{d}\boldsymbol{A}} = \boldsymbol{W} + \boldsymbol{W}^{\mathrm{T}} - \boldsymbol{W}\circ\boldsymbol{I} \quad (19)$$

$$\frac{\mathrm{d}(\mathrm{tr}(\boldsymbol{A}^{\mathrm{T}}\boldsymbol{A}))}{\mathrm{d}\boldsymbol{A}} = 2(\boldsymbol{A} + \boldsymbol{A}^{\mathrm{T}} - \boldsymbol{A}\circ\boldsymbol{I}) \quad (20)$$

The operator ∘ is the Hadamard product for matrices, and produces a matrix of equal size by element wise multiplication: if S = Q ∘ R then $S_{ij} = Q_{ij}R_{ij}$ for each i and j. More intuitively, Eqs. (19) and (20) represent a doubling of all the offdiagonal elements, leaving the diagonals unchanged.

Since A is symmetric we change notation for the derivatives and set

$$\hat{A} = A + A^{\mathrm{T}} - A \circ I \tag{21}$$

$$\hat{W} = W + W^{\mathrm{T}} - W \circ I \tag{22}$$

The final derivative can then be calculated as

$$\frac{\partial \ln p(\psi|X)}{\partial \psi} = -X^{\mathrm{T}} X\psi - \frac{1}{\sigma^2}\begin{bmatrix} \theta_{\hat{A}} \\ 0 \end{bmatrix} + \frac{1}{\sigma^2}\begin{bmatrix} \theta_{\hat{W}} \\ 0 \end{bmatrix} \tag{23}$$

In the final step we separate ψ from X and σ by distributing multiplication over addition. Here ψ depends on A and $\theta_{\hat{A}}$ on $\hat{A}$. Let the matrix J be a diagonal matrix which takes a value of 1 on the diagonal when the corresponding value of $\theta_{\hat{A}}$ is a diagonal element, and 2 otherwise. Setting the partials to zero it then follows that

$$\left(X^{\mathrm{T}} X + \frac{1}{\sigma^2}\begin{bmatrix} J & 0 \\ 0 & 0 \end{bmatrix}\right)\psi - \frac{1}{\sigma^2}\begin{bmatrix} \theta_{\hat{W}} \\ 0 \end{bmatrix} = 0 \tag{24}$$

Rearranging yields

$$\psi = \left(\sigma^2 X^{\mathrm{T}} X + \begin{bmatrix} J & 0 \\ 0 & 0 \end{bmatrix}\right)^{-1}\begin{bmatrix} \theta_{\hat{W}} \\ 0 \end{bmatrix} \tag{25}$$

CHOICE OF PRIOR

We use a matrix normal prior rather than a Wishart or inverse Wishart prior for the matrix A. The Wishart distributions are commonly used, well principled, and suitable for use with empirical data. However, computing a maximum a priori estimate in a similar fashion to that used in the previous section leads to a solution that must be determined by search, as opposed to the closed-form of Eq. (25). The requirement that Wishart and inverse Wishart matrices are symmetric positive definite is another limitation, particularly if cylinders or hyperboloids are necessary. Section 7 discusses how to construct a mean matrix W in order to favour different shapes of surface.

The prior is only cast over the matrix A since the parameters b and c contain information about the center and also the width of the resulting surface. These values are also coupled, in the sense that a change in center affects both b and c. Section 6 discusses how to extract the center and width from the parameters, but their relationship is pathological. Constraining A alone provides just enough information on the shape of the surfaceto ensure that it does not become a hyperboloid, without affecting the quality of the fit.

The matrices U and V determine the variance of the matrix A from the mean W. Smaller values more strongly favour the mean as opposed to the data. For our experiments we simply choose the parameter $\sigma \in \mathbb{R}^+$, which gives control over how close the matrix A is to the identity. The value σ is free, in the sense that it can be chosen arbitrarily or depending on a larger dataset. The results in Section 8 describe how it is chosen for our experiments; however, it could also be computed using a training set containing prior information about the variance of the observed quadric from the mean.

3D DATA

The results in Section 3 apply to data of arbitrary dimension. This section provides explicit formulae for 3 dimensions. We note that the same principles apply for 2 dimensional data, allowing us to model ellipses and hyperbola.

The parameters A and b can be written as follows:

$$\boldsymbol{A} = \begin{bmatrix} a_{11} & a_{12} & a_{13} \\ a_{12} & a_{22} & a_{23} \\ a_{13} & a_{23} & a_{33} \end{bmatrix}, \quad \boldsymbol{b} = \begin{bmatrix} b_1 \\ b_2 \\ b_3 \end{bmatrix} \tag{26}$$

In this case ψ is defined as

$$\boldsymbol{\psi} = \begin{bmatrix} \boldsymbol{\theta}_A \\ \boldsymbol{b} \\ c \end{bmatrix} \tag{27}$$

where

$$\boldsymbol{\theta}_A = [a_{11}, a_{22}, a_{33}, a_{12}, a_{13}, a_{23}]^{\mathrm{T}} \tag{28}$$

The matrix X is the quadratic data matrix:

$$\boldsymbol{X} = [\boldsymbol{x}_0, \ldots, \boldsymbol{x}_N] \tag{29}$$

where

$$\boldsymbol{x}_i = (z_{1i}^2, z_{1i}^2, z_{3i}^2, 2z_{1i}z_{2i}, \ldots, z_{1i}, z_{2i}, z_{3i}, 1)^{\mathrm{T}} \tag{30}$$

The matrix J is then taken to be

$$\boldsymbol{J} = \begin{bmatrix} \boldsymbol{I} & 0 \\ 0 & 2\boldsymbol{I} \end{bmatrix} \tag{31}$$

and

$$\theta_{\hat{W}} = \begin{bmatrix} 1 \\ 0 \end{bmatrix} \tag{32}$$

PARAMETRISATION

The quadric can be parameterised by finding the transformation between the canonical shape, such as the unit sphere, and the resulting surface.

Consider the following:

$$(x - \mu)^{\mathrm{T}} \Lambda (x - \mu) = \tau \tag{33}$$

Multiplying out and rearranging leads to

$$x^{\mathrm{T}} \Lambda x - \mu^{\mathrm{T}} (\Lambda + \Lambda^{\mathrm{T}}) x + \mu^{\mathrm{T}} \Lambda \mu - \tau = 0 \tag{34}$$

If we then take

$$A = \Lambda \tag{35}$$

$$b = -(\Lambda + \Lambda^{\mathrm{T}})\mu \tag{36}$$

$$c = \mu^{\mathrm{T}} \Lambda \mu - \tau \tag{37}$$

it can be seen that Eq. (1) takes on a canonical form of $y^{T}Wy = 1$ after an affine transformation of the data, where W is a diagonal matrix with entries from the set $\{1, 0, -1\}$.

For an ellipsoid, the matrix W is the identity. Faces and vertices on a unit sphere are generated using a spherical coordinate system, and we transform only the vertices. Letting v be an arbitrary vertex on the unit sphere, and $USU^{T} = A$ be the eigendecomposition of A (A is Hermitian, so $U^{T} = U^{-1}$), it follows that

$$\hat{v} = U \left(\frac{S}{\tau} \right)^{-\frac{1}{2}} v + \mu \tag{38}$$

where $\hat{v}$ is the required ellipsoid vertex. The same principle can be used for any canonical shape, including cylinders, ensuring that the principal axis is aligned with the zero entry of W.

Alternative methods exist for computing the solutions of Eq. (1): isosurface algorithms, for example. A parametric approach is beneficial for its speed; however, it then also requires algebraic methods for merging with other surfaces, which can become complicated.

EIGENDECOMPOSITION

The shape of the quadric can be determined by analysing the eigenvalues of A.

Since A is real and symmetric it is a special case of a Hermitian matrix, and so it follows that all of its eigenvalues are real. In the case of a symmetric positive definite matrix all of the eigenvalues are positive, and so the quadric forms an ellipsoid. If any of the eigenvalues are zero then the axis of the corresponding eigenvector is indeterminate; in the case of 2 positive eigenvalues it is a cylinder, for example. A single negative eigenvalue leads to a hyperboloid.

The final classifications can be obtained by studying the canonical form of the quadric introduced in Section 6, $y^T W y = 1$, where y is an affine transformation of the arbitrary data values z, and W is a diagonal matrix. Multiplying this equation out for different combinations of sign on the diagonal of W reduces to a scalar equation which can be compared to studied quadratic forms, as presented in Andrews and S´equin [3].

It is important to note that Eq. (1) is not only invariant to scale, but also negation, in the sense that multiplying the whole equation by a scalar $\alpha \in \mathbb{R}$ does not change the solution set. Consider the following expansion:

$$\boldsymbol{z}^{\mathrm{T}}\boldsymbol{A}\boldsymbol{z} + \boldsymbol{b}^{\mathrm{T}}\boldsymbol{z} + c = 0 \tag{39}$$

$$\Longleftrightarrow \alpha(\boldsymbol{z}^{\mathrm{T}}\boldsymbol{A}\boldsymbol{z} + \boldsymbol{b}^{\mathrm{T}}\boldsymbol{z} + c) = 0 \tag{40}$$

$$\Longleftrightarrow \boldsymbol{z}^{\mathrm{T}}(\alpha\boldsymbol{A})\boldsymbol{z} + \alpha\boldsymbol{b}^{\mathrm{T}}\boldsymbol{z} + \alpha c = 0 \tag{41}$$

Further, let $A = USU^T$ be the eigendecomposition of A. Then $\alpha A = U^T (\alpha S)U$. This means that a scaling and negation of the eigenvalues of A only affects the other parameters, but not the solutions, e.g., a full set of negative eigenvalues yields the same solution as a full set of positive eigenvalues. Since the signs of the eigenvalues of A are the same as the signs of the diagonal entries of W we can relate them to the shape of the surface.

A range of different surface shapes is attainable in this way. For example, a hyperboloid can be made from two negative and one positive eigenvalues, which is the same as two positive and one negative eigenvalues. A table showing the classification of the surface for different eigenvalues is shown in

Table 1, and represents only a sample of the possible combinations. We note that the parameters b and c also have an affect on the shape. For example, if b = 0 and we have one positive eigenvalue and two zero, the resulting shape is no longer a paraboloid but a pair of planes. Examples of these shapes can be seen in Fig. 1.

RESULTS

The results consider two scenarios. Firstly, it isshown that the method is able to perform well on data drawn from simulated quadrics, ellipsoids with known parameters.

Table 1: Relationship between eigenvalues and shape of the corresponding quadric

Shape	Positive	Zero	Negative
Ellipsoid	3	0	0
Cylinder	2	1	0
Paraboloid	2	1	0
Plane	0	0	0
Hyperboloid	2	0	1
Hyperbolic section	1	1	1

The algorithm is then shown to perform well on empirical data from a collection of 3D point clouds. The results are compare to quadric fitting without a prior and also against the work of Li and Griffiths [12], who solve a generalised eigenvalue problem similar to that of Ref. [11].

The work presented by Li and Griffiths is a candidate example of a modern ellipsoid fitting algorithm, from the larger set of algorithms available (e.g., Refs. [2–8]). Our method is a Bayesian model, which allows us to provide a parameter which expresses a measure of how close the shape should be to a canonical shape, such as a sphere. Rather than comparing to all other algorithms we simply show that a hard constraint to fit an ellipse does not always lead to the best fit to the data, particularly when there is a large percentage of missing or excess noise.

In the final section of the results, we give an example of a fit to a simulated hyperboloid, primarily to show that our algorithm also works for different surface types. It also demonstrates the value in choosing different diagonals on the mean matrix W.

Algorithmic Complexity

One of the most attractive attributes of a direct method for fitting, such as ours and Refs. [12] or [11], is its efficiency. A well known alternative is to compute the squared Euclidean (or geometric) distance between each point and the surface, and then minimise the sum of squares [9]. This isdifficult since computing the Euclidean distance is a non-linear problem (see Refs. [26, 27] for more details), and the complexity is dependent on a polynomial root finding algorithm.

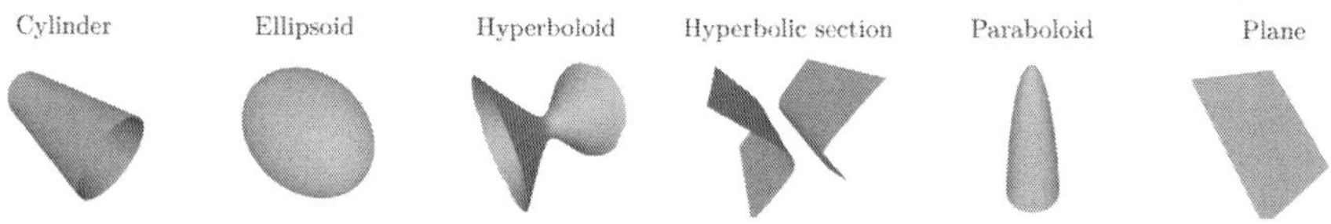

Figure 1: A collection of canonical quadric shapes.

Minimising the sum of squared errors in this way requires a nonlinear least-squares algorithm such as Levenberg– Marquardt. Computing the error alone has an order of O(NM3), which is a lower bound on the final fitting algorithm as it involves iteratively computing the error and parameter gradients at each step. It is possible, however, to approximate the distance function, as shown in Refs. [1] and [6], and fit using generalised eigenvalues or using a linear leastsquares estimate; these methods have a much lower complexity, but do not allow us to use a Bayesian prior when fitting.

The complexity of the algorithm presented in the paper arises in the computation of the scatter matrix X^TX in Eq. (25), which is $O(NM^2)$; the subsequent matrix operations have much lower complexity ($O(M^3)$ at worst). This makes ours an efficient approach, particularly since the scatter can be computed independently of the value σ. Faster algorithms exist for computing covariances, for example Ref. [28], at the expense of memory, although it may be generally quicker to use parallel computation, such as a GPU, to improve performance.

SIMULATED DATA

We first show results on a simulated dataset. Data is drawn from an ellipsoid of known dimensions, some of the data is removed, and spherical Gaussian noise is added. We show that with the correct choice of hyper-parameters, the Bayesian method fits best.

Figure 2 provides an example of fitting an ellipsoid to a simulated dataset, with missing data. In this experiment we only use 30% of the training data from the original ellipsoid and add Gaussian noise with a standard deviation of 0.01. It can be seen from the results that a value of σ = 10 produces a result most plausibly in agreement with the original, whereas the result of fitting with no prior produces a hyperboloid. It is clear from this example that lower values of σ produce a result closer to a sphere.

In order to determine the quality of the fit, we compare the vertices on the original quadric with the fitted quadric using the Hausdorff distance. Firstly, the distance between a point $x_i \in X$ and a point cloud $Y = \{y_0, \ldots, y_N\}$ is taken to be

$$\varrho(\boldsymbol{x}_i, Y) = \min_{j \in \{1,\ldots,N\}} (|\boldsymbol{x}_i - \boldsymbol{y}_j|) \tag{42}$$

The Hausdorff distance is then:

$$H(X, Y) = \max\Big(\max_{i \in \{1,\ldots,M\}} (\varrho(\boldsymbol{x}_i, Y)), \max_{j \in \{1,\ldots,N\}} (\varrho(\boldsymbol{y}_j, X)) \Big) \tag{43}$$

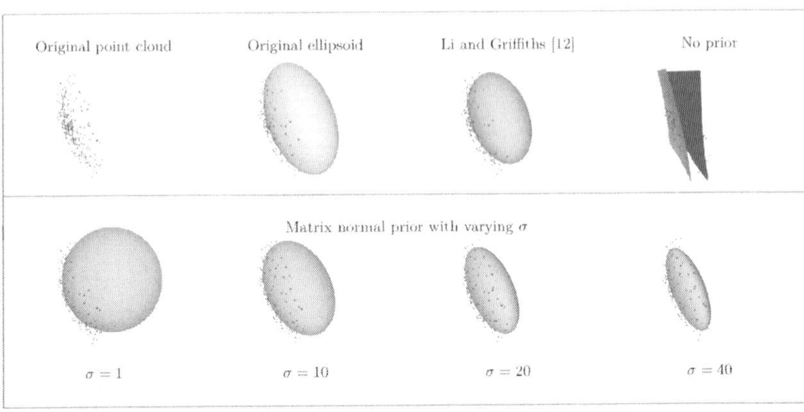

Figure 2: A collection of ellipses fitted to data. Bottom: results of fitting with a prior. Top: results of fitting with no prior. The first two images are the point cloud and the ellipsoid that they were generated from.

The first experiment considers how each of the algorithms perform when there is missing data. We generate an arbitrary ellipsoid and remove N% of the data, and add spherical Gaussian noise with 0.01 standard deviations. The parameters of the ellipsoid are drawn as follows:

$$\Lambda \sim \mathcal{W}(\omega^2 \boldsymbol{I}, 20) \quad (44)$$

$$\boldsymbol{\mu} \sim \mathcal{N}(\boldsymbol{0}, \omega^2 \boldsymbol{I}) \quad (45)$$

$$\tau \sim \mathcal{N}(1, \omega^2) \quad (46)$$

with $\omega^2 = 0.01$. The parameters A, b, and c are computed as in Section 6. The percentage of data removed is varied between 10% and 100%, and we choose the hyper-parameter $0.1 < \sigma < 30$ which produces the best Hausdorff distance. Using a multivariate matrix normal prior is compared with (1) using no prior at all and (2) the ellipsoid specific least-squares method of Li and Griffiths [12].

The results of the first experiment can be seen in Fig. 3. Our method consistently outperforms both methods, although the distance drops below the standard deviation in Gaussian noise in all methods after 60% of the data is used. Fitting without a prior also gives reasonable results if enough of the data is used.

In the second simulation experiment we demonstrate how the method is robust to noise by varying the amount of additive Gaussian noise added to each sampled ellipsoid. We denote the noise level by $\hat{\tau}$. In order to demonstrate the quality of fit over a range of values we sample 10 quadrics and compute error statistics. In this experiment we use 100% of the points, i.e., there is no missing data.

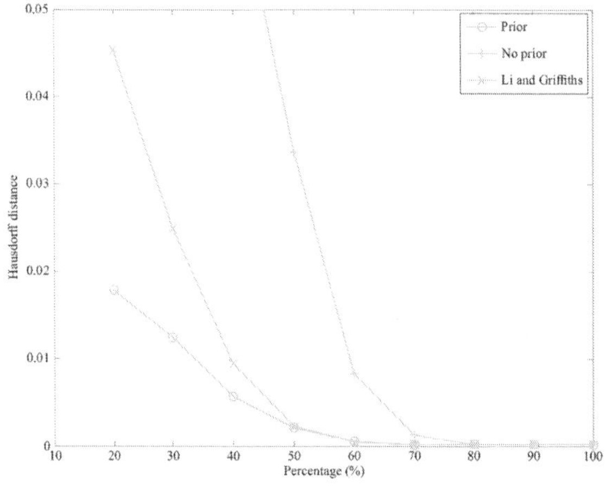

Figure 3: The Hausdorff distance for a varying amount of missing data.

The hyper-parameter is fixed at $\sigma = 10$.

The results from the second experiment can be seen in Fig. 4. Our method produces a Hausdorff distance which is comparable to Ref. [12] for lower values of $\hat{T}$, while as $\hat{T}$ increases the use of a prior proves to be beneficial. Although fitting without a prior gives reasonable results when $\hat{T}$ is small, it becomes very inaccurate when $\hat{T}$ is greater than 0.06. This tends to be due to negative eigenvalues in the matrix A which does not produce an ellipsoid.

In order to show that the algorithm works for other types of quadric surface, we simulated a hyperboloid and removed a percentage of the points. Three quadrics were fitted to the data using different priors for each of: an ellipsoid, no prior, and a hyperboloid. For the ellipsoid, W = I and $\sigma = 1e^{-6}$; for the hyperboloid W = Diag(−1, 1, 1) and $\sigma = 10$. Note that the variance for the ellipsoid needs to be low enough or it will favor a different shape. The hyperboloid hyper-parameter is aligned with the x-axis, as for the simulated data, based on our knowledge of the shape. Fitting without a prior chooses the hyperboloid's medial axis to be aligned with the z-axis. The results of this experiment can be seen in Fig. 5. The original quadric is in the top left of the figure. It is clear that using a hyperboloid prior is the best option for this data.

Empirical Data

We provide some examples which demonstrate the use of our algorithm on real data from point clouds extracted from a visual structure from a motion pipeline. In this scenario, regions of the point cloudcan be left without any reconstructed points.

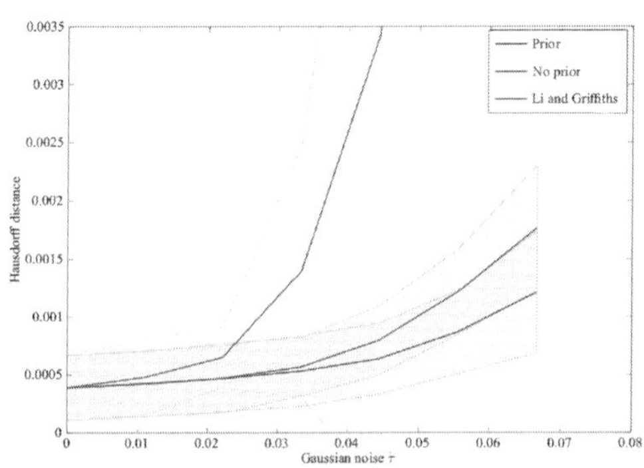

Figure 4: The Hausdorff distance for a varying amount of Gaussian noise. The shaded regions represent one standard deviation from the mean.

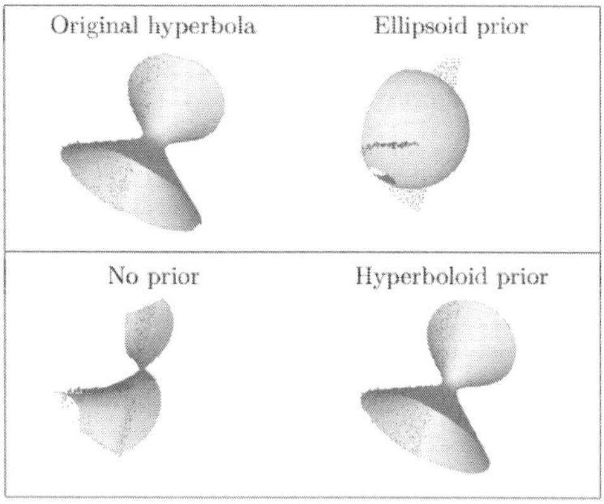

Figure 5: Fitting a hyperbola using a variety of different priors. The original points are shown as black dots.

An example is shown in Fig. 6. A smooth surface must be fitted to the roof of the phone box, presenting a challenge for most surface fitting algorithms due to the noisy and shapeless extracted points. Previous algorithms either do not fit at all, such as the one in Li and Griffiths [12], or do not allow control over the curvature of the data, which is the case when using no prior. The surface gradient can be chosen to match the roof curvature by varying σ and produces a final result as shown.

Some examples of the algorithm on point cloud data are shown in Fig. 7. The point clouds are computed from a collection of photographs comprising a post box, rugby ball, and roof. The surfaces are visually plausible even when there are large regions of missing data. This is particularly visible on the roof data, for which only the front is visible in each of the photographs.

CONCLUSIONS

We conclude by observing that state of the art methods for quadric fitting give reasonable results on noisy point clouds. Our algorithm provides a means to enforce a prior, allowing the algorithm to better fit the quadric that the points were drawn from, particularly when there is missing data or a large amount of noise. This has a practical use for real datasets, since it allows a user to specify the curvature of the surface where there are few points available.

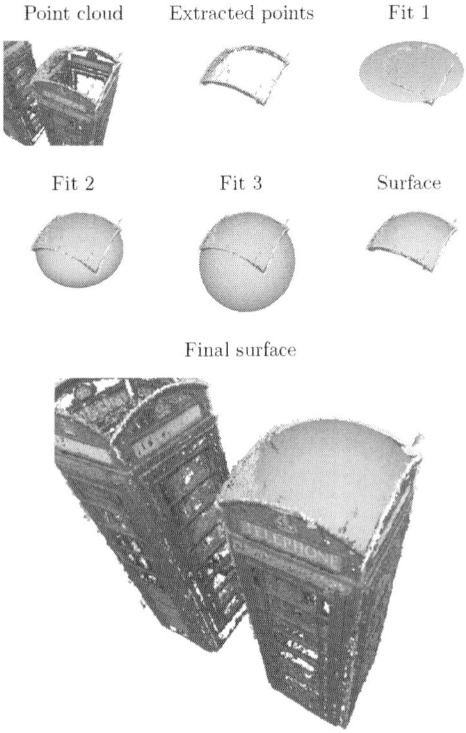

Figure 6: Phone box dataset. The top left shows the original point cloud and the extracted points that a quadric must be fitted to, followed by 3 examples with varying sigma values and the final surface at the bottom right.

Appendix A Symmetry of the Matrix A

The matrix A is assumed symmetric without loss of generality. This is possible since the function:

$$f : \mathbb{R}^{M\times M} \to \mathbb{R}^{M} \tag{47}$$

$$\boldsymbol{A} \mapsto \boldsymbol{x}^{\mathrm{T}}\boldsymbol{A}\boldsymbol{x} \tag{48}$$

is not injective for any $x \in \mathbb{R}^{M}$. For example, if we take

$$\boldsymbol{B} = \frac{1}{2}(\boldsymbol{A} + \boldsymbol{A}^{\mathrm{T}}) \tag{49}$$

which is a symmetric matrix, then

$$x^{T}Bx = x^{T}(\frac{1}{2}(A + A^{T}))x \tag{50}$$

$$= \frac{1}{2}(x^{T}Ax + x^{T}A^{T}x) \tag{51}$$

Since $x^{T}A^{T}x$ is a scalar, it is also symmetric, and so

$$x^{T}A^{T}x = (x^{T}A^{T}x)^{T} \tag{52}$$

$$= x^{T}(x^{T}A^{T})^{T} \tag{53}$$

$$= x^{T}Ax \tag{54}$$

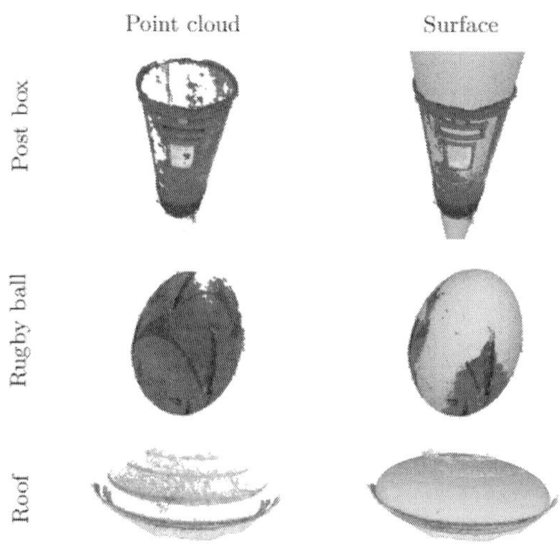

Figure 7: Some examples of the quadric fitting algorithm on point cloud data.

It then follows that

$$x^{T}Bx = x^{T}Ax \tag{55}$$

The choice of symmetry then follows, because if A, b, and c satisfy $x^{T}Ax + b^{T}x + c = 0$, then so does B, and B is symmetric.

REFERENCES

1. Taubin, G. Estimation of planar curves, surfaces, and nonplanar space curves defined by implicit equations with applications to edge and range image segmentation. IEEE Transactions on Pattern Analysis and Machine Intelligence Vol. 13, No. 11, 1115–1138, 1991.
2. Ahn, S. J.; Rauh, W.; Cho, H. S.; Warnecke, H.- J. Orthogonal distance fitting of implicit curves and surfaces. IEEE Transactions on Pattern Analysis and Machine Intelligence Vol. 24, No. 5, 620–638, 2002.
3. Andrews, J.; S′equin, C. H. Type-constrained direct fitting of quadric surfaces. Computer-Aided Design and Applications Vol. 11, No. 1, 107–119, 2014.
4. Atieg, A.; Watson, G. A. A class of methods for fitting a curve or surface to data by minimizing the sum of squares of orthogonal distances. Journal of Computational and Applied Mathematics Vol. 158, No. 2, 277–296, 2003.
5. Gander, W.; Golub, G. H.; Strebel, R. Leastsquares fitting of circles and ellipses. BIT Numerical Mathematics Vol. 34, No. 4, 558–578, 1994.
6. Luk′acs, G.; Martin, R.; Marshall, D. Faithful leastsquares fitting of spheres, cylinders, cones and tori for reliable segmentation. In: Lecture Notes in Computer Science, Vol. 1406. Burkhardt, H.; Neumann, B. Eds. Springer Berlin Heidelberg, 671–686, 1998.
7. Petitjean, S. A survey of methods for recovering quadrics in triangle meshes. ACM Computing Surveys (CSUR) Vol. 34, No. 2, 211–262, 2002.
8. Ruiz, O.; Arroyave, S.; Acosta, D. Fitting of analytic surfaces to noisy point clouds. American Journal of Computational Mathematics Vol. 3, No. 1A, 18–26, 2013.
9. Rouhani, M.; Sappa, A. D. Implicit polynomial representation through a fast fitting error estimation. IEEE Transactions on Image Processing Vol. 21, No. 4, 2089–2098, 2012.
10. Pratt, V. Direct least-squares fitting of algebraic surfaces. ACM SIGGRAPH Computer Graphics Vol. 21, No. 4, 145–152, 1987.
11. Fitzgibbon, A.; Pilu, M.; Fisher, R. B. Direct least square fitting of ellipses. IEEE Transactions on Pattern Analysis and Machine Intelligence Vol. 21, No. 5, 476–480, 1999.
12. Li, Q.; Griffiths, J. G. Least squares ellipsoid specific fitting. In: Proceedings of Geometric Modeling and Processing, 335–340, 2004.

13. Halir, R.; Flusser, J. Numerically stable direct least squares fitting of ellipses. In: Proceedings of the 6th International Conference in Central Europe on Computer Graphics and Visualization, Vol. 98, 125– 132, 1998.
14. Dai, M.; Newman, T. S.; Cao, C. Least-squares-based fitting of paraboloids. Pattern Recognition Vol. 40, No. 2, 504–515, 2007.
15. Subrahmonia, J.; Cooper, D. B.; Keren, D. Practical reliable Bayesian recognition of 2D and 3D objects using implicit polynomials and algebraic invariants. IEEE Transactions on Pattern Analysis and Machine Intelligence Vol. 18, No. 5, 505–519, 1996.
16. Cohen-Steiner, D.; Alliez, P.; Desbrun, M. Variational shape approximation. ACM Transactions on Graphics Vol. 23, No. 3, 905–914, 2004.
17. Wu, J.; Kobbelt, L. Structure recovery via hybrid variational surface approximation. Computer Graphics Forum Vol. 24, No. 3, 277–284, 2005.
18. Yan, D.-M.; Liu, Y.; Wang, W. Quadric surface extraction by variational shape approximation. In: Lecture Notes in Computer Science, Vol. 4077. Kim, M.-S.; Shimada, K. Eds. Springer Berlin Heidelberg, 73–86, 2006.
19. Schnabel, R.; Wahl, R.; Klein, R. Efficient RANSAC for point-cloud shape detection. Computer Graphics Forum Vol. 26, No. 2, 214–226, 2007.
20. Li, Y.; Wu, X.; Chrysathou, Y.; Sharf, A.; Cohen-Or, D.; Mitra, N. J. GlobFit: Consistently fitting primitives by discovering global relations. ACM Transactions on Graphics Vol. 30, No. 4, Article No. 52, 2011.
21. Ohtake, Y.; Belyaev, A.; Alexa, M.; Turk, G.; Seidel, H.-P. Multi-level partition of unity implicits. In: Proceedings of ACM SIGGRAPH 2005 Courses, Article No. 173, 2005.
22. Hamann, B. Curvature approximation for triangulated surfaces. In: Computing Supplementum, Vol. 8. Farin, G.; Noltemeier, H.; Hagen, H.; Kndel, W. Eds. Springer Vienna, 139–153, 1993.
23. Vona, M.; Kanoulas, D. Curved surface contact patches with quantified uncertainty. In: Proceedings of IEEE/RSJ International Conference on Intelligent Robots and Systems, 1439–1446, 2011.
24. Powell, M. W.; Bowyer, K. W.; Jiang, X.; Bunke, H. Comparing curved-surface range image segmenters. In: Proceedings of the 6th

International Conference on Computer Vision, 286–291, 1998.

25. Petersen, K. B.; Pedersen, M. S. The Matrix Cookbook. Technical University of Denmark. Version: November 15, 2012. Available at https://www.math.uwaterloo. ca/vhwolkowi/matrixcookbook. pdf.
26. Lott, G. K. Direct orthogonal distance to quadratic surfaces in 3D. IEEE Transactions on Pattern Analysis and Machine Intelligence Vol. 36, No. 9, 1888–1892, 2014.
27. Eberly, D. Distance from a point to an ellipse, an ellipsoid, or a hyperellipsoid. Geometric Tools, LLC, 2011. Available at http://www. geometrictools.com/ Documentation/DistancePointEllipseEllipsoid. pdf.
28. Kwatra, V.; Han, M. Fast covariance computation and dimensionality reduction for sub-window features in images. In: Lecture Notes in Computer Science, Vol. 6312. Daniilidis, K.; Maragos, P.; Paragios, N. Eds. Springer Berlin Heidelberg, 156–169, 2010.

14

Multivariate Rational Response Surface Approximation of Nodal Displacements of Truss Structures

Shan Chai[1], **Xiang-Fei Ji**[2], **Li-Jun Li**[1] **& Ming Guo**[3]

[1]School of Transportation and Vehicle Engineering, Shandong University of Technology, Zibo, 255049, China
[2]Key Laboratory of Electronic Equipment Structure Design, Xidian University, Xi'an, 710071, China
[3]Shandong Linglong Tire Co., Ltd., Zhaoyuan, 265400, China

ABSTRACT

Polynomial-basis response surface method has some shortcomings for truss structures in structural optimization, concluding the low fitting accuracy and the great computational effort. Based on the theory of approximation, a response surface method based on Multivariate Rational Function basis (MRRSM) is proposed. In order to further reduce the computational workload of MRRSM, focusing on the law between the cross-sectional area and the nodal displacements of truss structure, a conjecture that the determinant of the stiffness matrix and the corresponding elements of adjoint matrix involved in displacement determination are polynomials with the

Citation: Chai, S., Ji, X. F., Li, L. J., & Guo, M. (2018). Multivariate rational response surface approximation of nodal displacements of truss structures. Chinese Journal of Mechanical Engineering, 31(1), 1. (14 pages)

Copyright: © Open Access. This article is distributed under the terms of the Creative Commons Attribution 4.0 International (CC BY 4.0) License.

same order as their respective matrices, each term of which is the product of cross-sectional areas, is proposed. The conjecture is proved theoretically for statically determinate truss structure, and is shown corrected by a large number of statically indeterminate truss structures. The theoretical analysis and a large number of numerical examples show that MRRSM has a high fitting accuracy and less computational effort. Efficiency of the structural optimization of truss structures would be enhanced.

Keywords: Multivariate rational function, Response surfaces method, Truss structures, Structure optimization

INTRODUCTION

The response surface methodology (RSM) explores the relationships between several explanatory variables and one or more response variables. The method was introduced by Box and Wilson in 1951. The main idea of RSM is to use a sequence of designed experiments to obtain an optimal response. Box and Wilson suggest using a second-degree polynomial model to do this [1].

Because response surface methodology can establish the unknown function between design variables and response variables, it has been widely applied to optimization. Choon-Man Jang and Ka-Ram Choi [2] carried out the optimization of the blower impeller by using the response surface method (RSM). Heng Jiang et al. [3] carried out a dynamic and static multi-objective optimization of a vertical machining center based on response surface method. Quadratic polynomials are employed to construct response surface (RS) model, which reflects the relationship between design inputs and structural response outputs, according to the response outputs of these samples obtained by analyzing the dynamic and static characteristics of the machining centre at these samples with the software ANSYS. In the field of vehicle optimization, RSM also get a widely application [4,5,6]. Chun-Lin Wang et al. [7] carried out a variable curvature blade of fire pump based on experimental design theory and response surface approximation. Dong-Sheng Jia et al. [8] established the mass function of impeller of water valve-controlled hydrodynamic coupling with RSM and optimized the impeller key structural parameters. Response surface method is applied in the field of robust optimization design [9, 10]. Reliability analysis has an improvement based on response surface method [11, 12].

Generally speaking, in the mathematical models of structural optimization, the objective function is the explicit function of the design variables and the constraint function is the implicit function of the design variables. This characteristic is also the important difference between the structure optimization problem and the mathematical program problem and it greatly increases the complexity of the structure optimization problem [13]. In order to solve this problem, the implicit function must be made explicit and the response surface methodology, is the most commonly used method [14]. Hong-Wu et al. [15] used response surface methodology to establish the function expressions of deflection and structural weight,and optimized the truss structure. Roux et al. [16] carried out RS-based optimization of truss structures and concluded that RS accuracy depends on the range of the design area and the form of the explicit function. Zhang et al. [17] establish a fitted regression model for the dynamic transmission error(DTE) fluctuations to quantify the relationship between modification amounts and DTE fluctuations by using response surface method.

In order to increase the accuracy and efficiency, Fan, et al. [18] derived a criterion for judging the existence of cross terms and proposed an adaptive response surface method. Houten et al. [19] considered fifth-order polynomials to check if higher-order polynomials improved RSM accuracy. Venter et al. [20] solved plate optimization problems using higher order RS approximations.

Sui and his collaborators [21,22,23,24,25] developed the central point accurate response surface method where the response value at the center point of the approximate model is equal to the true response value. This method was successfully applied to optimization of membrane structures, 2D-continuum structures, plate and shell structures subject to frequency constraints, and multidisciplinary optimization. Using the property of the Kreisselmerier-Steinhauser function, RS approximation was built with function derivation, Taylor expansion near experimental points and numerical iteration, trying to minimize the largest difference between response function values and true response values.

In the finite element displacement method, the nodal displacement is the basic unknown quantity. The stress, dynamic response and so on can be derived from the displacement, so this paper is devoted to discuss the response surface of truss node displacement, its research method and results can be easily extended to other types of structures. It should be noted that the analytical form of polynomial-basis response surface approximation is often very different from the functions describing the actual displacement field. As this may result in low quality approximation, it is important to assess the real analytical form of the displacement field. This study will show that the nodal displacement field determined by finite element method for truss structures can be expressed as multivariate rational functions. In view of this, a response surface methodology based on multivariate rational functions will be developed. It will be illustrated that the determinant of the stiffness matrix and the corresponding elements of adjoint matrix involved in displacement determination are polynomials depending on cross-sectional areas, with the same order as their respective matrices, each term of which is the product of cross-sectional areas. The validity of the proposed approach is verified for many design examples of statically determinate and statically indeterminate truss structures.

POLYNOMIAL-BASIS RESPONSE SURFACE APPROXIMATION FOR NODAL DISPLACEMENT OF TRUSS STRUCTURES

Polynomial-basis RSM

The commonly used forms of polynomial-basis response surface methodology are as follows.

Linear form

$$\tilde{y} = \beta_0 + \sum_{i=1}^{k} \beta_i x_i, \tag{1}$$

separable quadratic form

$$\tilde{y} = \beta_0 + \sum_{i=1}^{k} \beta_i x_i + \sum_{i=1}^{k} \beta_{ii} x_i^2, \tag{2}$$

and the complete quadratic form (containing cross terms)

$$\tilde{y} = \beta_0 + \sum_{i=1}^{k} \beta_i x_i + \sum_{i=1}^{k} \beta_{ii} x_i^2 + \sum_{i=1}^{k-1} \sum_{j=i+1}^{k} \beta_{ij} x_i x_j, \tag{3}$$

where $\tilde{y}$ is the response function to be constructed, and β are the unknown RS coefficients.

RSM Evaluation Standards

The most commonly used parameters used to evaluate accuracy of response surface approximation are as follows.

The multiple fitting coefficient

$$R^2 = 1 - \frac{\delta}{\gamma}, \quad (4)$$

and the modified multiple fitting coefficient

$$R^2_{\text{adj}} = 1 - \left(\frac{m-1}{m-k}\right)\frac{\delta}{\gamma}, \quad (5)$$

where δ is the sum of the square of the difference between the true response value and the response estimated value; γ is the sum of the square of the difference between the response estimated value and the response mean value; m is the total number of experimental points; and k is the number of unknown parameters:

$$\gamma = Y^{\mathrm{T}}Y - \left(\sum_{i=1}^{m} y_i\right)^2 / m. \quad (6)$$

R^2 can reflect the extent to which the response surface matches the data given. It ranges from 0 to 1, and the closer it is to 1, the smaller the influence of various errors.

For the approximation accuracy of check points, the most commonly used parameters are as follows.

The maximum relative error

$$\zeta_{\max} = \max\{\zeta_r | r = 1, 2, \cdots, S\}, \quad (7)$$

minimum relative error

$$\zeta_{\min} = \min\{\zeta_r | r = 1, 2, \cdots, S\}, \quad (8)$$

average relative error

$$\zeta_{\text{avg}} = \frac{\sum_{r=1}^{M} \zeta_r}{S}, \quad (9)$$

where S is the total number of check points. The relative error is $\zeta_r = |\delta_r - \delta'_r| / \delta'r \times 100\%$, δ_r' is the true response value of the rth check ponit, δ_r is the response estimated value of the rth check point.

Evaluation of Polynomial-Basis RSM Accuracy

In the following, fitting accuracy of polynomial-basis RS method is analyzed for some simple truss structures. The cross-sectional area of the truss structure is selected as the design variable. The cross-sectional area ranges from 0.01 m^2 to 0.05 m^2, and all cross sections are circular. The elastic modulus is $E = 2.1 \times 10^{11}$ Pa, and Poisson's ratio is 0.3. The nodal displacement is taken as the structural response. The value of the ith design variable of design points P_D is $AP_{D,i} \in \{0.01,0.02,0.03,0.04,0.05\}$ and the value of the ith design variable of check points P_E is $AP_{E,i} \in \{0.015,0.025,0.035,0.045\}$.

Example 1: Planar 2-bar Truss Structure

The first example is the very simple two-bar truss structure shown in Figure 1.

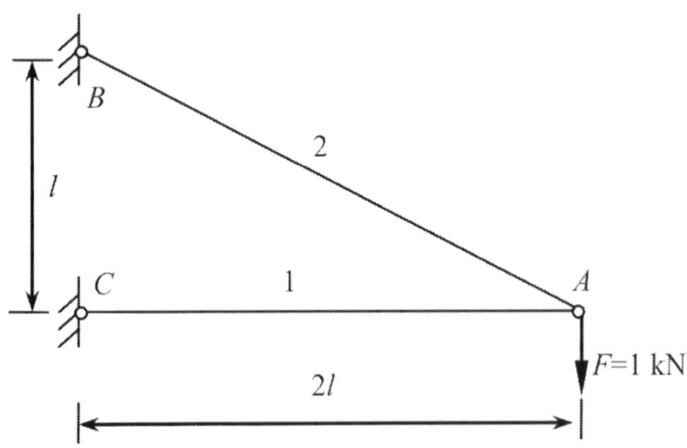

Figure 1. Schematics of the planar 2-bar truss structure.

The given length of bar 1 is 2 l, and the length of bar 2 is 5–√15l. Let l = 1 m.

1. Analytic expression of nodal displacement

Using the analytic method of structural mechanics, the displacement of node A along the force F direction is as follows:

$$\delta_F = \frac{8F}{EA_1} + \frac{5\sqrt{5}F}{EA_2}$$
$$= 1.0 \times 10^{-7} \times \left(0.380\,952\,381\frac{1}{A_1} + 0.532\,397\,138\frac{1}{A_2}\right), \quad 10$$

2. Nodal displacement computed with polynomial-basis RSM

Trial values of cross-sectional areas are given in input to Eq. (10) to determine the displacement of node A in the direction of applied force. Design points form the matrix Xwhile the corresponding displacements form the response matrix Y. The RS coefficient vector β is determined for various types of polynomial-basis approximation of displacement δ_F.

Linear form:

$$\tilde{y} = 9.240\,1 \times 10^{-6} - 7.047\,6 \times 10^{-5} A_1 - 9.849\,3 \times 10^{-5} A_2. \quad 11$$

Separable quadratic form:

$$\tilde{y} = 1.373\,1 \times 10^{-5} - 2.310\,2 \times 10^{-4} A_1 - 3.228\,6 \times 10^{-4} A_2 + 0.002\,7 A_1^2 + 0.003\,7 A_2^2. \quad 12$$

Complete quadratic form:

$$\tilde{y} = 1.373\,1 \times 10^{-5} - 2.310\,2 \times 10^{-4} A_1 - 3.228\,6 \times 10^{-4} A_2 + 0.002\,7 A_1^2 + 0.003\,7 A_2^2 - 8.456\,8 \times 10^{-18} A_1 A_2. \quad 13$$

The solving accuracy of polynomial-basis response surface method and the selection of the response surface forms are shown in Table 1.

Table 1. Accuracy of linear, separable and complete quadratic RS models for the 2-bar truss structure problem

Evaluation standards	**Linear form**	**Separable quadratic form**	**Complete quadratic form**
Multiple fitting coefficient R^2	0.7962	0.9727	0.9713
Modified multiple fitting coefficient R^2_{adj}	0.7962	0.9727	0.9713
Maximum relative error ζ_{max}/%	37.29	15.56	15.56
Minimum relative error ζ_{min}/%	10.13	6.56	6.56
Average relative error ζ_{avg}/%	23.56	11.59	11.59

Using the MATLAB software, the surface comparison figure of the analytic value and the response surface estimated value and relative error surface figure of the check points are shown in Figure 2.

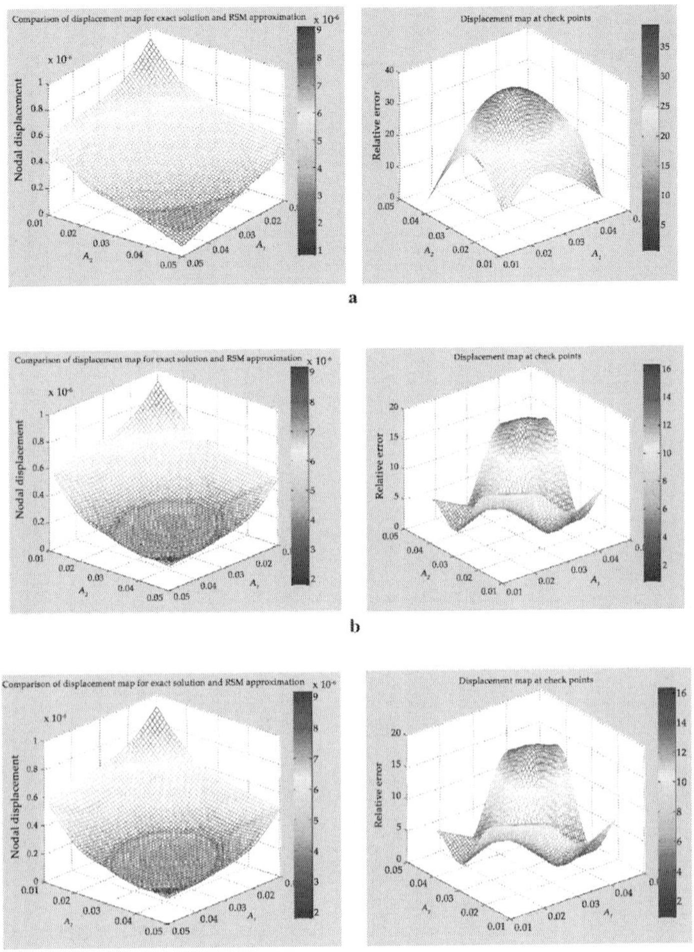

Figure 2. Evaluation of RS approximation accuracy for the 2-bar structure problem. **a** Linear RS model. **b**Separable quadratic RS model. **c** Full quadratic RS model.

The accuracy of these RS models is evaluated in Table 1 by means of the performance indicators described in Section 2.2. Figure 2 compares the variation of nodal displacement over the design space determined from the analytical expression (10) or by means of the three RS models. In particular, the MATLAB plots show the displacement map and the relative error map. The following conclusions can be drawn from the results obtained for this example.

- The analytical form of the polynomial-basis RS is completely different from the analytic solution (10) which depends on the inverse of cross-sectional areas.

- The coefficients of the linear RS model are small. The quadratic RS models are much more accurate than the linear RS model although relative errors at check points remain rather high. Adding cross-terms in the quadratic model does not yield substantial improvements.

Example 2: Planar 8-bar Truss Structure

In order to check the validity of the above analysis for more complicated structures, the statically determinate planar 8-bar truss structure shown in Figure 3 was considered. The length of all bars is 2 m, except for bars 6 and 8 which are 22–√22 m long. Loads and kinematic constraints for this structure are shown in Figure 3.

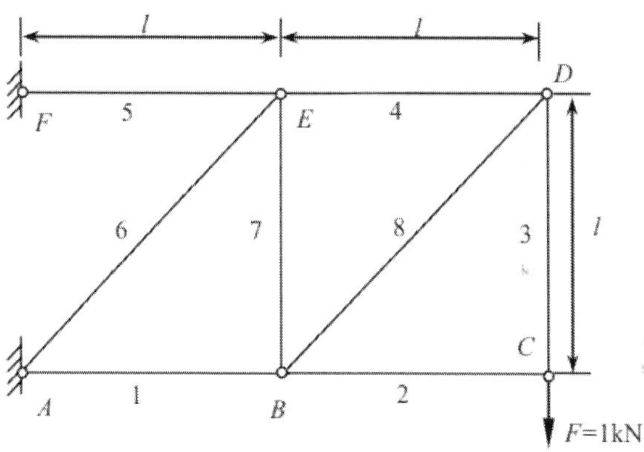

Figure 3. Schematics of the planar 8-bar truss structure.

The analytical solution for the displacement of node C along the direction of applied force F is as follows:

$$\begin{aligned}\delta_y &= \frac{8F}{EA_1} + \frac{2F}{EA_2} + \frac{2F}{EA_3} + \frac{2F}{EA_5} + \frac{4\sqrt{2}F}{EA_6} + \frac{2F}{EA_7} + \frac{4\sqrt{2}F}{EA_8} \\ &= 1.0 \times 10^{-7} \times \left[0.380952381\frac{1}{A_1} + 0.09523809524 \right. \\ &\quad \left. \times \left(\frac{1}{A_2} + \frac{1}{A_3} + \frac{1}{A_5} + \frac{1}{A_7}\right) + 0.269340119 \times \left(\frac{1}{A_6} + \frac{1}{A_8}\right)\right] \qquad 14\end{aligned}$$

Polynomial-basis response surface approximations of displacement give the following results.

Linear form:

$$\tilde{y} = 1.315\,81 \times 10^{-5} - 1.0 \times 10^{-5} \times (7.047\,6A_1 + 1.761\,9 \times (A_2 + A_3 + A_5 + A_7) + 4.982\,8 \times (A_6 + A_8)). \quad 15$$

Separable quadratic form:

$$\tilde{y} = 1.955\,2 \times 10^{-5} - 1.0 \times 10^{-4} \times (2.310\,2A_1 + 5.775\,5 \times (A_2 + A_3 + A_5 + A_7) + 1.633\,4 \times (A_6 + A_8)) + 0.002\,7 \times A_1^2 + 6.689\,3 \times 10^{-4} \times (A_2^2 + A_3^2 + A_5^2 + A_7^2) + 0.001\,9 \times (A_6^2 + A_8^2). \quad 16$$

For the sake of brevity, the expression of complete quadratic RS is not reported in the article. Performance indicators for these RS models are listed in Table 2. The quadratic RS models are again much more accurate than the linear RS model but relative errors at check points remain rather high, thus not satisfying the requirements on overall accuracy. Adding cross-terms in the quadratic model again does not yield substantial improvements. As expected, the analytical form of nodal displacement (14) is completely different from the expressions of RS models.

Table 2. Accuracy of linear, separable and complete quadratic RS models for the 8-bar truss structure problem

Evaluation standards	Linear form	Separable quadratic form	Complete quadratic form
Multiple fitting coefficient R^2	0.8132	0.9773	0.9773
Modified multiple fitting coefficient R^2_{adj}	0.8132	0.9773	0.9773
Maximum relative error ζ_{max}/%	37.292	15.563	15.563
Minimum relative error ζ_{min}/%	0.039 6	2.292×10^{-4}	2.292×10^{-4}
Average relative error ζ_{avg}/ %	16.096	4.747	4.747

Since the two truss examples refer to statically determinate structures, a statically indeterminate structure is now analyzed to draw more general conclusions.

Example 3: Planar 6-bar Truss Structure

Figure 4 shows the schematic of a statically indeterminate 6-bar truss structure, in which l is the length and l = 1 m. The displacement to be computed is that of node B along the direction of applied force F.

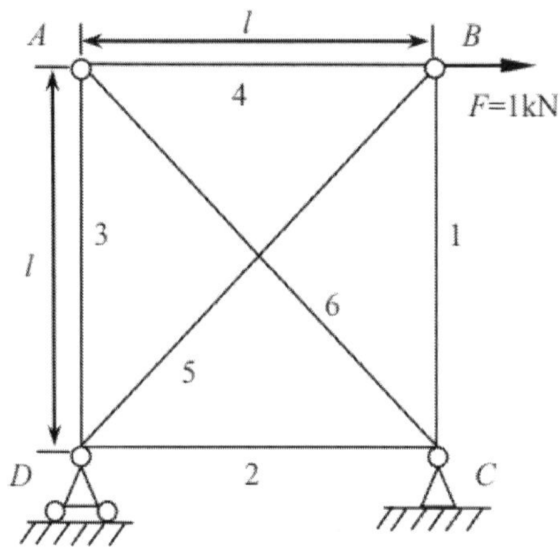

Figure 4. Schematics of the planar 6-bar truss structure.

The accuracy of linear, separable quadratic and full quadratic RS models is listed in Table 3.

Table 3. Accuracy of linear, separable and complete quadratic RS models for the 6-bar truss structure problem

Evaluation standards	**Linear form**	**Separable quadratic form**	**Complete quadratic form**
Multiple fitting coefficient R^2	0.8026	0.8991	0.9996
Modified multiple fitting coefficient R^2_{adj}	0.8026	0.8990	0.9996
Maximum relative error ζ_{max}/%	28.68	22.66	14.43
Minimum relative error ζ_{min}/%	2.76	3.98	5.01
Average relative error ζ_{avg}/%	18.93	11.82	8.96

The data listed in the table lead to similar conclusions as in the case of statically determinate structures. However, introducing cross-terms in the quadratic RS model now yields a more clear improvement over separable quadratic RS model.

The above examples demonstrate that nodal displacements of both statically determinate and indeterminate truss structures are multivariate rational functions of cross-sectional areas. This model is totally different from the analytical form of polynomial-basis response surfaces which cannot hence describe correctly the nodal displacement field of truss structures.

Evaluation of the Accuracy of Polynomial-Basis RSM with Reciprocal Variables

The displacement of a certain node in a given direction computed using Mohr's theorem can be expressed as

$$\delta_i = \sum_{j=1}^{M} \frac{N_j^0 N_j l_j}{EA_j}, \tag{17}$$

where N_j is the internal force of jth element under the action of true load, N^0_j is the internal force of the jth element under the action of unit force with the computed displacement direction, l_j is the element length, and M is the number of elements.

Let $x_j=1/A_j$, consequently, Eq. (17) can be expressed as

$$\delta_i = \sum_{j=1}^{n} \frac{N_j^0 N_j l_j}{E} x_j. \tag{18}$$

The most commonly used RS approximation function of nodal displacements of truss structures can be obtained by taking the reciprocal variables x_j as design variables and substituting them into Eqs. (1), (2) and (3).

In order to analyze the accuracy of the polynomial-basis response surface method with reciprocal variables, the accuracy analysis is carried out by reusing Example 1–3 as the examples.

Example 4: Planar 2-bar Truss Structure

For the 2-bar truss structure studied in Example 1, the computing accuracy of polynomial-basis RSM with reciprocal variables is listed in Table 4.

Table 4. Accuracy of linear, separable and complete quadratic reciprocal variables RSM models for the 2-bar truss structure problem

Evaluation standards	Linear form	Separable quadratic form	Complete quadratic form
Multiple fitting coefficient R^2	1	1	1
Modified multiple fitting coefficient R^2_{adj}	1	1	1
Maximum relative error ζ_{max}/%	5.78×10^{-14}	1.68×10^{-12}	1.13×10^{-12}
Minimum relative error ζ_{min}/%	1.14×10^{-14}	1.09×10^{-12}	5.22×10^{-13}
Average relative error ζ_{avg}/%	3.38×10^{-14}	1.40×10^{-12}	8.21×10^{-13}

Example 5: Planar 8-bar Truss Structure

For the planar 8-bar truss structure studied in Example 2, accuracy of polynomial-basis RSM with reciprocal variables is listed in Table 5.

Table 5. Accuracy of linear, separable and complete quadratic reciprocal variables RS models for the 8-bar truss structure problem

Evaluation standards	Linear form	Separable quadratic form	Complete quadratic form
Multiple fitting coefficient R^2	1	1	1
Modified multiple fitting coefficient R^2_{adj}	1	1	1
Maximum relative error ζ_{max}/%	2.76×10^{-11}	2.70×10^{-9}	4.72×10^{-11}
Minimum relative error ζ_{min}/%	9.90×10^{-12}	4.15×10^{-14}	5.23×10^{-13}
Average relative error ζ_{avg}/%	1.90×10^{-11}	6.70×10^{-10}	2.10×10^{-11}

Example 6: Planar 6-bar Truss Structure

For the planar 6-bar truss structure studied in Example 3, accuracy of polynomial-basis RSM with reciprocal variables is listed in Table 6.

Table 6. Accuracy of linear, separable and complete quadratic reciprocal variables RS models for the 6-bar truss structure problem

Evaluation standards	**Linear form**	**Separable quadratic form**	**Complete quadratic form**
Multiple fitting coefficient R^2	0.897	0.908	0.998
Modified multiple fitting coefficient R^2_{adj}	0.897	0.908	0.998
Maximum relative error ζ_{max}/%	23.194	27	5.967
Minimum relative error ζ_{min}/%	7.07×10^{-4}	3.36×10^{-3}	3.167×10^{-4}
Average relative error ζ_{avg}/%	5.096	4.065	1.024

The following conclusions can be drawn from the results of the above these examples.

- For the statically determinate structure, the evaluation coefficients are very good. The all relative errors are less than 10^{-10}, therefore the approximation accuracy is very high. This is mainly because the polynomial-basis RSM with reciprocal variables equation have a high consistency in the form with the analytic solution, and the response surface equation reflects the true relationship between displacement and design variables.
- For the statically indeterminate structure, the approximation accuracy is very low. This is mainly because internal forces are changing as the changes of design variables, and these changes don't be reflected in RSM equation, which lead to the difference in the form between RSM approximation with reciprocal variables and analytic solution. In order to improve the approximation accuracy, a new method of high accuracy response surface should be developed.

DETERMINATION OF NODAL DISPLACEMENTS OF TRUSS STRUCTURES

According to the finite element method theory, the balance equation is as

$$K\delta = F. \tag{19}$$

The displacement equation is as

$$\delta = K^{-1}F. \tag{20}$$

Where the inverse matrix can be obtained by the formula as

$$K^{-1} = \frac{\text{adj}\mathbf{K}}{|\mathbf{K}|}. \tag{21}$$

According to the theory of determinants, the n-order determinant is a number firmed by n^2elements of a_{ij}.

$$|\mathbf{K}| = \begin{vmatrix} a_{11} & a_{12} & \cdots & a_{1n} \\ a_{21} & a_{22} & \cdots & a_{2n} \\ \vdots & \vdots & & \vdots \\ a_{n1} & a_{n2} & \cdots & a_{nn} \end{vmatrix} = \sum_{j_1 \cdots j_n} (-1)^{\tau(j_1 j_2 \cdots j_n)} a_{1j_1} a_{2j_2} \cdots a_{nj_n}, \tag{22}$$

where $\tau(j_1 j_2 \cdots j_n)$ is the inverted sequence number of natural numbers 1,2,…,n, and $\sum_{j1 \cdots jn}$ is the sum of all permutations of natural numbers 1,2,…,n.

Since elements a_{ij} of stiffness matrix include products of linear terms corresponding to cross-sectional areas, the determinant can be simplified as

$$|\mathbf{K}| = f(A^n), \tag{23}$$

where $f(A^n)$ is an n-order polynomial of cross-sectional area variables A, and n is the order of the stiffness matrix, equal to the DOF of the structure system.

Elements of the adjoint matrix adj**K** are polynomials $g_{ij}(A^{n-1})$ which are one order lower than the determinant of the stiffness matrix. Thus, the form of the adjoint matrix is

$$\text{adj}\mathbf{K} = \begin{bmatrix} g_{11}(A^{n-1}) & \cdots & g_{1n}(A^{n-1}) \\ \vdots & & \vdots \\ g_{n1}(A^{n-1}) & \cdots & g_{nn}(A^{n-1}) \end{bmatrix} \tag{24}$$

Combining all of the above formulas, it can be obtained the equation to analytically determine nodal displacements of truss structures:

$$\delta = \mathbf{K}^{-1}\mathbf{F} = \frac{\text{adj}\,\mathbf{K}}{|\mathbf{K}|}\mathbf{F}. \tag{25}$$

Consequently, the displacement of a certain node in a given direction can be expressed as

$$\delta_i = \frac{1}{|\mathbf{K}|}\sum_{j=1}^{n} g_{ij}(A^{n-1})F_j = \frac{1}{f(A^n)}\sum_{j=1}^{n} g_{ij}(A^{n-1})F_j. \tag{26}$$

CONJECTURE ABOUT THE DETERMINANT EXPRESSION OF STIFFNESS MATRIX AND ELEMENTS OF ADJOINT MATRIX

The previous derivation indicate that the determinant of the stiffness matrix of the truss structure and the elements of adjoint matrix involved in the determination of nodal displacements are respectively n-order and (n–1)-order polynomials. Stiffness matrix determinant is an algebraic sum including n! terms each of which is the product of n factors belonging to different rows and columns. The sign of each term depends on the corresponding inverted sequence number. Therefore, there are at least n! coefficients. Similarly, determinants of adjoint matrix elements are algebraic sums of (n–1)! terms each of which corresponds to the product of (n–1) factors belonging to different rows and different columns. The adjoint matrix is an n-order square matrix. Equation (26) implies that 2n! function evaluations must be done. However, if matrix order changes, the number of coefficients also will change in a factorial way. Hence, if the equation order gets large, too many combinations should be considered. Variable linking and relationships between cross-sectional area variables may greatly reduce computational complexity of combinations and multivariate rational function response surfaces can be built based on these relationships. The above mentioned relationships will be illustrated by the following examples.

1. For the 2-bar truss structure considered in Example 1, displacement of node A in the direction of applied load F has the following analytical expression:

$$\delta_{AF} = \frac{1}{|\mathbf{K}|}\sum_{j=1}^{n} g_{1j}(A^{n-1})F_j = \frac{1}{f(A^2)}g_{12}(A)F, \tag{27}$$

where,

$$f(A^2) = \frac{E^2}{50l^2}\sqrt{5}A_1A_2, \quad g_{12}(A) = \frac{E}{l}\left(\frac{A_1}{2} + \frac{4\sqrt{5}A_2}{25}\right)$$

2. For the statically indeterminate 6-bar truss structure considered in Example 3, the analytical expression for the nodal displacement of node B along the direction of applied force F is as follows:

$$\delta_{BF} = \frac{1}{|\mathbf{K}|}\sum_{j=1}^{n} g_{1j}(A^{n-1})F_j = \frac{1}{f(A^5)}g_{33}(A^4)F, \qquad 28$$

where

$$f(A^5) = \frac{E^5}{16l^5}(2A_1A_2A_3A_5A_6 + 2A_1A_2A_4A_5A_6 + A_1A_3A_4A_5A_6 + A_2A_3A_4A_5A_6 + 4\sqrt{2}A_1A_2A_3A_4A_5 + 4\sqrt{2}A_1A_2A_3A_4A_6),$$

$$g_{33}(A^4) = \frac{E^4}{16l^4}(16A_1A_2A_3A_4 + 4\sqrt{2}A_1A_2A_3A_6 + 4\sqrt{2}A_1A_2A_4A_6 + 4\sqrt{2}A_1A_3A_4A_5 + 2A_1A_3A_5A_6 + 2A_1A_4A_5A_6 + 4\sqrt{2}A_2A_3A_4A_5 + 2A_2A_3A_5A_6 + 2A_2A_4A_5A_6).$$

The following conjecture can be made from the previous relationships. The determinant of stiffness matrix and the adjoint matrix term entailed in the determination of a certain nodal displacement are polynomials of the same order as the corresponding matrices. Each term of these polynomials is the product of first power terms corresponding to cross-sectional areas.

$$\begin{cases} f(A^n) = \sum_{i=1}^{m_1} \alpha_i A_{i1}A_{i2}\cdots A_{in}, \\ g(A^{n-1}) = \sum_{i=1}^{m_2} \beta_i A_{i1}A_{i2}\cdots A_{i(n-1)}, \end{cases} \qquad 29$$

where

$$i1 \neq i2 \neq i3 \neq \cdots \neq i(n-1) \neq in.$$

The number of DOF n is the same as the order of stiffness matrix; m is the number of design variables. A^n are the possible combinations of cross-sectional areas A_i selected from the set of m design variables; m_1 is the number of these combinations, namely, $m_1 = C_m^n = m!/[(m-n)!n!]$. Similarly, A^{n-1} are other possible combinations of cross-sectional areas A_i; m_2 is the number of these combinations, namely, $m_2 = C_m^{n-1} = m!/[(m-n+1)!\cdot(n-1)!]$; α and β are unknown coefficients to be determined.

The number of terms included in matrix stiffness determinant decreases from n! to m!/[(m−n)!n!]; while the number of terms in the corresponding adjoint matrix decreases from (n-1)! to m!/[(m−n+1)!·(n−1)!], thus reducing the computational cost of the fitting process.

Let's now consider a statically determinate structure with m bars and n DOFs. Since there are no redundant bars, it holds m = n. Hence, in Eq. (29), it holds $m_1 = C_m^n = C_n^n = 1,\ m_2 = C_m^{n-1} = C_n^{n-1} = n$ and the equation simplifies to

$$\begin{cases} f(A^n) = b_0 A_1 A_2 \cdots A_n, \\ g_{ij}(A^{n-1}) = a_1 A_1 A_2 \cdots A_{n-1} + a_2 A_1 A_3 \cdots A_n + \cdots + \\ \qquad a_i A_1 A_i \cdots A_n + \cdots + a_n A_2 A_3 \cdots A_n. \end{cases} \tag{30}$$

By substituting Eq. (30) into Eq. (26), it follows

$$\delta_i = \frac{C_1}{A_1} + \frac{C_2}{A_2} + \cdots + \frac{C_n}{A_n} = \sum_{j=1}^{n} \frac{C_j}{A_j}, \tag{31}$$

where $C_j = a_{n-j+1}/b_0$.

For statically determinate structures, if the external force P is known, internal forces N_j developed in each bar are constant. If the generalized displacement δ_i in the direction i is computed using Mohr's theorem, a unit generalized force N^0_i is applied in the direction i and the corresponding internal forces N_j in each bar also are constant. Element lengths l_j and cross-sectional areas A_j are constant values as well. Therefore, nodal displacements of a truss structure can be expressed as

$$\delta_i = \sum_{j=1}^{n} \frac{N_j N_j^0 l_j}{EA_j} = \sum_{j=1}^{n} \frac{C_j}{A_j}, \tag{32}$$

where the constant C is equal to $C_j = N_j N^0_j l_j / E$.

The consistency of Eqs. (31) and (32) proves that it may be reasonable to approximate nodal displacements of statically determinate truss structures with multivariate rational response surfaces. The following examples will prove the validity of this conclusion also for statically indeterminate trusses.

Example 7: Planar 10-bar Truss Structure

For the planar 10-bar truss structure shown in Figure 5, the general form of the determinant of the stiffness matrix is for this structure is as follows:

$$\begin{aligned} f(A^n) = \frac{E^8}{l^8}(&8A_1A_{10}A_2A_3A_4A_5A_6A_8 \\ &+ 8A_1A_{10}A_2A_3A_4A_5A_7A_9 \\ &+ A_1A_{10}A_2A_3A_4A_6A_7A_9 + \cdots \\ &+ \beta A_aA_bA_cA_dA_eA_fA_gA_h + \cdots). \end{aligned} \quad 33$$

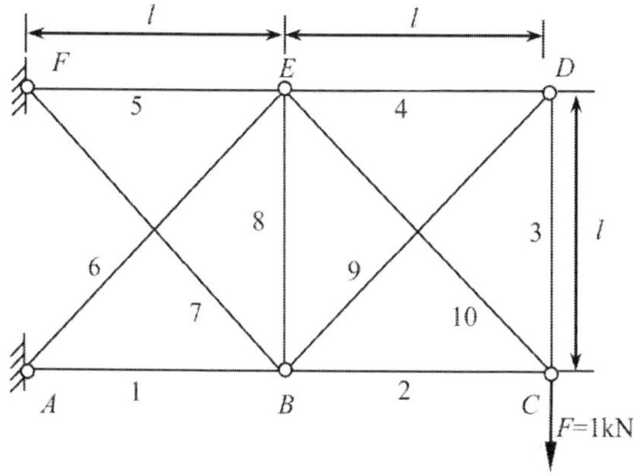

Figure 5. Schematics of the planar 10-bar truss structure.

The corresponding adjoint matrix element involved in the determination of the displacement of node C in the direction of applied load F is as follows:

$$\begin{aligned} g_{67}(A^{n-1}) = \frac{E^7}{l^7}(&8A_1A_{10}A_2A_3A_4A_5A_7 \\ &+ 8A_1A_{10}A_2A_3A_4A_5A_9 \\ &+ 8A_1A_{10}A_2A_3A_4A_8A_7 + \cdots \\ &+ \alpha A_aA_bA_cA_dA_eA_fA_g + \cdots). \end{aligned} \quad 34$$

where {a, b, c, d, e, f, g, h} ∈ {1, 2, 3, 4, 5, 6, 7, 8, 9, 10}, α and β are the unknown coefficients of the products of design variable A_i, and a ≠ b ≠ c ≠ d ≠ e ≠ f ≠ g ≠ h.

Example 8: Spatial 18-bar Truss Structure

For the spatial 18-bar truss structure shown in Figure 6, determinant of stiffness matrix is as follows:

$$f(A^n) = \frac{E^{12}}{l^{12}}(2175A_1A_2A_{17}A_{18}A_5A_9A_{11}A_{12}A_{15}A_{16}A_{13}A_{14}$$
$$+ 76\sqrt{2}A_1A_{17}A_{18}A_4A_5A_9A_{10}A_{12}A_{15}A_{16}A_{11}A_{14}$$
$$+ 128\sqrt{2}A_1A_3A_6A_8A_2A_9A_{10}A_4A_{11}A_{12}A_{15}A_{16} + \cdot$$
$$+ \alpha A_aA_bA_cA_dA_eA_fA_gA_hA_iA_jA_kA_l + \cdots + \cdots). \quad 35$$

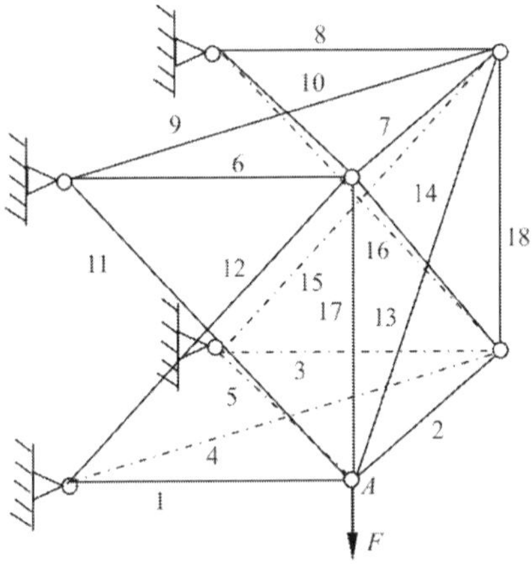

Figure 6. Schematics of the Spatial 18-bar truss structure.

The corresponding adjoint matrix element involved in the determination of the displacement of node A in the direction of applied load F is as follows:

$$g_{12}(A^{n-1}) = \frac{E^{11}}{l^{11}}(4096\sqrt{2}A_1A_3A_6A_2A_7A_{17}A_5A_{10}A_{11}A_{12}A_{15}$$
$$+ 2048A_1A_3A_6A_2A_7A_5A_4A_{11}A_{16}A_{13}A_{14}$$
$$+ 768A_1A_3A_6A_{17}A_{18}A_5A_4A_9A_{10}A_{13}A_{16}$$
$$+ 6144\sqrt{2}A_1A_3A_2A_7A_{17}A_{18}A_5A_9A_{10}A_{11}A_{16} + \cdots$$
$$+ \alpha A_aA_bA_cA_dA_eA_fA_gA_hA_iA_jA_k + \cdots), \quad 36$$

where {a, b, c, d, e, f, g, h, i, j, k, l, m, n, o, p, q, r} ∈ {1, 2, 3, 4, 5, 6, 7, 8, 9, 10, 11, 12, 13, 14, 15, 16, 17, 18}, α and β are the unknown coefficients of the products of design variable A_i, and a ≠ b ≠ c ≠ d ≠ e ≠ f ≠ g ≠ h ≠ i ≠ j ≠ k ≠ l ≠ m ≠ n ≠ o ≠ p ≠ q ≠ r.

Other examples not documented in this article for the sake of brevity confirm the validity of the conjecture made on the mutual relationship between the stiffness matrix of the truss structure and adjoint matrix elements.

MULTIVARIATE RATIONAL RESPONSE SURFACE APPROXIMATION OF TRUSS NODAL DISPLACEMENTS

Response surface approximation of truss nodal displacements can be built as follows:

$$\begin{aligned}\delta_i &= \frac{1}{f(A^n)}\sum_{j=1}^{n} g_{ij}(A^{n-1})F_j \\ &= \frac{\sum_{j=1}^{n}\left(\sum_{i=1}^{m_2}\beta_{ij}A_{i1}A_{i2}\cdots A_{in-1}\right)F_j}{\sum_{i=1}^{m_1}\alpha_i A_{i1}A_{i2}\cdots A_{in}} \\ &= \frac{\sum_{F_j\neq 0}\sum_{i=1}^{m_2}\beta'_{ij}A_{i1}A_{i2}\cdots A_{in-1}}{\sum_{i=1}^{m_1}\alpha_i A_{i1}A_{i2}\cdots A_{in}},\end{aligned} \tag{37}$$

where α_i and β'_{ij} are the RS fitting coefficients to be determined with $\beta'_{ij}=\beta_{ij}F_j$; m_1 and m_2, respectively, are the number of terms included in polynomials $f(A^n)$ and $g_{ij}(A^{n-1})$.

Suitable design points are defined including the cross-sectional areas of the M bars forming the truss. For the generic design "C", it holds $A_k = A^C_k$ with $k = 1, 2,\ldots, M$.

After the displacement δ^C_i corresponding to design points is determined from finite element analysis, Eq. (37) is rewritten as follows:

$$\delta_i^C\left(\sum_{i=1}^{m_1}\alpha_i A_{i1}^C A_{i2}^C\cdots A_{in}^C\right) = \sum_{F_j\neq 0}^{n}\sum_{i=1}^{m_2}\beta'_{ij}A_{i1}^C A_{i2}^C\cdots A_{i(n-1)}^C \tag{38}$$

or

$$\delta_i^C\left(\sum_{i=1}^{m_1}\alpha_i A_{i1}^C A_{i2}^C\cdots A_{in}^C\right) - \sum_{F_j\neq 0}^{n}\sum_{i=1}^{m_2}\beta'_{ij}A_{i1}^C A_{i2}^C\cdots A_{i(n-1)}^C = 0. \tag{39}$$

The above expression is a homogeneous equation. Since the ratio between numerator and denominator in Eq. (37) stays the same if both terms are multiplied by the same quantity, RS model coefficients can be normalized with respect to any fitting coefficient. In order to keep generality, let us set $\alpha_1 = 1$ and rewrite Eq. (39) as follows:

$$\sum_{F_j \neq 0}^{n} \sum_{i=1}^{m_2} \beta'_{ij} A_{i1}^{C} A_{i2}^{C} \cdots A_{in-1}^{C} - \delta_i^{C} \left(\sum_{i=2}^{m_1} \alpha_i A_{i1}^{C} A_{i2}^{C} \cdots A_{in}^{C} \right) = \delta_i^{C} A_{i1}^{C}. \qquad 40$$

Selecting M design points ($M \geq m_1 + Km_2$, where K is the number of loads F_j that are not equal to zero), a system of M linear equations in the unknown coefficients α_i and β'_{ij} is obtained:

$$\begin{cases} \sum_{F_j \neq 0}^{n} \sum_{i=1}^{m_2} \beta'_{ij} A_{i1}^{C_1} A_{i2}^{C_1} \cdots A_{i(n-1)}^{C_1} - \delta_i^{C_1} \left(\sum_{i=2}^{m_1} \alpha_i A_{i1}^{C_1} A_{i2}^{C_1} \cdots A_{in}^{C_1} \right) = \delta_i^{C_1} A_{i1}^{C_1}, \\ \sum_{F_j \neq 0}^{n} \sum_{i=1}^{m_2} \beta'_{ij} A_{i1}^{C_2} A_{i2}^{C_2} \cdots A_{i(n-1)}^{C_2} - \delta_i^{C_2} \left(\sum_{i=2}^{m_1} \alpha_i A_{i1}^{C_2} A_{i2}^{C_2} \cdots A_{in}^{C_2} \right) = \delta_i^{C_2} A_{i1}^{C_2}, \\ \vdots \\ \sum_{F_j \neq 0}^{n} \sum_{i=1}^{m_2} \beta'_{ij} A_{i1}^{C_M} A_{i2}^{C_M} \cdots A_{i(n-1)}^{C_M} - \delta_i^{C_M} \left(\sum_{i=2}^{m_1} \alpha_i A_{i1}^{C_M} A_{i2}^{C_M} \cdots A_{in}^{C_M} \right) = \delta_i^{C_M} A_{i1}^{C_M}. \end{cases} \qquad 41$$

The determination of these coefficients by solving this system can be done in two ways.

- If $M > m_1 + Km_2$, the number of design points is larger than the number of unknown coefficients. Solving accuracy is maximized by using the least square method.
- If $M = m_1 + Km_2$, the number of design points is equal to the number of unknown coefficients and classical linear algebraic solvers can be utilized.

According to best uniform approximation theory, quality of RS approximation improves as the fitting model resembles the real structural response. Since the multivariate rational RS model developed in this study reproduces the actual analytical model behind determination of nodal displacements of truss structures, the present model can give results very close to the real structural response of a truss structure.

VERIFICATION OF MULTIVARIATE RATIONAL RS MODEL FOR TRUSS DISPLACEMENT APPROXIMATION

The accuracy of the multivariate rational response surface approximation model developed in this study is verified in some design examples. The cross-sectional area of truss elements is selected as the design variable which can range between 0.01 and 0.05 m^2; cross sections of all elements are supposed to be circular. The Young's modulus is 2.1×10^{11} Pa while Poisson's ratio is 0.3. Node displacement is taken as the structural response to be approximated. Value of design variables selected for design points P_D is $AP_{D,i} \in \{0.01,0.02,0.03,0.04,0.05\}$ while value of design variables selected for check points P_E is $AP_{E,i} \in \{0.015,0.025,0.035,0.045\}$.

Example 9: Planar 2-bar Truss Structure

For the truss studied in Example 1, accuracy of multivariate rational RS model is evaluated in Table 7. Figure 7(a) compares the displacement maps corresponding to the true structural response and the approximate model. The relative errors on nodal displacements evaluated at check points are shown in Figure 7(b).

Table 7. Accuracy of multivariate rational RS model for the 2-bar truss structure problem

Evaluation standards	**Node A**
Multiple fitting coefficient R^2	1
Modified multiple fitting coefficient R^2_{adj}	1
Maximum relative error ζ_{max}/%	6.58×10^{-14}
Minimum relative error ζ_{min}/%	1.15×10^{-14}
Average relative error ζ_{avg}/%	2.05×10^{-14}

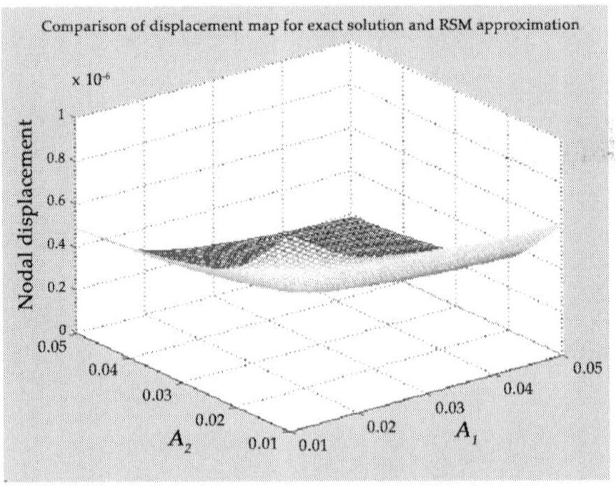

a

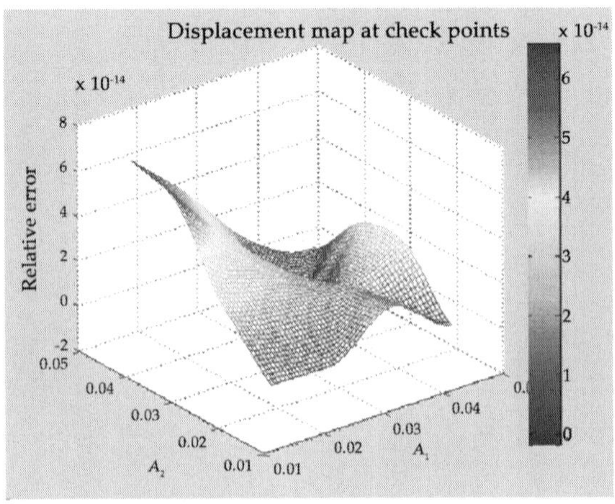

b

Figure 7. Evaluation of multivariate rational RS approximation accuracy for the 2-bar structure problem. **a**Comparison of displacement map for exact solution and RSM approximation. **b** Displacement error map at check points.

Example 10: Planar 6-bar Truss Structure

The structural layout, kinematic constraints and loads are the same as for Example 3. Table 8lists the values of performance indicators for the multivariate rational RS model.

Table 8. Accuracy of multivariate rational RS model for the 6-bar truss structure problem

Evaluation standards	**Node B**
Multiple fitting coefficient R^2	1
Modified multiple fitting coefficient R^2_{adj}	1
Maximum relative error ζ_{max}/%	3.63×10^{-4}
Minimum relative error ζ_{min}/%	1.51×10^{-7}
Average relative error ζ_{avg}/%	7.46×10^{-5}

Example 11: Planar 12-bar Truss Structure

The planar truss shown in Figure 8 is a statically indeterminate structure constrained by fixed-hinged supports around the edge. There are only two degrees of freedom for this design example. Performance indicators for the multivariate rational RS model, evaluated with respect to the solution provided by the commercial finite element software ANSYS, are listed in Table 9.

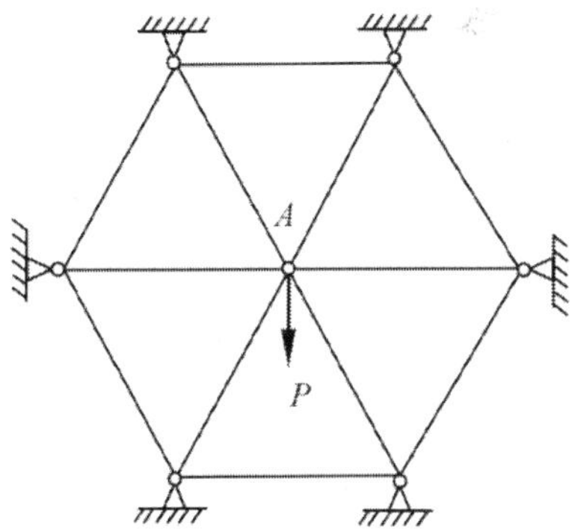

Figure 8. Schematics of the planar 12-bar truss structure.

Table 9. Accuracy of multivariate rational RS model for the 12-bar truss structure problem

Evaluation standards	**Node A**
Multiple fitting coefficient R^2	1

Modified multiple fitting coefficient R^2_{adj}	1
Maximum relative error ζ_{max}/%	2.03×10^{-3}
Minimum relative error ζ_{min}/%	1.54×10^{-9}
Average relative error ζ_{avg}/%	1.67×10^{-4}

Example 12: Monolithic Tower

For the monolithic tower schematized in Figure 9, the lateral load P_2 simulates the action of the wind while the vertical load P_1 simulates the weight of the tower body and other concentrated loads. Two loading conditions are considered: (i) $P_1 = 0$, $P_2 = 1$ kN; (ii) $P_1 = 1$ kN, $P_2 = 1$ kN. The displacement to be computed is the lateral displacement of node A and B respectively. Performance indicators for the multivariate rational RS model, evaluated with respect to the solution provided by the commercial finite element software ANSYS, are listed in Table 10.

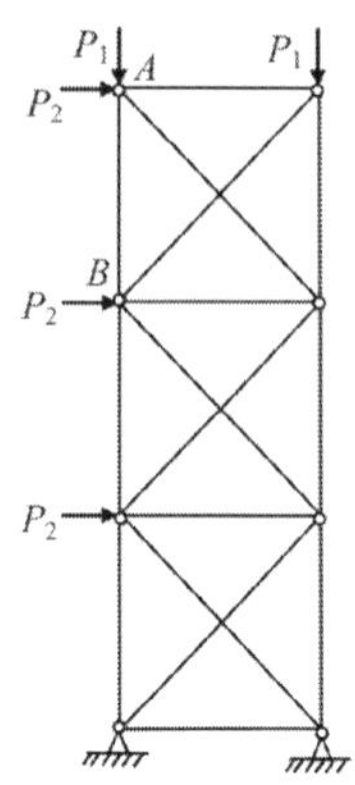

Figure 9. Schematics of the monolithic tower.

Table 10. Accuracy of multivariate rational RS model for the monolithic tower problem

Evaluation standards	**Condition 1**		**Condition 2**	
	Node A	**Node B**	**Node A**	**Node B**
Multiple fitting coefficient R^2	1	1	1	1

Modified multiple fitting coefficient R^2_{adj}	1	1	1	1
Maximum relative error ζ_{max}/%	3.67×10^{-4}	3.63×10^{-2}	4.83×10^{-3}	7.08×10^{-3}
Minimum relative error ζ_{min}/%	2.10×10^{-3}	7.30×10^{-8}	7.65×10^{-8}	2.76×10^{-8}
Average relative error ζ_{avg}/%	5.58×10^{-4}	2.45×10^{-3}	6.72×10^{-4}	5.18×10^{-4}

Example 13: Secondary Mirror Supporting Structure

The secondary mirror supporting structure schematized in Figure 10 is a highly lightweight triangular structure in which six supporting bars form a closed bar system in an end to end way between the primary and secondary mirror plates. The weight of the objects on the top of the supporting structure is equivalent to the concentrated loads on the three top nodes (see Figure 10). The displacement to be computed is the vertical displacement of node A. Performance indicators for the multivariate rational RS model, evaluated with respect to the solution provided by the commercial finite element software ANSYS, are listed in Table 11.

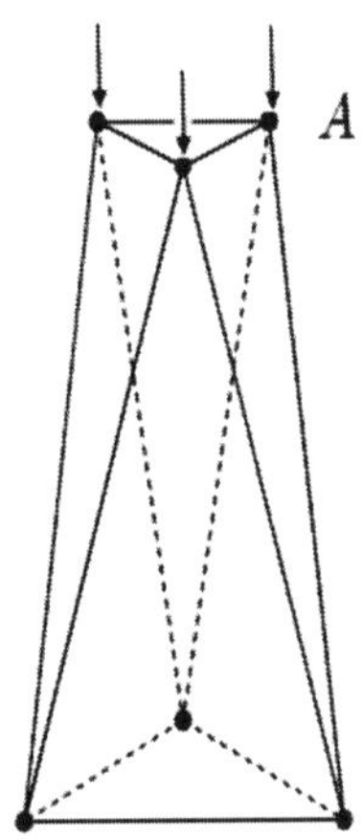

Figure 10. Schematics of the secondary mirror supporting structure.

Table 11. Accuracy of multivariate rational RS model for the secondary mirror supporting structure problem

Evaluation standards	**Node A**
Multiple fitting coefficient R^2	1
Modified multiple fitting coefficient R^2_{adj}	1
Maximum relative error ζ_{max}/%	1.24×10^{-4}
Minimum relative error ζ_{min}/%	7.37×10^{-6}
Average relative error ζ_{avg}/%	4.48×10^{-5}

It can be seen that relative error between response results obtained by multivariate rational RS approximation and ANSYS is very small for all design examples. Evaluation coefficients are always very high, thus satisfying engineering requirements. The same conclusion holds true for both statically determinate and indeterminate truss structures.

CONCLUSIONS

- The polynomial-basis response surface approximation is different from the real function between cross-sectional area and nodal displacement, relative error ranging from 4.747% to 23.56% which does not meet the engineering requirement.
- Polynomial-basis RSM with reciprocal variables has a great improvement on the statically determinate structure, with relative errors less than 10^{-10}. However, for the statically indeterminate structure, the approximation accuracy is still low.
- Relationship between the nodal displacement and the cross-sectional area is derived, following the conjecture about the determinant expression of stiffness matrix and elements of adjoint matrix.
- On the basis of previous conjecture, multivariate rational response surface approximation of truss nodal displacements is established. The fitting accuracy of both statically determinate structure and statically indeterminate structure is greatly improved, with relative errors less than 10^{-3}.

REFERENCES

1. André I Khuri, Siuli Mukhopadhyay. Response surface methodology. Wiley Interdisciplinary Reviews Computational Statistics, 2010, 2(2): 128-149.
2. Choon-Man Jang, Ka-Ram Choi. Optimal design of splitters attached to turbo blower impeller by RSM. Journal of Thermal Science, 2012, 21(3): 215-222.
3. Heng Jiang, Yi-Sheng Guan, Zhi-Cheng Qiu, et al. Dynamic and static multi-objective optimization of a vertical machining center based on response surface method. Journal of Mechanical Engineering, 2011, 47(11): 125-133. (in Chinese)
4. Hui Lu, De-Jie Yu, Zhan Xie, et al. Optimization of vehicle disc brakes stability based on response surface method. Journal of Mechanical Engineering, 2013, 49(9): 55-60. (in Chinese)
5. Yong Zhang, Guang-Yao Li, Zhi-Hua Zhong. Design optimization on lightweight of full vehicle based on moving least square response surface method. Journal of Mechanical Engineering, 2008, 44(11): 192-196. (in Chinese)
6. Guang-Yong Sun, Guang-Yao Li, Zhi-Hua Zhong, et al. Optimization design of multi-objective particle swarm in crashworthiness based on sequential response surface method. Journal of Computational Mechanics, 2009, 45(2): 224-230. (in Chinese)
7. Chun-Lin Wang, Hai-Bo Peng, Jian Ding et al. Optimization for S-type blade of fire pump based on response surface method. Journal of Mechanical Engineering, 2013, 49(10): 170-177. (in Chinese)
8. Dong-Sheng Jia, Guo-Ping Li, Xiao-Hui Bai. Structural optimization design of impeller of hydrodynamic coupling based on method of response surface. Coal Mine Machinery, 2010, 31(12): 3-5. (in Chinese)
9. Xiao-Jing Wu, Wei-Wei Zhang, Hua Xiao et al. A Robust Aerodynamic Design for Airfoil Based on Response Surface Method. Engineering Mechanics, 2015, 32(2): 250-256. (in Chinese)
10. Wei Wang, Wen-Hui Fan, Tian-Qing Chang, et al. Robust collaborative optimization method based on dual-response surface. Chinese Journal of Computational Mechanics, 2009, 22(2): 169-176.
11. Min-Jiu Zhang, Ru-Jia Gu, Hong-Shuang Li, et al. Application of response surface method on structural reliability analysis of pin joint

in composites. Journal of Computational Mechanics, 2016, 33(5): 711-716. (in Chinese)

12. Guang-Chen Bai, Cheng-Wei Fei. Distributed collaborative response surface method for mechanical dynamic assembly reliability design. Chinese Journal of Computational Mechanics, 2013, 26(6): 1160-1168.
13. Shan Chai, Xiao-Jiang Shang, Xian-Yue Gang. Optimization design method and application of engineering structure. Beijing: China Railway Publishing House, 2015. (in Chinese).
14. Yun-Kang Sui, Hui-Ping Yu. Improvement of response surface method and its application to engineering optimization. Beijing: Science Press, 2011. (in Chinese)
15. Hong-Hu Chen, Cheng Tian, Gong-Gong Peng, et al. Truss structure optimization design based on response surface method. Machine Design and Research, 2016, 32(1): 48-54. (in Chinese)
16. W J Roux, N Stander, R T Haftka. Response surface approximations for structural optimization. International Journal of Numerical Methods in Engineering, 1998, 42(3): 517-534.
17. Jun Zhang, Fan Guo. Statistical modification analysis of helical planetary gears based on response surface method and Monte Carlo simulation. Chinese Journal of Computational Mechanics, 2015, 28(6): 1194-1203.
18. Wen-Liang Fan, Chun-Tao Zhang, Zheng-Liang Li, et al. An adaptive response surface method with cross terms. Engineering Mechanics, 2013, 30(4): 68-72. (in Chinese)
19. M H V Houten, A J G Schoofs, D H V Campen. Response surface techniques in structural optimization. 1st,WCSMO World Congress of Structural and Multidisciplinary Optimization, Goslar, Germany, 28 May-2 June, 1995: 89-95.
20. G Venter, R T Haftka, J H Starnes. Construction of response surface approximations for design optimization. 6th AIAAA/NASA/ISSMO Symposium on Multidisciplinary Analysis and Optimization, Bellevue, Washington, September 4-6, 1996: 548-564.
21. Yun-Kang Sui, Hai-Bo Bai. Plate shell structural optimization based on central point's accurate response surface method. Journal of Machine Design, 2005, 22(11): 10-13. (in Chinese)
22. Nai-Mei Tang, Yun-Kang Sui. Membrane structure cross-sectional optimization based on response surface method. Academic Conference

of Chinese Society of Theoretical and Applied Mechanics, Beijing, China, August 26-28, 2005: 1157-1158. (in Chinese)

23. Yun-Kang Sui, Shan-Po Li. The application of improved RSM in shape optimization of two-dimension continuum. Engineering Mechanics, 2006, 23(10): 1-6. (in Chinese)

24. Yan-Fei Zhao, Yun-Kang Sui, Hong-Ling Ye. Optimum design of shell structures with frequency constraints based on response surface methodology. Journal of Beijing University of Technology, 2006, 32(s1): 19-23. (in Chinese)

25. Hui-Ping Yu, Yun-Kang Sui, Li Ding, et al. Min-Max distance response surface and Kreisselmerier-Steinhauser function. Science & Technology Review, 2010, 28(1): 63-68. (in Chinese)

15

Visual Data Mining Based on Differential Topology: A Survey

Osamu Saeki[1] & Shigeo Takahashi[2]

[1]Institute of Mathematics for Industry, Kyushu University, 744, Motooka, Nishi-ku, 819-0395, Fukuoka, Japan
[2]Graduate School of Information Science and Technology, The University of Tokyo, 7-3-1 Hongo, Bunkyo-ku, Tokyo, 133-8565, Japan

ABSTRACT

In this article, we describe techniques for visual data mining based on differential topology. Data scientists have been working long on the analysis of data obtained from a wide variety of sources. The data is often represented as discrete sample points of a function $R^n \rightarrow R^m$, while the dimensions of the data domain and range have rapidly increased due to recent advancement in computational power and measurement technology. Mathematical formulations of differential topology effectively help us to analyze such data in a hierarchical fashion and to visually extract significant features from it. We present new algorithms and application examples as well as existing ones, including the authors' recent results, so that we can fully elucidate the potential power of this approach especially in data visualization.

Citation: Saeki, O., & Takahashi, S. (2014). Visual data mining based on differential topology: a survey. Pacific Journal of Mathematics for Industry, 6(1), 4. (10 pages

Copyright: © Open Access. This article is distributed under the terms of the Creative Commons Attribution 4.0 International (CC BY 4.0) License.

INTRODUCTION

Recent advancement of supercomputers provides us with easy access to highly parallelized and efficient computational environments, and thus the associated computer simulation usually produces large-sized high-dimensional data. This so-called big data usually helps us to reproduce exact and detailed behaviors of target phenomena, while it may hide its important features in the sense that we cannot easily locate such features locally due to its huge size. This often leads us to a negative spiral of data size and understanding of the target phenomena. Actually, techniques for visually identifying important features within such big data have been demanded, since they allow us to elucidate such important features as visualization images effectively.

As the demand increases, the concept of differential topology has attracted considerable attention of many researchers since 1990s. Indeed, this mathematical framework has a unique capability to analyze such big data in a hierarchical fashion by extracting its topological structure as a higher layer over the entire data. Recently, the advantage has encouraged more and more researchers to go into this research topic because of high potential for the analysis of complicated data. In particular, the data representation based on differential topology is thought of as one of top ten innovative techniques in the visualization community. In this article, we present approaches for extracting features of differential topology from discrete samples of a function $R^n \rightarrow R^m$, including existing and new algorithms together with their application examples.

OVERVIEW

In general, scientific simulation data can be represented as a set of discrete points obtained by sampling a function $f:R^n \rightarrow R^m$, where $\mathbf{R}^n$ and $\mathbf{R}^m$ indicate the data domain and range, respectively. Data analysis techniques based on differential topology basically provide an effective means of encoding topological changes in the inverse image $f^{-1}(c)(\subset R^n)$ of faccording to the change of $\mathbf{c}(\in \mathbf{R}^m)$. One of the significant advantages of this type of data analysis is its ability to extract not only singular points where local topological changes arise in the inverse image but also their connectivity over the entire data for understanding its global structure.

Consider a discrete elevation model, which represents a set of discrete samples on a terrain surface as shown in Figure 1. In this case, we can denote the latitude and longitude of each sample by x and y coordinates, respectively, and its corresponding height as z=f(x,y). Note that the singular point of a scalar function is defined as a point where the cross-sectional contour produces a topological change, and thus it corresponds to a peak, a pit, or a pass as shown in Figure 2. Furthermore, we can also track the connectivity among the singular points to precisely locate the splitting and merging in the contour. In practice, the topological change in the inverse image of some scalar value is often described as a tree structure called a contour tree[1] as shown in Figure 3, which has successfully been applied to various situations such as analyzing contour topology of terrain surfaces [17], designing visualization parameters for volume rendering [19], and extracting spatial embeddings of contours in volumes [20].

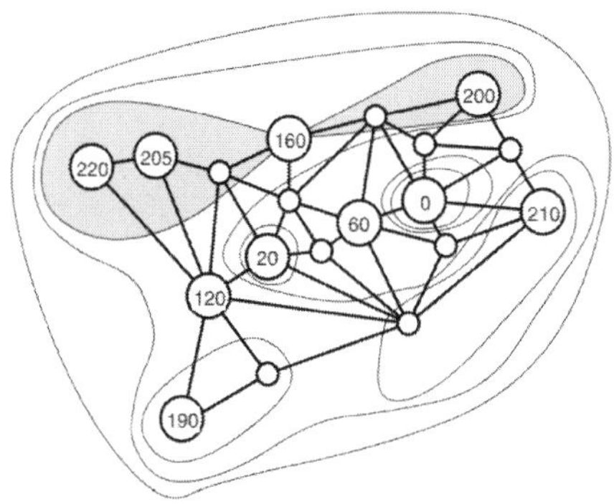

Figure 1. Analyzing a discrete elevation model. Features of differential topology are extracted by tracking the topological change in cross-sectional contours with respect to the height. The boundary of the yellow region represents one of the cross sections of the terrain surface at height 160: the region itself corresponds to height ≥160 and corresponds to the subgraph in the join tree marked yellow in Figure 3.

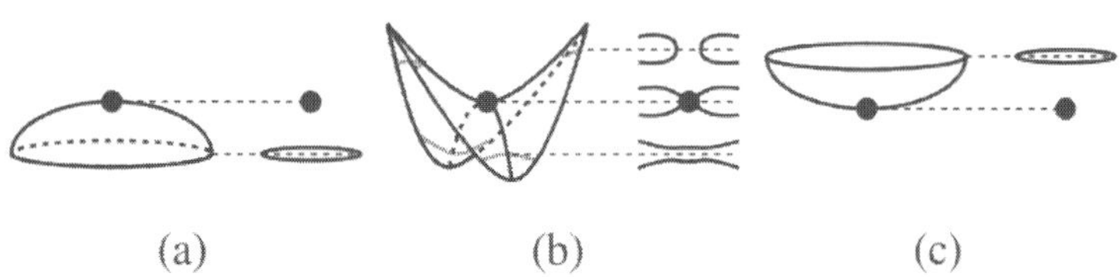

Figure 2. Singular points of a 2-dimensional scalar field. (a) A peak (index 2). (b) A pass (index 1). (c)A pit (index 0).

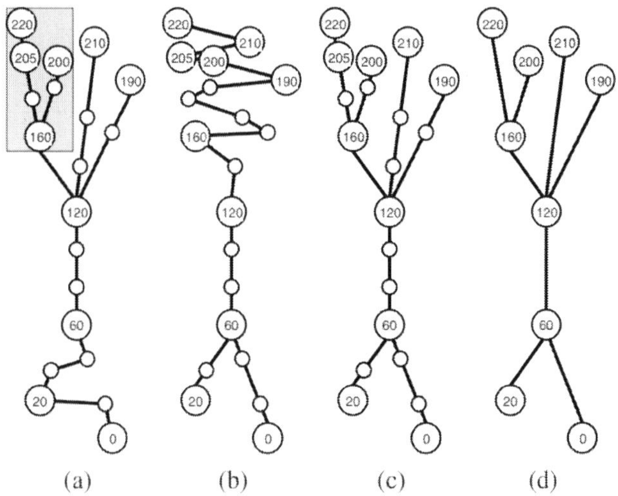

Figure 3. Extracting a contour tree. (a) A join tree. (b) A split tree. (c) An augmented contour tree. (d) A final contour tree.

Constructing algorithms for computing such contour trees of scalar functions $R^n \rightarrow R$ started from the mid 1990s, and early in 2000s Carr et al. [4] presented an excellent algorithm for constructing contour trees, which is fully sophisticated both in simplicity and in computational complexity and thus has been commonly employed so far. Although this algorithm intrinsically has no limitation on the dimension n of the data domain, it still suffers from practical implementation issues when n is equal to 4 or more due to the troubles in composing connectivity among discrete samples.

This technical problem was alleviated by algorithms that were developed in the late 2000s, which aggressively incorporated dimensionality reduction approaches from the field of machine learning. These new algorithms successfully eliminate the limit on the number of data dimensions, and thus enabled computation of contour trees even from time-varying and higher-dimensional data samples.

On the other hand, the dimension of the data range has long been limited to 1, since it is considerably difficult to track the change in the inverse image in terms of multiple function values simultaneously. Of course, we can construct a contour tree for each of the multiple function values individually while this scheme does not provide any information about relationships among the multiple function values. For example, it is preferable to extract some coherent relationships between temperature and pressure in some space when we try to extract features of differential topology from data samples of a multivariate function $R^3 \rightarrow R^2$ in this case. In this article, we also show that recent technical challenges can solve this problem by extracting the topological change in a fiber, which is defined to be the intersection among the inverse images of the given multiple function values.

These three types of approaches will be described in the following sections.

ANALYZING SAMPLES OF A FUNCTION $R^2 \rightarrow R$ OR $R^3 \rightarrow R$

As an example of the first type of data, we consider discrete samples of a scalar function $f: R^2 \rightarrow R$, which represents a height field of a terrain surface as shown in Figure 1. In the following, we assume that f is differentiable of class C^∞. A point $\mathbf{p} \in \mathbf{R}^2$ is called a singular point of the function f if

$$\frac{\partial f}{\partial x}(\boldsymbol{p}) = \frac{\partial f}{\partial y}(\boldsymbol{p}) = 0. \quad (1)$$

We also define the number of negative eigenvalues of the Hessian matrix at **p**, i.e.,

$$\boldsymbol{H}(\boldsymbol{p}) = \begin{pmatrix} \frac{\partial^2 f}{\partial x^2}(\boldsymbol{p}) & \frac{\partial^2 f}{\partial y \partial x}(\boldsymbol{p}) \\ \frac{\partial^2 f}{\partial x \partial y}(\boldsymbol{p}) & \frac{\partial^2 f}{\partial y^2}(\boldsymbol{p}) \end{pmatrix} \quad (2)$$

to be the index of the singular point **p**. In the following, we assume that the singular points are all non-degenerate, i.e., the Hessian matrix is always non-degenerate. A scalar function whose singular points are all non-degenerate is called a Morse function, and it is known that the space of Morse functions is dense in the space of all functions: any given function can be approximated arbitrarily well by a Morse function (for example, see [8]).

In the case of such a height field, the singular points are classified into three categories as shown in Figure 2. Note that in each category in Figure 2 the left hand side illustration depicts a terrain shape around each type of a singular point while the right hand side illustration shows the topological transition of the corresponding contour with respect to the height. As the reader can observe, a singular point corresponds to a point where a topological change arises in the corresponding contour.

Now we move on to the case of one dimension higher, i.e., a set of discrete samples of a 3D scalar field $f:R^3 \rightarrow R$. This case covers 3D volume data such as 3D medical images provided by measurement equipments (i.e. CT, MRI, etc.) and 3D spatial data obtained through computer simulations. In this case, a contour corresponds to an isosurface on which the scalar field values are all equal, and the singular points are classified into four different groups according to their indices, as long as the singular points are all non-degenerate.

As described earlier, the primary advantage of data analysis based on differential topology is the capability to extract not only local features such as singular points but also the global structure of the entire data as the mutual connection among the local features, which easily allows us to represent the data in a hierarchical fashion. In particular, a tree structure called contour tree[1] serves as an effective tool for encoding topological changes in the inverse image according to the scalar function value changes, and thus has been employed in many visualization problems.

For a given function $f:\mathbf{R}^n \rightarrow \mathbf{R}$, by contracting each connected component of the inverse images to a point, we get a space R_f, which inherits the quotient topology from $\mathbf{R}^n$. It is known that R_f has the structure of a graph in general and is often called the Reeb graph[12] of f. In some contexts, it is a tree and is called a contour tree.

The standard version of an algorithm for constructing contour trees has been invented by Carr et al. [4], and consists of the following steps (Figure 3):

- Constructing a join tree and a split tree,
- Constructing an augmented contour tree,
- Composing a final contour tree.

The first step is to construct a join tree, which describes how connected components in the inverse image joins as the function value decreases. In the case of discrete elevation samples shown in Figure 1, we first triangulate the sample points to interpolate the height field over the 2D data domain first, and then compose a tree by incorporating the discrete samples in the order of the corresponding function values. Suppose that we construct the contour tree of the sample points contained in the yellow region as shown in Figure 1. We first pick up the highest sample point at the height of 220 and add the corresponding vertex to the join tree. We then insert the second highest sample point at the height of 205 as a vertex that is connected with the previously inserted vertex, since it is also adjacent to that vertex in the triangulation of the terrain data. The next highest point at the height of 200 will be incorporated into the tree as a disjoint vertex, since it has no direct connection with the already registered vertices. Finally, when we insert the point at the height of 160, two disjoint sets of vertices will be merged into one. Figure 3(a) shows a join tree of the entire set of discrete terrain data.

A split tree can be constructed in the same way if we reorder the discrete samples in an ascending order with respect to the function value, as shown in Figure 3(b). As demonstrated in Figure 3(c), a preliminary version of a contour tree called an augmented contour tree can be constructed, by identifying the topological branches in the inverse image by tracking the join tree from the top and the split tree from the bottom. Figure 3(d) presents the final version of the contour tree, which has been obtained by removing non-branching vertices from the augmented contour tree.

As additional options, we can introduce steps for decomposing degenerate singular points into non-degenerate ones [17],[19] to better figure out the spatial configuration of the inverse image [20], for simplifying the contour trees by pruning minor edges for noise removal [5],[18], and for extracting change in genus of the inverse image especially for the case of 3D volumes [11].

In this way, especially for functions $R^2 \rightarrow R$ and $R^3 \rightarrow R$. , extracting features of differential topology can be accomplished by composing a proximity graph over the discrete samples first and then constructing a tree that represents the splitting and merging of the inverse images according to the function value changes. Figure 4 provides examples of contour trees extracted from 2D and 3D scalar fields using the algorithm described above.

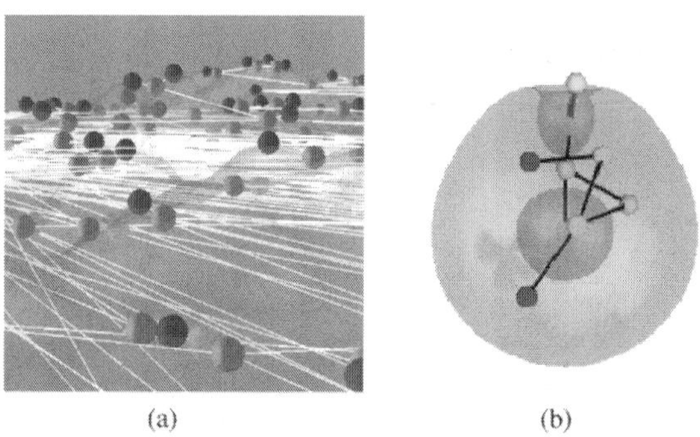

Figure 4. Examples of contour trees extracted from discrete samples on functions $R^2 \rightarrow R$ and $R^3 \rightarrow R$. (a) A digital elevation model around Hakone area,

which contains a crater lake (Lake Ashi) surrounded by mountains [17]. (b) Voxel samples at a 3D grid in a volume data, which is obtained through a simulation of the probability distribution of a nucleon in the atomic nucleus ^{16}O[18].

Remark 1. In practical applications, we are lead to analyze functions $V \to \mathbf{R}$, where V is a bounded domain in $\mathbf{R}^2$ or $\mathbf{R}^3$. In such a case, the function often takes extreme values near the boundary: consequently, by adding a virtual point outside of V that takes a still more extreme value, we can eliminate the domain boundary so that we get a function on a manifold without boundary. Usually, this kind of a technique is useful for analyzing scalar functions, while this usually does not make sense for multivariate functions.

ANALYZING SAMPLES OF A FUNCTION $R^N \to R$

The aforementioned algorithm is quite effective for discrete samples of 2D/3D scalar fields, since we can linearly interpolate the function value in the data domain easily by decomposing it into triangles/tetrahedra. Nonetheless, implementing the algorithm incurs other problems due to the complexity for partitioning the data domain especially when its dimension is equal to 4 or more.

This problem has recently been tackled by approaches to projecting the high-dimensional data samples onto screen space using dimensionality reduction techniques [16], which have been available in the community of machine learning. The basic idea behind this approach is to introduce different metrics among the discrete data samples so that we can effectively locate each sample point on the contour tree to be constructed. This has been accomplished by constructing a proximity graph over the data samples to infer its manifold connectivity and then projecting the samples onto the screen space to stipple the contour tree.

In practice, we first construct the proximity graph by searching for the nearest k-neighbors of each sample using the new metric based on their spatial relationship and function values, and then projecting the samples onto the contour tree by approximating the distance between every pair of sample points. As shown in Figure 5, the dimensionality reduction process effectively permits us to construct the contour tree from a set of discrete samples embedded in the high-dimensional data domain. Figure 6 shows another example where the features of differential topology are extracted from time-varying volume data.

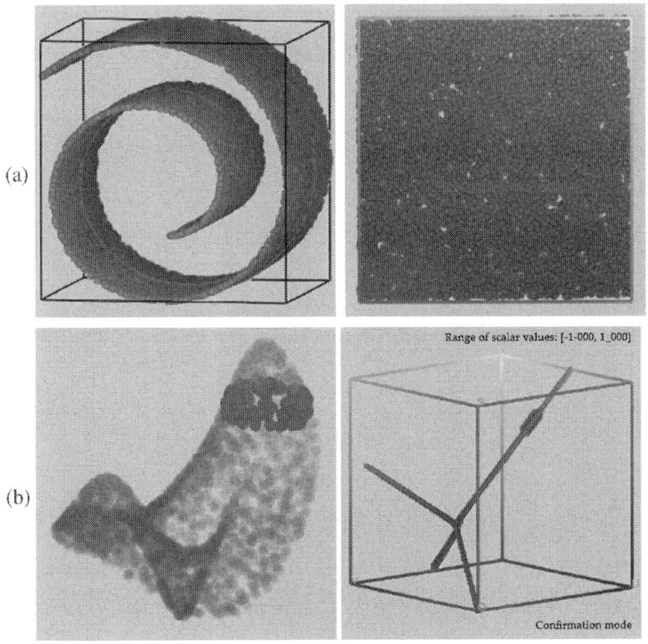

Figure 5. Constructing contour trees using manifold learning techniques [[16]]. (a) Discrete samples on the swiss roll data in 3D (left) and their projections onto 2D plane via nonlinear dimensionality reduction. (b) Discrete samples in the height field and their projection onto a contour tree.

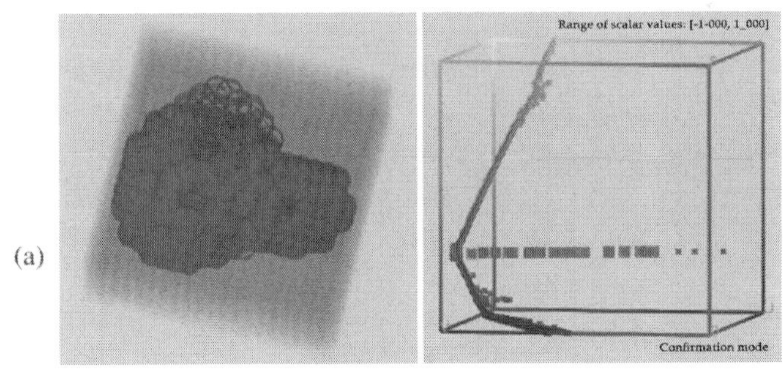

Figure 6. Simulation data of proton - hydrogen atom collision [[16]]. (a) Contour tree extracted from the corresponding time-varying volume data. (b) Snapshots of the feature time-varying volume data corresponding to the branch

of the contour tree (highlighted in red).

We can also use the proximity graphs to directly define the partial orders among data samples in terms of the scalar function value. Nonetheless, it is still hard to select appropriate types of graphs for constructing the manifold connectivity, since we have to keep the graph as sparse as possible to minimize the associated computational cost. In practice, we can refer to [10] for several possible types of vertex connectivities. The work also successfully constructed a visual metaphor called topological landscape[21], which is a variant of a contour tree for a terrain surface and is currently often employed for the visual analysis of high dimensional data.

ANALYZING SAMPLES OF A FUNCTION $R^N \rightarrow R^M$

Data analysis based on differential topology has long been primarily focused on data samples on univariate (scalar-valued) functions $f:R^n \rightarrow R$. Nonetheless, with the recent advent of high-performance computers and sensors of high resolution, we are more likely to tackle data samples of multivariate functions $f:R^n \rightarrow R^m$, $m>1$. For example, illuminating the global structure of 3D fields of multiple scalar values such as temperature, pressure, humidity, etc. poses a very important technical problem for weather forecast, and the visual data mining based on differential topology is again expected to help us to tackle the challenging problems.

A natural extension of the previous approach for multivariate data is to extract the intersection among inverse images of multiple function values first, and then track the topological changes inherent in that intersection with respect to the multivariate function value changes.

For a multivariate function $f:\mathbf{R}^n \rightarrow \mathbf{R}^m$ and a point $\mathbf{c} \in \mathbf{R}^m$, the inverse image

$$f^{-1}(c) = \{x \in R^n \,|\, f(x) = c\} \qquad 3$$

is called a fiber[13]. A more rigorous mathematical definition goes as follows. For two multivariate functions f_0 and f1:Rn→Rmf1:Rn→Rm and points $\mathbf{y}_0$ and $\mathbf{y}_1$ in the range $\mathbf{R}^m$, we say that the fibers of f_0 and f_1 over the points $\mathbf{y}_0$ and $\mathbf{y}_1$, respectively, are equivalent if for some open neighborhoods U_i of $\mathbf{y}_i$ in $\mathbf{R}^m$, i=0,1, there exist C∞C∞ diffeomorphisms $\Phi:f^{-1}(U_0) \rightarrow f^{-1}(U_1)$ and $\varphi:U_0 \rightarrow U_1$ with $\varphi(\mathbf{y}_0)=\mathbf{y}_1$ which make the following diagram commutative:

$$\begin{array}{ccc} \left(f_0^{-1}(U_0), f_0^{-1}(y_0)\right) & \xrightarrow{\Phi} & \left(f_1^{-1}(U_1), f_1^{-1}(y_1)\right) \\ f_0 \downarrow & & \downarrow f_1 \\ (U_0, y_0) & \xrightarrow{\varphi} & (U_1, y_1). \end{array}$$

Therefore, a fiber of a differentiable multivariate function over a point in the range refers to such an equivalence class. Note that this information encodes the semi-local behavior of the function around the whole inverse image of a point, and not just the inverse image as a set.

A point $\mathbf{p}\in\mathbf{R}^n$ is a singular point if the rank of the Jacobian matrix

$$\left(\frac{\partial f_i}{\partial x_j}(\boldsymbol{p})\right)_{1\leq i\leq m,\, 1\leq j\leq n} \quad 4$$

is strictly less than min{n,m}min{n,m}, where $f=(f_1,f_2,\ldots,f_m)$ and $(x_1,x_2,\ldots,x_n)$ are the coordinates of $\mathbf{R}^n$. In general, we can characterize singular points as the points in the domain where a topological change occurs in the fibers. A fiber is a singular fiber if it contains a singular point. Furthermore, the set of all singular points is called the Jacobi set of f and is often denoted by J(f). As the reader can easily imagine, recognizing singular fibers is very important in visualizing a given set of large data.

In practice, the analysis of multivariate data samples begins with extracting the Jacobi set, which has been tackled by Edelsbrunner et al. [6],[7]. They successfully developed an algorithm for extracting such Jacobi sets from functions to $\mathbf{R}^2$ by identifying sample points where the gradients of the two corresponding scalar functions are parallel to each other. However, the algorithm extracts singular points only individually from the given data samples, and cannot identify the topological changes in fibers; in particular, we cannot get any information on the types of the singular points. Thus, identifying only the Jacobi set does not help us very much to identify global structures inherent in the entire data. Therefore, it remained as an important problem to identify the global structures of a given large data by seeking the underlying connectivity among the singular points or singular fibers.

In order to represent a given set of data in a hierarchical fashion, an extension of the Reeb graph is useful. For a function $f:\mathbf{R}^n\rightarrow\mathbf{R}^m$, by contracting each connected component of its inverse images to a point, we get a space R_f

, which is called the Reeb space[7] of f. This is expected to play an important role similar to that of a contour tree.

Quite recently, the above long lasting problem has been solved again by Carr et al. [2],[3], where they invented an algorithm for constructing a joint contour net as a variant of a Reb space for data samples of a multivariate function $f:R^n \rightarrow R^m$, m>1. The basic idea of this algorithm is to quantize the image of function samples in the m-dimensional range space $\mathbf{R}^m$into a set of small blocks in terms of the m coordinate axes, and then seek the connectivity of the fibers between every pair of adjacent blocks along each coordinate axis, which finally allows us to compose a joint contour net as a network structure over the range space. Figure 7shows such an example, where two different function values are defined over a 3D polygonal surface to characterize its shape with the corresponding joint contour net. Here, the two function values are the integral of geodesic distance over the polygonal shape [9] and surface curvature, where the integral of geodesic distance represents how much the sample point is far away from the object center. Figure 7(a) shows an ellipsoid and its corresponding joint contour net projected to the 2D range space. Since the surface curvature becomes large as the sample point moves away from the object center, the joint contour net is projected to a narrow region between the left bottom corner and the right top corner. Figure 7(b) presents 3D spectacles where the associated joint contour net becomes more complicated, since the 3D shape model has variation in surface curvature. On the other hand, singular points of the multivariate function can be easily detected as the branches of the joint contour net. An example is demonstrated in Figure 8, where a certain explicit function $V \rightarrow \mathbf{R}^2$ is analyzed with V being a bounded domain in $\mathbf{R}^3$. For each case, the left window exhibits a singular fiber at some function value that is marked in the right window.

(a)

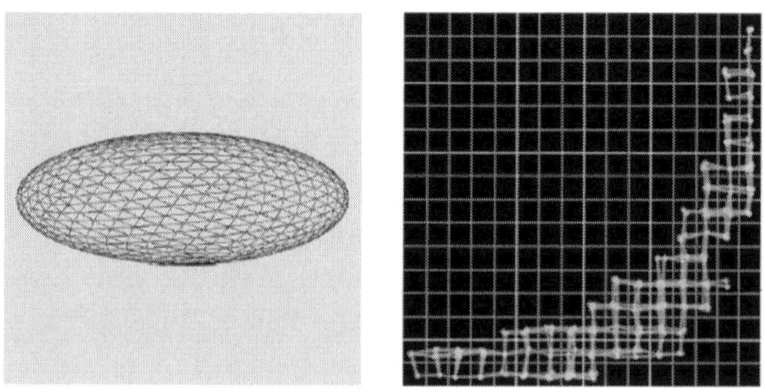

(b)

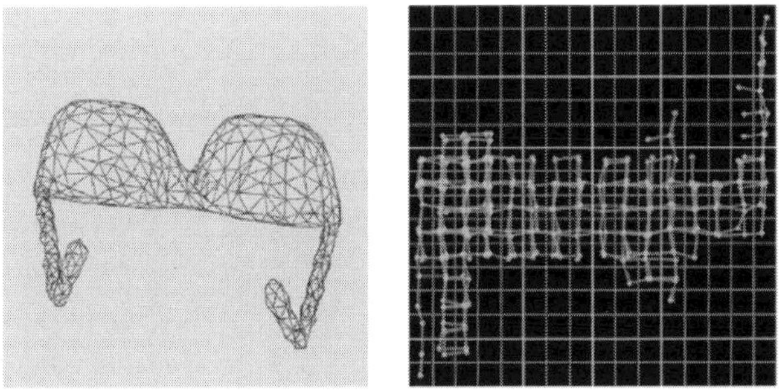

Figure 7. Topological changes in the intersection of two isolines extracted from a function of two variables defined over 3D polygonal surfaces M→ R^2. The images on the left and right show the 3D polygonal surface M and range R^2, respectively. Here, 16×16 resolution is employed to quantize the 2D range space, which is spanned by the two function values, i.e., the integral of geodesic distance from the shape center and surface curvature. (a) An ellipsoid surface and its corresponding joint contour net. (b) A spectacles polygonal surface and its corresponding joint contour net.

(a)

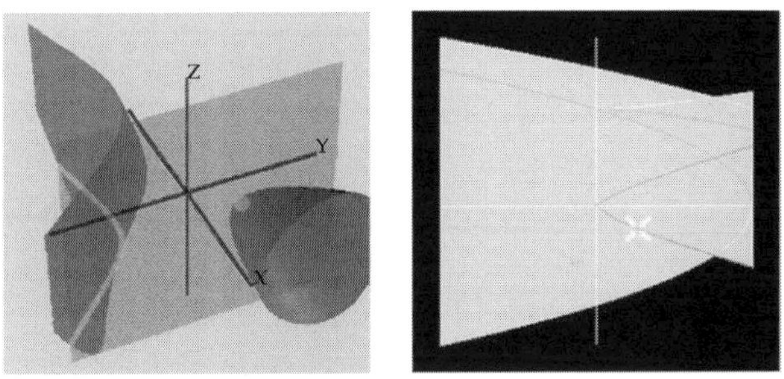

(b)

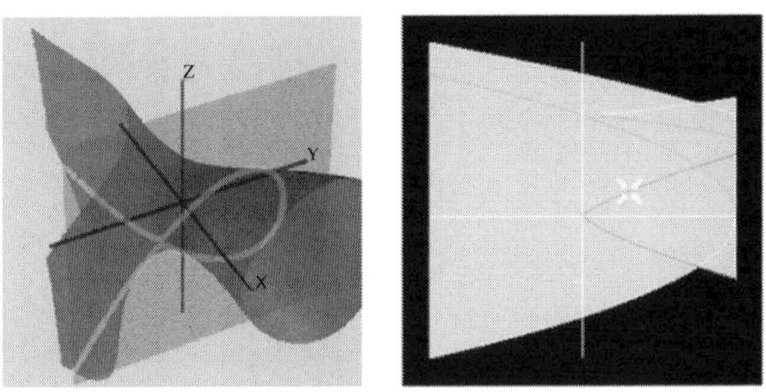

Figure 8. Topological changes in the intersection of two isosurfaces extracted from a function V→ R^2, where V is a bounded domain in R^3. The images on the left and right show the function domain in R^3 and the range R^2, respec-

tively. Here, images of singular fibers are represented as a line in the range [14]. (a) A single connected component in a fiber appears/disappears. (b) A pair of connected components merges/splits within a fiber.

The joint contour nets indeed allow us to track the connectivity among the singular fibers, while precisely identifying the topological type of each singular fiber still remains to be tackled.

In fact, some classification results for singular fibers have been obtained in singularity theory. For scalar functions on 2D domains, we have seen a classification of local topological changes of contour lines in Figure 2, under the assumption that all the singular points are non-degenerate (see Section 3). For functions $V \to \mathbf{R}^2$, where V is a bounded domain in 3D space, recently a classification of singular fibers and their topological changes has been obtained in [15], under the assumption that the functions are stable. A function $f: V \to \mathbf{R}^2$ is stable if for any of its C^∞ approximation g (in the sense of the Whitney C^∞ topology), there exist diffeomorphisms $\Phi: V \to V$ and $\varphi: \mathbf{R}^2 \to \mathbf{R}^2$ such that $g = \varphi \circ f \circ \Phi^{-1}$ (for details, see [8]). Note that the space of stable functions is open and dense in the space of all smooth functions $V \to \mathbf{R}^2$.

The list of singular fibers for such stable functions is as in Figures 9 and 10. Note that this classifies the singular fibers up to equivalence, and therefore describes completely the topological change of the fibers around the singular fibers. Such a classification clearly helps us to identify the type of a singular fiber for a given set of data. By combining such classification results with the technique of joint contour net, we can visualize a given set of data in an efficient and robust way (for details, see [14]). Figure 8 shows such a visualization result: each curve in the right windows corresponds to a specific singular fiber type in Figure 9according to its color.

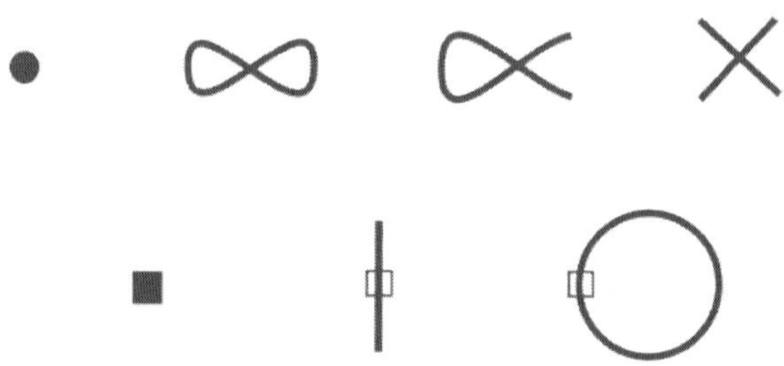

Figure 9. First list of singular fibers for stable functions $V \to R^2$for a 3D bounded domain V : those in this list appear along curves in the range [[15]]. Both the black dot and the black square represent a point: however, the former one is in

the interior of V, while the latter one appears on the boundary of V. Void squares represent tangencies with the boundary ∂ V.

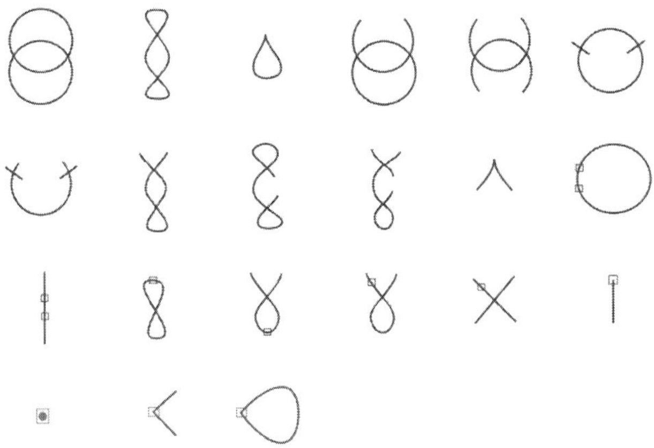

Figure 10. Second list of singular fibers for stable functions V → $R^2$for a 3D bounded domain V : those in this list appear discretely [[15]]. Red squares correspond to the intersection points of the Jacobi set with the boundary ∂ V.

Nevertheless, identifying the topological type of each singular fiber is considerably hard, especially for higher dimensional cases, where theoretical investigation has been still devoted for more detailed classification of singular fibers according to their topological types [13]. We are also working on this problem so that we can provide learners of differential topology with an interface for visually inspecting singular fibers of multivariate functions on the screen space. Furthermore, by referring to the topological types of the singular fibers embedded in the data domain, we can effectively extract meaningful features from multivariate data samples with minimal cost.

CONCLUSION

It is naturally expected that extracting differential topological features of multivariate data will provide us with various new information. It might not be so simple or straightforward to interpret real multivariate data by using differential topological features, compared with the case of scalar functions. Nevertheless, with the help of singularity theoretical progress in recent years, it is expected that visualization of large multivariate data featuring singular fibers will play essential role in visual data mining.

REFERENCES

1. Bajaj, CL., Pascucci, V., Schikore, DR.: The contour spectrum. In: Proc. IEEE Vis.'97, pp. 167–173, (1997)
2. Carr, H., Duke, D.: Joint contour nets. Accepted for publication in IEEE Transactions on Visualization and Computer Graphics (2013)
3. Carr, H., Duke, D.: Joint contour nets: Computation and properties. In: Proceedings of the 6th IEEE Pacific Visualization Symposium (PacificVis 2013), pp. 161–168, (2013)
4. Carr, H., Snoeyink, J., Axen, U.: Computing contour trees in all dimensions. Comput. Geometry: Theory Appl. 24(2), 75–94 (2003)
5. Carr, H., Snoeyink, J., van de Panne, M.: Simplifying flexible isosurfaces using local geometric measures. In: Proc. IEEE Vis. 2004, pp. 497–504, (2004)
6. Edelsbrunner, H., Harer, J.: Jacobi sets of multiple Morse functions. In: Cucker F., DeVore R., Olver P., Süli E. (eds.) Foundations of Computational Mathematics, pp. 37–57. Cambridge University Press, (2002)
7. Edelsbrunner, H., Harer, J., Patel, AK.: Reeb spaces of piecewise linear mappings. In: Proceedings of the Twenty-Fourth Annual Symposium on Computational Geometry, pp. 242–250, (2008)
8. Golubitsky, M., Guillemin, V.: Stable Mappings and their Singularities. Grad. Texts in Math, vol 14. Springer (1973)
9. Hilaga, M., Shinagawa, Y., Kohmura, T., Kunii, TL.: Topology matching for fully automatic similarity estimation of 3d shapes. In: Computer Graphics (Proceedings of Siggraph 2001), pp. 203–212, (2001)
10. Oesterling, P., Heine, C., Jänicke, H., Scheuermann, G., Heyer, G.: Visualization of high-dimensional point clouds using their density distribution's topology. IEEE Trans. Vis. Comput. Graph. 17(11), 1547–1559 (2011)
11. Pascucci, V., Cole-McLaughlin, K.: Efficient computation of the topology of level sets. In: Proc. IEEE Vis. 2002, pp. 187–194, (2002)
12. Reeb, G.: Sur les points singuliers d'une forme de Pfaff complètement intégrable ou d'une fonction numérique. Comptes Rendus Acad. Sci. Paris. 222, 847–849 (1946)

13. Saeki, O.: Topology of Singular Fibers of Differentiable Maps. Lecture Notes in Mathematics. vol. 1854. Springer (2004)
14. Saeki, O., Takahashi, S., Sakurai, D., Wu, Hsiang-Yun, Kikuchi, K., Carr, H., Duke, D., Yamamoto, T.: Visualizing multivariate data using singularity theory. In: Wakayama M., Anderssen RS., Cheng J., Fukumoto Y., McKibbin R., Polthier K., Takagi T., Toh K-C. (eds.) The Impact of Applications on Mathematics, Proceedings of the Forum of Mathematics for Industry 2013. Mathematics for Industry, vol. 1, pp. 51–65. Springer, (2013)
15. Saeki, O., Yamamoto, T.: Singular fibers of stable maps of 3-manifolds with boundary into surfaces and their applications. preprint (2014)
16. Takahashi, S., Fujishiro, I., Okada, M.: Applying manifold learning to plotting approximate contour trees. IEEE Trans. Vis. Comput. Graph. 15(6), 1185–1192 (2009)
17. Takahashi, S., Ikeda, T., Shinagawa, Y., Kunii, TL., Ueda, M.: Algorithms for extracting correct critical points and constructing topological graphs from discrete geographical elevation data. Comput. Graph. Forum. 14(3), 181–192 (1995)
18. Takahashi, S., Nielson, GM., Takeshima, Y., Fujishiro, I.: Topological volume skeletonization using adaptive tetrahedralization. In: Proc Geometric Modeling and Processing 2004, pp. 227–236, (2004)
19. Takahashi, S., Takeshima, Y., Fujishiro, I.: Topological volume skeletonization and its application to transfer function design. Graphical Models. 66(1), 22–49 (2004)
20. Takeshima, Y., Takahashi, S., Fujishiro, I., Nielson, GM.: Introducing topological attributes for objective-based visualization of simulated datasets. In: Proc. Volume Graphics 2005, pp. 137–236, (2005)
21. Weber, GH., Bremer, P-T., Pascucci, V.: Topological landscapes: a terrain metaphor for scientific data. IEEE TVCG. 13(6), 1416–1423 (2007)

16

The Method of Finding Solutions of Partial Dynamic Equations on Time Scales

Hsuan-Ku Liu

Department of Mathematics and Information Education, National Taipei University of Education, Taipei, Taiwan

ABSTRACT

On time scales, one area lacking of development is the method of finding solutions on partial dynamic equations. This paper proposes a method for finding the exact solution of linear partial dynamic equations on arbitrage time scales. We modify the variational iteration method on $\mathbb{R}$ to find an approximation of the nonlinear partial dynamic equation on $\overline{q^{\mathbb{N}}}$. As an example, the modified variational iteration method is applied to q-Berger equations and to q-Fisher equations. Their numerical results reveal that the proposed method is very effective.

Citation: Liu, H. K. (2013). The method of finding solutions of partial dynamic equations on time scales. Advances in Difference Equations, 2013(1), 141. (16 pages)

Copyright: © Open Access. This article is distributed under the terms of the Creative Commons Attribution 2.0 Generic (CC BY 2.0) License.

INTRODUCTION

A time scale is a nonempty closed subset of real numbers. On time scale calculus, notations and theorems have been well established for the univariate case [1]. Solutions of ordinary differential equations, such as initial value problems and boundary value problems, have been studied and published during the past two decades on time scales. In recent years, Hoffacker [2] and Ahlbrandt and Morian [3] demonstrated the related ideas to the multivariate case and studied partial dynamic equations on time scales. Notations and definitions on multivariate time scales calculus can be found in Bohner and Guseinov [4, 5]. Jackson [6] extended the existing ideas of the time scales calculus [3] to the multivariate case. The method of generalized Laplace transform on time scales is applied to find solutions of the homogeneous and nonhomogeneous heat and wave equations. Recent developments in the method of finding solutions have aroused further interest in the discussion of partial dynamic equations on time scales.

For the nonlinear cases, methods of finding solutions are not mentioned for partial dynamic equations on time scales. One of the difficulties for developing a theory of series solutions to nonlinear equations on time scales is that the formula for multiplications of two generalized polynomials is not easily found. If a time scale has constant graininess, Haile and Hall [7] provided an exact formula for the multiplication of two generalized polynomials. Using the obtained results, the series solutions for linear dynamic equations are proposed on the time scales $\mathbb{R}$ and T=hZ (difference equations with step size h). On generalized time scales, Mozyrska and Pawtuszewicz [8] presented a formula for the multiplication of generalized polynomials of degree one and degree n∈ N. Liu [9] provided a product rule of two generalized polynomials on the time scale $q^{\mathbb{Z}} = \{q^n \mid n \in \mathbb{N}\} \cup \{0\}$.

The variational iteration method proposed by He [10] is a powerful mathematical tool in analyzing the nonlinear problems on $\mathbb{R}$ (the set of real numbers). Over the last few years, the variational iteration method (VIM) has been widely applied to analyze the nonlinear boundary value problems [11], the nonlinear heat diffusion equations [12] and the nonlinear reaction-diffusion equations [13]. An advantage of the VIM is that there is no need to make the assumption of the small parameters. On nonlinear partial dynamic equations, approximate solutions obtained by the variational iteration method are not found yet.

In this paper, we first explore a simple method to find the exact solution of linear partial dynamic equations on time scales. For the nonlinear cases, we derive a product rule of two generalized polynomials on $\overline{q^{\mathbb{Z}}}$, which provides an idea for developing a series solutions on q-calculus. Applying the product rules, we extend the variational iteration method from the set of real numbers ℝ to the time scales $\overline{q^{\mathbb{Z}}}$. The extension provides a method to find an approximate solution on the nonlinear partial dynamic equation on $\overline{q^{\mathbb{Z}}}$. Moreover, the VIM is applied to find an approximation of the q-Berger equation and the q-Fisher equation. By the numerical results, we found that the modified VIM is very effective. The VIM can be applied to other time scales when the multiplication rule of two generalized polynomials on these time scales is obtained.

This paper is organized as follows. In Section 2, the basic ideas of partial dynamic equations on time scales are introduced. In Section 3, a method is explored to find an exact solution of linear initial value problems on time scales. In Section 4, a product rule of two generalized polynomials at 0 is derived on $\overline{q^{\mathbb{Z}}}$ and the variational iteration method is applied to find an approximate solution of the Burger equation and the Fisher equation. In Section 5, numerical examples reveal that the proposed method is very effective. Finally, a concise conclusion and future directions are provided in Section 6.

BASIC CONCEPTS ON TIME SCALES

A time scale is an arbitrary nonempty closed subset of the real numbers. The calculus of time scales was introduced by Hilger [14] in order to create a theory that can unify discrete and continuous analysis.

An introduction to Time Scales

In this subsection, we first define the forward and backward jump operators on time scales and then introduce the delta derivative and the integration.

Definition 1 Let T be a time scale. For $t \in T$ the forward jump operator $\sigma: T \to T$ is defined by

$$\sigma(t) := \inf\{s > t \mid s \in \mathbb{T}\}$$

and the backward jump operator $\rho: T \to T$ is defined by

$$\rho(t) := \sup\{s < t \mid s \in \mathbb{T}\}.$$

The gain function $\mu: T \to [0,\infty)$ is defined by

$\mu(t) := \sigma(t) - t.$

According to the forward jump operator and the gain function, the delta derivative on the time scale T is given as follows.

Definition 2 Assume that f:T→ R is a function and let t∈T. If σ(t)> t, the delta derivative of f(t) at t on the time scale T is defined as

$$f^{\Delta}(t) = \frac{f(\sigma(t)) - f(t)}{\mu(t)}.$$

A function f(t) on TT is said to be differentiable at t if its derivative exists at t, ∀t∈T.

Integration on a time scale can be viewed as an anti-derivative.

Definition 3 If we have delta derivative g(t)=fΔ(t) on the time scale T, then the anti-derivative is

$$f(t) = \int_a^t g(s)\Delta s + \text{constant}, \quad a, t \in \mathbb{T}$$

and the definite integral on the time scale T follows as

$$\int_a^b g(s)\Delta s = f(b) - f(a), \quad a, b \in \mathbb{T}.$$

Following the delta derivative and integration, we define the generalized polynomials as follows.

Definition 4 On the time scale T, the generalized polynomials $h_k(\cdot, t0)$:T→R are defined recursively as follows:

$$h_0(t,s) = 1, \qquad h_{k+1}(t,s) = \int_s^t h_k(\tau, s)\Delta\tau.$$

Hence, for each fixed s, the delta derivative of h_k with respect to t satisfies

$$h_k^{\Delta^j}(t,0) = \begin{cases} h_{k-j}(t,0) & \text{if } k \geq j, \\ 0 & \text{if } k < j. \end{cases}$$

An Introduction to q-calculus

Let

$$q^{\mathbb{N}} = \{q^n \mid n \in \mathbb{N}\} \quad \text{and} \quad \overline{q^{\mathbb{N}}} = q^{\mathbb{N}} \cup \{0\},$$

where $\mathbb{N}$ denotes the set of positive integers.

If a and q are real numbers such that $0<q<1$, then the q-shift factorial [15] is defined by

$$(a;q)_0 = 1 \quad \text{and} \quad (a;q)_n = \prod_{k=0}^{n-1}(1-aq^k), \quad n \in \mathbb{N}.$$

Following Definition 2, the q-derivative is given as follows.

Definition 5 Assume that $f : \overline{q^{\mathbb{N}}} \to \mathbb{R}$ is a function and $t \in \overline{q^{\mathbb{N}}}$. The q-derivative [1] at t is defined by

$$f^{\Delta}(t) = \frac{f(qt) - f(t)}{(q-1)t}, \quad t \neq 0$$

and

$$f^{\Delta}(0) = \lim_{n\to\infty} \frac{f(q^n) - f(0)}{q^n}.$$

By computing the recurrence relation, the q-polynomials are represented as

$$h_k(t,s) = \prod_{\nu=0}^{k-1} \frac{t - sq^{\nu}}{\sum_{j=0}^{\nu} q^j}$$

on $\overline{q^{\mathbb{N}}}$ [1].

A q-difference equation is an equation that contains q-derivatives of a function defined on $\overline{q^{\mathbb{N}}}$.

Multivariable Calculus on Time Scales

The differentiation and integrations are introduced for functions of two variables on time scales [3]. Definitions on multivariate calculus on time scales can be found in Bohner and Guseinov [4, 5]. Following the line of ideas, the dynamic equations on time scales are extended to partial dynamic equations on time scales [2, 3, 14].

Let T_1 and T_2 be any two time scales. Consider the 'rectangle' $T=T_1\times T_2$. For any $t\in T$, the jump operator of $t=(t_1,t_2)$ for $t_1\in T_1$ and $t_2\in T_2$ is given as follows:

- The forward jump operators σ:T→T by $\sigma(t)=(\sigma(t_1),\sigma(t_2))$ are defined as $\sigma(t_1)=\inf(s\in T_1|s>t_1)$ and $\sigma(t_2)=\inf(s\in T_2|s>t2)$.
- The backward jump operators τ:T→T by $\tau(t)=(\tau(t_1),\tau(t_2))$ are defined as $\tau(t_1)=\sup(s\in T_1|s<t_1)$ and $\tau(t_2)=\sup(s\in T_2|s<t_2)$.

To use the notations for partial derivatives with respect to time scale variables t_1 and t_2, respectively, we employ lexicographic ordering for consistency. Let $f\Delta_1$ denote the time scale partial derivative with respect to t1t1 and let $f\Delta_2$ denote the time scale partial derivative with respect to t_2. Definitions of these partial derivatives are given as below [2, 3].

Definition 6 Let f be a real-valued function on T. At $(t_1,t_2)\in T=T_1\times T_2$ we say f has a Δ_1-partial derivative $f\Delta_1(t_1,t_2)$ if for each 0ε>0, there exists a neighborhood U of t_1, with $U=(t_1-\delta,t_1+\delta)\cap T_1$ for 0δ>0, such that

$$\left|f(\sigma(t_1),t_2)-f(s,t_2)-f^{\Delta_1}(t_1,t_2)(\sigma(t_1)-s)\right|\leq\left|\sigma(t_1)-s\right|$$

for all s∈U. On the other hand, we say f has a Δ_2-partial derivative $f\Delta_2(t_1,t_2)$ if for each ε>0, there exists a neighborhood V of t_2, with $V=(t_2-\delta,t_2+\delta)\cap T_2 V =(t_2-\delta,t_2+\delta)\cap T_2$ for δ>0, such that

$$\left|f(t_1,\sigma(t_2))-f(t_1,s)-f^{\Delta_2}(t_1,t_2)(\sigma(t_2)-s)\right|\leq\left|\sigma(t_2)-s\right|$$

for all s∈V.

Using the ideas of time scale partial derivatives, notations of mixed partial and high order partial derivatives are given as follows:

- $f^{\Delta_{ij}}(\mathbf{t})$ (if this value exists) denotes first taking the partial derivative with respect to titiand then taking the partial derivative with respect to tj, so that $f^{\Delta_{ij}}=(f^{\Delta_i})^{\Delta_j}$, i,j=1,2.
- $f^{\Delta_i^n}(\mathbf{t})$ (if this value exists) denotes taking the partial derivative of f(t) with respect to t_i n times.

The details and examples can be found in [3] and [6].

THE EXACT SOLUTION OF LINEAR INITIAL VALUE PROBLEMS ON TIME SCALES

Let $h_k(t_1,0)$ and $g_k(t_2,0)$ be the generalized polynomials on T_1 and T_2, respectively. In this section, the variational iteration method on ℝ is extended to provide a method of finding the exact solution of linear partial dynamic equations on time scales. The introduction and the details of the variational iteration method can be found in the Appendix and in [10].

The Exact Solution of the First-Order Linear Partial Dynamic Equations

We first consider the first-order linear partial dynamic equations as the form

$$\begin{cases} u^{\Delta_1} = c^2 u^{\Delta_2} & \text{on } \mathbb{T}_1 \times \mathbb{T}_2, \\ u(0,t_2) = f(t_2) & \text{on } \mathbb{T}_2, \end{cases} \quad 1$$

where $f(t_2) = \sum_{i=0}^{K} a_i g_i(t_2, 0)$ on T_2 and aiai, i=1,…,K are real numbers.

The basic character of the variational iteration method is to construct a correction functional for the system, which reads

$$u_{n+1}(t_1,t_2) = u_n(t_1,t_2) + \int_{t_0}^{t_1} \lambda\{Lu_n(s,t_2) + N\tilde{u}_n(s,t_2)\}\Delta s,$$

where L is a linear operator on T_1, N is a linear (or nonlinear) operator on T_2 (or $T_1 \times T_2$), λis a Lagrange multiplier which can be identified optimally by variational theory, u_n is the n th approximation, and $\tilde{u}_n$ denotes a restricted variation, that is, $\delta\tilde{u}_n = 0$.

The linear operator L is selected as

$$Lu = u^{\Delta_1}$$

and the other operator N is selected as

$$Nu = -c^2 u^{\Delta_2}.$$

Make the above correction functional stationary with respect to u_n

$$\delta u_{n+1}(t_1,t_2) = \delta u_n(t_1,t_2) + \delta \int_0^{t_1} \lambda\{u^{\Delta_1}(s,t_2) + N\tilde{u}_n(s,t_2)\}\Delta s$$

$$= (1 + \lambda(t_1))\delta u_n(t_1,t_2) + \int_0^{t_1} \lambda^{\Delta}(s)\delta u_n(\sigma(s),t_2)\Delta s.$$

We, therefore, have the following stationary conditions:

$$1 + \lambda(t_1) = 0,$$

$$\lambda^{\Delta}(s) = 0.$$

The Lagrange multiplier can be readily identified

$$\lambda(s) = -1.$$

As a result, the variational iteration formula is obtained

$$u_{n+1}(t_1, t_2) = u_n(t_1, t_2) - \int_0^t \left\{u_n^{\Delta_1}(s, t_2) + Nu_n(s, t_2)\right\} \Delta s.$$

2

Using the initial condition u0=f(t2)u0=f(t2) and the iteration formula (2), we have the following equations:

$$u_1(t_1, t_2) = f(t_2) - \int_0^{t_1} f^{\Delta_2}(t_2) \Delta s$$

$$= f(t_2) + c^2 f^{\Delta_2}(t_2) h_1(t_1, 0),$$

$$u_2(t_1, t_2) = u(t_1, t_2) - \int_0^{t_1} c^2 f^{\Delta_2^2}(t_2) h_1(s, 0) \Delta s$$

$$= u_1(t_1, t_2) + c^4 f^{\Delta_2^2}(t_2) h_2(t_1, 0)$$

$$= f(t_2) + c^2 f^{\Delta_2}(t_2) h_1(t_1, 0) + c^4 f^{\Delta_2^2}(t_2) h_2(t_1, 0)$$

and

$$u_k(t_1, t_2) = \sum_{j=0}^{k} c^{2j} f^{\Delta_2^j}(t_2) h_j(t_1, 0).$$

As k is large enough such that $f^{\Delta_2^k}$ equals to zero, the series solution u_k is the exact solution of (1).

Example 1 Consider the initial value problem

$$\begin{cases} u^{\Delta_1} = c^2 u^{\Delta_2} & \text{on } \mathbb{T}_1 \times \mathbb{T}_2, \\ u(0, t_2) = g_k(t_2, 0) & \text{on } \mathbb{T}_2, \end{cases}$$

3

where $g_k(t_2,0)$ is a generalized polynomial on T_2. The function

$$u_k(t_1,t_2) = \sum_{j=0}^{k} c^{2j} g_{k-j}(t_2,0)h_j(t_1,0)$$

is the exact solution of (3).

Proof. We now verify that the obtained function u_k actually solves the initial value problem (3). First, we show that the obtained function satisfies the initial condition. Since h0(t,s)≡ 1for all t, s and h_j (0,0)≡1 for j>0j>0, we have

$$u_k(0,t_2) = g_k(t_2,0).$$

Second, we display the obtained function u_k satisfying the equation by

$$u_k^{\Delta_1}(t_1,t_2) = \sum_{j=1}^{k} c^{2j} g_{k-j}(t_2,0)h_{j-1}(t_1,0),$$

$$c^2 u_k^{\Delta_2}(t_1,t_2) = \sum_{j=0}^{k-1} c^{2(j+1)} g_{k-j-1}(t_2,0)h_j(t_1,0) = \sum_{j=1}^{k} c^{2j} g_{k-j}(t_2,0)h_{j-1}(t_1,0).$$

This implies that $u_k^{\Delta_1}(t_1,t_2) - c^2 u_k^{\Delta_2}(t_1,t_2) = 0$ on $T_1 \times T_2$.

The exact solution of the second-order linear partial dynamic equations

Consider the second-order partial dynamic equation as the form

$$\begin{cases} u^{\Delta_1} = c^2 u^{\Delta_2^2} & \text{on } \mathbb{T}_1 \times \mathbb{T}_2, \\ u(0,t_2) = f(t_2) & \text{on } \mathbb{T}_2, \end{cases} \qquad 4$$

where $f(t_2) = \sum_{i=0}^{K} a_i g_i(t_2,0)$ on T_2 and a_i, i=1,…,K are real numbers.

In this work, the linear operator L is selected as

$$Lu = u^{\Delta_1}$$

and the other operator N is selected as

$$Nu = -c^2 u^{\Delta_2^2}.$$

Using the initial condition $u_0(t_1,t_2)=f(t_2)$ and the iteration formula (2), we have the following equations:

$$u_1(t_1,t_2) = f(t_2) - \int_0^{t_1} c^2 f^{\Delta_2^2}(t_2)\Delta s$$

$$= f(t_2) + c^2 f^{\Delta_2^2}(t_2) h_1(t_1,0),$$

$$u_2(t_1,t_2) = u_1(t_1,t_2) - \int_0^{t_1} c^2 f^{\Delta_2^4}(t_2) h_1(s,0)\Delta s$$

$$= u_1(t_1,t_2) + f^{\Delta_2^4}(t_2) h_2(t_1,0)$$

$$= f(t_2) + c^2 f^{\Delta_2^2}(t_2) h_1(t_1,0) + c^4 f^{\Delta_2^4}(t_2) h_2(t_1,0)$$

and

$$u_k(t_1,t_2) = \sum_{j=0}^{k} c^{2j} f^{\Delta_2^{(2j)}}(t_2) h_j(t_1,0).$$

As k is large enough such that $f^{\Delta_2^{(2k)}}$ equals to zero, the series solution ukuk is the exact solution of (4).

Example 2 Consider the IVP

$$\begin{cases} u^{\Delta_1} = c^2 u^{\Delta_2^2} & \text{on } \mathbb{T}_1 \times \mathbb{T}_2, \\ u(0,t_2) = g_k(t_2,0) & \text{on } \mathbb{T}_2, \end{cases} \qquad 5$$

where $g_k(t_2,0)$ is a generalized polynomial of T_2. The function

$$u_{\lfloor k/2 \rfloor}(t_1,t_2) = \sum_{j=0}^{\lfloor k/2 \rfloor} c^{2j} g_{k-2j}(t_2,0) h_j(t_1,0)$$

is the exact solution of (5).

The exact solution of Example 2 is also obtained by Jackson [6]. He transformed the IVP into an ODE and obtained the exact solution as

$$u(t_1,t_2) = \sum_{j=0}^{\lfloor k/2 \rfloor} c^{2j} g_{k-2j}(t_2,0) h_j(t_1,0),$$

where $\lfloor k/2 \rfloor$ denotes the floor of k/2.

When the initial condition can be represented as a finite series of generalized polynomials, we have proposed a useful method of finding the exact solution of partial dynamic equations on time scales. When the initial condition is represented as an infinite series of generalized polynomials, the approximate solution can be obtained by the same manner. In the following section, we consider the nonlinear partial dynamic equation on the specific time scales.

APPROXIMATION SOLUTIONS OF NONLINEAR Q-PARTIAL DYNAMIC EQUATIONS

In this section, we extend the variational iteration method to find an approximate solution of nonlinear initial value problems on the time scale $\overline{q^{\mathbb{N}}}$.

To extend the variational iteration method, we first display a production rule [9] of two q-polynomials at 0 which will be used to derive an approximate solution in the following discussion.

Theorem 1 Let $h_i(t,0)$ and $h_j(t,0)$ be two q-polynomials at zero. We have

$$h_i(t,0)h_j(t,0) = \frac{(q^{i+1};q)_j}{(q;q)_j} h_{i+j}(t,0).$$

Proof Since

$$h_{i+j}(t,0) = \prod_{\nu=0}^{i+j-1} \frac{t}{\sum_{\mu=0}^{\nu} q^{\mu}},$$

we have

$$h_{i+j}(t,0) = \left(\prod_{\nu=0}^{i-1} \frac{t}{\sum_{\mu=0}^{\nu} q^{\mu}}\right)\left(\prod_{\nu=i}^{i+j-1} \frac{t}{\sum_{\mu=0}^{\nu} q^{\mu}}\right)$$

$$= h_i(t,0)\left(\frac{\prod_{\nu=0}^{j-1}\sum_{\mu=0}^{\nu} q^{\mu}}{\prod_{\nu=0}^{j-1}\sum_{\mu=0}^{\nu} q^{\mu}}\right) t^j \left(\prod_{\nu=i}^{i+j-1} \frac{1}{\sum_{\mu=0}^{\nu} q^{\mu}}\right)$$

$$= h_i(t,0)\left(\prod_{\nu=0}^{j-1} \frac{t}{\sum_{\mu=0}^{\nu} q^{\mu}}\right)\left(\prod_{\nu=0}^{j-1}\sum_{\mu=0}^{\nu} q^{\mu}\right)\left(\prod_{\nu=i}^{i+j-1} \frac{1}{\sum_{\mu=0}^{\nu} q^{\mu}}\right)$$

$$= h_i(t,0)h_j(t,0)\left(\prod_{\nu=0}^{j-1} \frac{\sum_{\mu=0}^{\nu} q^{\mu}}{\sum_{\mu=0}^{\nu+i} q^{\mu}}\right).$$

This implies that

$$h_i(t,0)h_j(t,0) = \left(\prod_{\upsilon=0}^{j-1}\frac{\sum_{\mu=0}^{\upsilon+i}q^{\mu}}{\sum_{\mu=0}^{\upsilon}q^{\mu}}\right)h_{i+j}(t,0) = \prod_{\upsilon=0}^{j-1}\frac{(1-q^{\upsilon+i+1})}{(1-q^{\upsilon+1})}h_{i+j}(t)$$

$$= \frac{(q^{i+1};q)_j}{(q;q)_j}h_{i+j}(t,0).$$

Proposition 1 Let $h_i(t,0)$ and $h_j(t,0)$ be any two q-polynomials. We have $h_i(t,0)h_j(t,0) = h_j(t,0)h_i(t,0)$.

Proof It suffices to show that

$$\frac{(q^{i+1};q)_j}{(q,q)_j} = \frac{(q^{j+1};q)_i}{(q,q)_i}.$$

Suppose i >j, we have

$$\frac{(q^{i+1};q)_j}{(q,q)_j} - \frac{(q^{j+1};q)_i}{(q,q)_i}$$

$$= \frac{(1-q^{j+1})\cdots(1-q^{i+j})}{(1-q)\cdots(1-q^{i})} - \frac{(1-q^{i+1})\cdots(1-q^{i+j})}{(1-q)\cdots(1-q^{j})}$$

$$= \frac{(1-q^{j+1})\cdots(1-q^{i+j})}{(1-q)\cdots(1-q^{i})} - \frac{(1-q^{i+1})\cdots(1-q^{i+j})(1-q^{j+1})\cdots(1-q^{i})}{(1-q)\cdots(1-q^{j})(1-q^{j+1})\cdots(1-q^{i})} = 0.$$

Let h_k and g_k be generalized polynomials of $\overline{q_1^{\mathbb{N}}}$ and $\overline{q_2^{\mathbb{N}}}$. The variational iteration method is now applied to find an approximate solution of the nonlinear partial dynamic equations as the form

$$\begin{cases} u^{\Delta_1} = Nu & \text{on } \overline{q_1^{\mathbb{N}}} \times \overline{q_2^{\mathbb{N}}}, \\ u(0,t_2) = g_k(t_2,0) & \text{on } \overline{q_2^{\mathbb{N}}}. \end{cases}$$

When the linear operator L is selected as

$$Lu = u^{\Delta_1}$$

and the other operator N is selected as −Nu−Nu, the variational iteration formula is obtained as

$$u_{n+1}(t_1,t_2) = u_n(t_1,t_2) - \int_0^{t_1}\left\{u_n^{\Delta_1}(s,t_2) - Nu_n(s,t_2)\right\}\Delta s$$

with the initial approximation

$$u_0 = g_k(t_2,0).$$

Example 3 Consider the partial dynamic equations as the form

$$\begin{cases} u^{\Delta_1} = uu^{\Delta_2} & \text{on } \mathbb{T}_1 \times \mathbb{T}_2, \\ u(0,t_2) = g_k(t_2,0) & \text{on } \mathbb{T}_2. \end{cases}$$

With the variational iteration formula, we obtain the first few components of $u_n(t_1,t2)$:

$$\begin{aligned} u_1(t_1,t_2) &= u_0(t_1,t_2) - \int_0^{t_1} \left\{ u_0^{\Delta_1}(s,t_2) - u_0(s,t_2)u_0^{\Delta_2}(s,t_2) \right\} \Delta s \\ &= g_k(t_2,0) + \int_0^{t_1} \left[g_k(t_2,0) g_{k-1}(t_2,0) \right] \Delta s \\ &= g_k(t_2,0) + H_2(k,k-1) g_{2k-1}(t_2,0) h_1(t_1,0), \end{aligned}$$

$$\begin{aligned} u_2(t_1,t_2) &= u_1(t_1,t_2) - \int_0^{t_1} \left\{ u_1^{\Delta_1}(s,t_2) - u_1(s,t_2)u_1^{\Delta_2}(s,t_2) \right\} \Delta s \\ &= g_k(t_2,0) + H_2(k,k-1) g_{2k-1}(t_2,0) h_1(t_1,0) \\ &\quad - \int_0^{t_1} \left[H_2(k,k-1) g_{2k-1}(t_2,0) - \left(g_k(t_2,0) + H_2(k,k-1) g_{2k-1}(t_2,0) h_1(s,0) \right) \right. \\ &\quad \left. \times \left(g_{k-1}(t_2,0) + H_2(k,k-1) g_{2k-2}(t_2,0) h_1(s,0) \right) \right] \Delta s \\ &= g_k(t_2,0) + H_2(k,k-1) g_{2k-1}(t_2,0) h_1(t_1,0) \\ &\quad + H_2(k,k-1) H_2(k,2k-2) g_{3k-2}(t_2,0) h_2(t_1,0) \\ &\quad + H_2(k,k-1) H_2(k-1,2k-1) g_{3k-2}(t_2,0) h_2(t_1,0) \\ &\quad + H_2(k,k-1) H_2(k,k-1) H_2(2k-1,2k-2) g_{4k-3}(t_2,0) H_1(1,1) h_3(t_1,0), \end{aligned}$$

where $H_1(k,l) = \frac{(q_1^{k+1};q_1)_l}{(q_1;q_1)_l}$ and $H_2(k,l) = \frac{(q_2^{k+1};q_2)_l}{(q_2;q_2)_l}$.

In the same manner, the rest of components of the iteration formula are obtained iteratively.

Applications to the q-Burger equation and the Fisher equation

q-Burger equation

First of all, we consider the q-Burger equation as the form

$$\begin{cases} u^{\Delta_1} - uu^{\Delta_2} - \alpha u^{\Delta_2^2} = 0 & \text{on } \overline{q_1^{\mathbb{N}}} \times \overline{q_2^{\mathbb{N}}}, \\ u(0,t_2) = g_k(t_2,0) & \text{on } \overline{q_2^{\mathbb{N}}}. \end{cases}$$

When the linear operator and the nonlinear operator are selected as $Lu = u^{\Delta_1}$ and $-Nu = -uu^{\Delta_2} - \alpha u^{\Delta_2^2}$, respectively, the variational iteration formula is obtained as

$$u_{n+1}(t_1,t_2) = u_n(t_1,t_2) - \int_0^{t_1} \left\{u_n^{\Delta_1}(s,t_2) - u_n(s,t_2)u_n^{\Delta_2}(s,t_2) - \alpha u_n^{\Delta_2^2}(s,t_2)\right\}\Delta s.$$ 6

Let

$$G(t_2) = H_2(k,k-1)g_{2k-1}(t_2,0) + \alpha g_{k-2}(t_2,0) \text{ and } H_2(k,l) = \frac{(q_2^{k+1};q_2)_l}{(q_2;q_2)_l}.$$

With the initial condition

$$u_0(t_1,t_2) \equiv u(0,t_2) = g_k(t_2,0),$$

we have

$$\begin{aligned}
u_1(t_1,t_2) &= u_0(t_1,t_2) - \int_0^{t_1} \left\{u_0^{\Delta_1}(s,t_2) - u_0(s,t_2)u_0^{\Delta_2}(s,t_2) - \alpha u_0^{\Delta_2^2}(s,t_2)\right\}\Delta s \\
&= g_k(t_2,0) - \int_0^{t_1} \left\{-g_k(t_2,0)g_{k-1}(t_2,0) - \alpha g_{k-2}(t_2,0)\right\}\Delta s \\
&= g_k(t_2,0) + \left(H_2(k,k-1)g_{2k-1}(t_2,0) + \alpha g_{k-2}(t_2,0)\right)h_1(t_1,0) \\
&= g_k(t_2,0) + G(t_2)h_1(t_1,0),
\end{aligned}$$

$$\begin{aligned}
u_2(t_1,t_2) &= u_1(t_1,t_2) - \int_0^{t_1} \left\{u_1^{\Delta_1}(s,t_2) - u_1(s,t_2)u_1^{\Delta_2}(s,t_2) - \alpha u_1^{\Delta_2^2}(s,t_2)\right\}\Delta s \\
&= g_k(t_2,0) + G(t_2)h_1(t_1,0) \\
&\quad - \int_0^{t_1} \left\{G(t_2) - \left(g_k(t_2,0) + G(t_2)h_1(s,0)\right)\right. \\
&\quad \times \left(g_{k-1}(t_2,0) + g^{\Delta_2}(t_2)h_1(s,0)\right) \\
&\quad \left. - \alpha\left(g_{k-2}(t_2,0) + g^{\Delta_2^2}(t_2)h_1(s,0)\right)\right\}\Delta s \\
&= g_k(t_2,0) + H_2(k,k-1)g_{2k-1}(t_2,0)h_1(t_1,0) \\
&\quad + G(t_2)g_{k-1}(t_2,0)h_2(t_1,0) + g^{\Delta_2}(t_2)g_k(t_2,0)h_2(t_1,0) \\
&\quad + G(t_2)g^{\Delta_2}(t_2)H_1(1,1)h_3(t_1,0) \\
&\quad + \alpha g_{k-2}(t_2,0)h_1(t_1,0) + \alpha g^{\Delta_2^2}(t_2)h_2(t_1,0) \\
&= g_k(t_2,0) + \left[H_2(k,k-1)g_{2k-1}(t_2,0) + \alpha g_{k-2}(t_2,0)\right]h_1(t_1,0) \\
&\quad + \left[G(t_2)g_{k-1}(t_2,0) + g^{\Delta_2}(t_2)g_k(t_2,0) + \alpha g^{\Delta_2^2}(t_2)\right]h_2(t_1,0) \\
&\quad + G(t_2)g^{\Delta_2}(t_2)H_1(1,1)h_3(t_1,0).
\end{aligned}$$

In the same manner, the rest of components of the iteration formula are obtained iteratively.

q-Fisher equation

Secondly, we consider the q-Fisher equation, which is a nonlinear reaction diffusion equation, as the form

$$\begin{cases} u^{\Delta_1} - \alpha u^{\Delta_2^2} - \beta u(1-u) = 0 & \text{on } \overline{q_1^{\mathbb{N}}} \times \overline{q_2^{\mathbb{N}}}, \\ u(0,t_2) = g_k(t_2,0) & \text{on } \overline{q_2^{\mathbb{N}}}. \end{cases}$$

The variational iteration formula is obtained as

$$u_{n+1}(t_1,t_2) = u_n(t_1,t_2) - \int_0^{t_1} \left\{ u_n^{\Delta_1}(s,t_2) - \alpha u_n^{\Delta_2^2}(s,t_2) \right. \\ \left. - \beta u_n(s,t_2) + \beta u_n(s,t_2) u_n(s,t_2) \right\} \Delta s. \quad 7$$

Let

$G(t_2) = \alpha h_{k-2}(t_2) + \beta g_k(t_2,0) - \beta H_2(k,k) g_{2k}(t_2,0)$ and $H_2(k,l) = \frac{(q_2^{k+1};q_2)_l}{(q_2;q_2)_l}$.

With the initial condition

$$u_0(t_1,t_2) \equiv u(0,t_2) = g_k(t_2,0),$$

we have

$$\begin{aligned} u_1(t_1,t_2) &= u_0(t_1,t_2) - \int_0^{t_1} \left\{ u_0^{\Delta_1}(s,t_2) - \alpha u_0^{\Delta_2^2}(s,t_2) - \beta u_0(s,t_2) + \beta u_0(s,t_2) u_0(s,t_2) \right\} \Delta s \\ &= g_k(t_2,0) + \int_0^{t_1} \alpha g_{k-2}(t_2,0) + \beta g_k(t_2,0) - \beta H_2(k,k) g_{2k}(t_2,0) \Delta s \\ &= g_k(t_2,0) + \left(\alpha h_{k-2}(t_2) + \beta g_k(t_2,0) - \beta H_2(k,k) g_{2k}(t_2,0) \right) h_1(t_1,0) \\ &= g_k(t_2,0) + G(t_2) h_1(t_1,0), \end{aligned}$$

$$\begin{aligned} u_2(t_1,t_2) &= u_1(t_1,t_2) - \int_0^{t_1} \left\{ u_1^{\Delta_1}(s,t_2) - \alpha u_1^{\Delta_2^2}(s,t_2) - \beta u_1(s,t_2) + \beta u_1(s,t_2) u_1(s,t_2) \right\} \Delta s \\ &= g_k(t_2,0) + G(t_2) h_1(t_1,0) \\ &\quad - \int_0^{t_1} \left\{ G(t_2) - \alpha \left(g_{k-2}(t_2,0) + g^{\Delta_2^2}(t_2) h_1(s,0) \right) - \beta \left(g_k(t_2,0) + G(t_2) h_1(s,0) \right) \right. \\ &\quad \left. + \beta \left(g_k(t_2,0) + G(t_2) h_1(s,0) \right) \left(g_k(t_2,0) + G(t_2) h_1(s,0) \right) \right\} \Delta s \end{aligned}$$

$$= g_k(t_2,0) + \alpha g_{k-2}(t_2,0)h_1(t_1,0) + \alpha g^{\Delta_2^2}(t_2)h_2(t_1,0)$$
$$+ \beta g_k(t_2,0)h_1(t_1,0) + \beta G(t_2)h_2(t_1,0)$$
$$- \int_0^{t_1} \beta\big(H_2(k,k)g_{2k}(t_2,0) + 2G(t_2)g_k(t_2,0)h_1(s,0)$$
$$+ g^2(t_2)H_1(1,1)h_2(s,0)\big)\Delta s$$
$$= g_k(t_2,0) + \left[\alpha g_{k-2}(t_2,0) + \beta g_k(t_2,0) - \beta H_2(k,k)g_{2k}(t_2,0)\right]h_1(t_1,0)$$
$$+ \left(\alpha g^{\Delta_2^2}(t_2) + \beta G(t_2) - 2\beta G(t_2)g_k(t_2,0)\right)h_2(t_1,0)$$
$$- \beta g^2(t_2)H_1(1,1)h_3(t_1,0).$$

In the same manner, the rest of components of the iteration formula are obtained iteratively.

NUMERICAL RESULTS

The approximate solutions introduced in the previous sections will be illustrated with some examples.

Let $\mathbb{T}_1 = \mathbb{T}_2 = \overline{0.9^{\mathbb{N}}} = \{0.9, 0.81, 0.729, \ldots, 0\}$, where 0 is the cluster point of $\overline{q^{\mathbb{N}}}$. The q-shift factorial with q=0.9 is given as

$$(a;0.9)_0 = 1 \quad \text{and} \quad (a;0.9)_n = \prod_{k=0}^{n-1}(1 - 0.9^k a), \quad n \in \mathbb{N}$$

and the q-polynomials are represented as

$$h_k(t,0) = \prod_{\nu=0}^{k-1} \frac{t}{\sum_{j=0}^{\nu} 0.9^j} = \frac{t^k}{\prod_{\nu=0}^{k-1}\sum_{j=0}^{\nu} 0.9^j} = \frac{t^k}{\prod_{\nu=1}^{k-1} 9(1-0.9^{\nu})},$$

$t \in \{0.9^n, n \in \mathbb{N}\} \cup \{0\}$ and $h_k(0,0) = 1$. The multiplication of two generalized polynomials $h_k(t,0)$ and $h_l(t,0)$ is obtained as

$$h_k(t,0)h_l(t,0) = H(k,l)h_{k+l}(t,0),$$

where $H(k,l) = \frac{(0.9^{k+1};0.9)_l}{(0.9;0.9)_l}$. For example, $H(1,1) = \frac{(0.9^2;0.9)_1}{(0.9;0.9)_1} = \frac{(1-0.9^2)}{(1-0.9)} = \frac{1.9}{0.1} = 1.9$.

Example 4 Consider the nonlinear partial dynamic equation as the form

$$\begin{cases} u^{\Delta_1} - uu^{\Delta_2} - u^{\Delta_2^2} = 0 & \text{on } \overline{0.9^{\mathbb{N}}} \times \overline{0.9^{\mathbb{N}}}, \\ u(0,t_2) = g_1(t_2,0) & \text{on } \overline{0.9^{\mathbb{N}}}. \end{cases}$$

8

The initial approximation can be given as

$$u_0(t_1,t_2) = g_1(t_2,0)$$

according to the initial condition. By the variational iteration formula (2), the first two components of $u_n(t_1,t_2)$ are obtained:

$$u_1(t_1,t_2) = \big(1 + h_1(t_1,0)\big)g_1(t_2,0),$$

$$u_2(t_1,t_2) = \big(1 + h_1(t_1,0) + 2h_2(t_1,0) + 1.9h_3(t_1,0)\big)g_1(t_2,0).$$

The rest of components of the iteration formula are obtained in the same manner. The responses of $u(t_1,t_2)$ are shown in Figure 1.

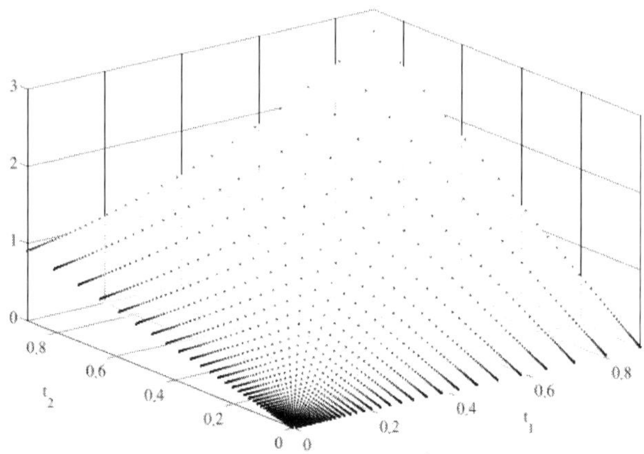

Figure 1. Response for (8) with 2 VIM iterations.

Example 5 Consider the Fisher equation as the form

$$\begin{cases} u^{\Delta_1} - u^{\Delta_2^2} - u(1-u) = 0 & \text{on } \overline{0.9^{\mathbb{N}}} \times \overline{0.9^{\mathbb{N}}}, \\ u(0,t_2) = g_1(t_2,0) & \text{on } \overline{0.9^{\mathbb{N}}}. \end{cases}$$

9

With the initial condition $u_0(t_1,t_2)=g_1(t_2,0)$, the first two components are obtained

$$
\begin{aligned}
u_1(t_1,t_2) &= g_1(t_2,0) + \left(g_1(t_2,0) - 1.9g_2(t_2,0)\right)_1(t_2,0) = g_1(t_2,0) + G(t_2)g_1(t_2,0),\\
u_2(t_1,t_2) &= g_1(t_2,0) + G(t_2)h_1(t_1,0) + \left[g^{\Delta_2^2}(t_2) + G(t_2) - 2G(t_2)g_1(t_2,0)\right]h_2(t_1,0)\\
&\quad - g^2(t_2)H_1(1,1)h_3(t_1,0)\\
&= g_1(t_2,0) + \left[g_1(t_2,0) - 1.9g_2(t_2,0)\right]h_1(t_1,0)\\
&\quad + \left[-1.9 + g_1(t_2,0) - 1.9g_2(t_2,0) - 2\left(g_1(t_2,0) - 1.9g_2(t_2,0)\right)g_1(t_2,0)\right]h_2(t_1,0)\\
&\quad - 1.9\left(g_1(t_2,0) - 1.9g_2(t_2,0)\right)^2 h_3(t_1,0)\\
&= g_1(t_2,0) + \left[g_1(t_2,0) - 1.9g_2(t_2,0)\right]h_1(t_1,0)\\
&\quad + \left[-1.9 + g_1(t_2,0) - 5.7g_2(t_2,0) + 10.298g_3(t_2,0)\right]h_2(t_1,0)\\
&\quad - 1.9\left[1.9g_2(t_2,0) - 10.298g_3(t_2,0) + 4.9051g_4(t_2,0)\right]h_3(t_1,0),
\end{aligned}
$$

where $G(t_2) = g_1(t_2,0) - 1.9g_2(t_2,0)$. The rest of components of the iteration formula are obtained in the same manner. The responses of $u(t_1,t_2)$ are shown in Figure 2.

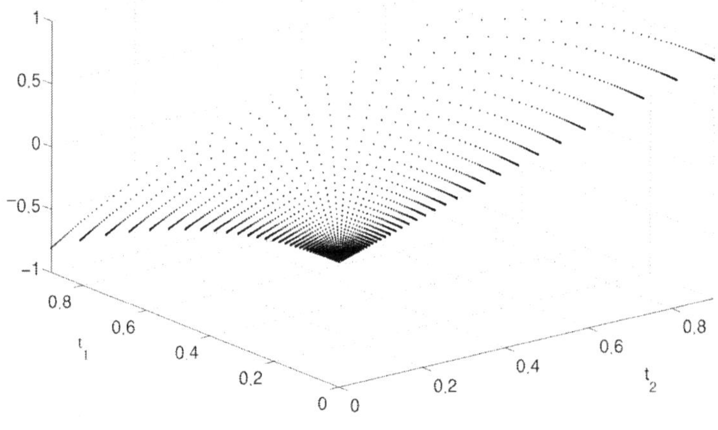

Figure 2. Response for (9) with 2 VIM iterations.

Now, we have demonstrated a method for finding an approximate solution of nonlinear partial dynamic equations on $\overline{q_1^{\mathbb{N}}} \times \overline{q_2^{\mathbb{N}}}$. The proposed tool could also be applied to other nonlinear q-partial dynamic equations.

In future studies, we intend to derive the multiplication rule of two generalized polynomials and extend the application of the variational iteration method to nonlinear partial dynamic equations on other time scales.

CONCLUSION AND FUTURE DIRECTION

In this paper, we have propose a method to find the exact solution of the linear partial dynamic equation on time scales and to find an approximate solution of the nonlinear q-partial dynamic equations. Moreover, this method is applied to provide an approximate solution of the q-Berger equations and the q-Fisher equations.

To extend the method to other time scales, it is important to derive a multiplication rule of two generalized polynomials on the other time scales. On the other hand, approximate solutions as well as their properties of the nonlinear partial dynamic equations, such as Benjamin-Ono equations and the Benjamin-Bona-Mahony equations, are not found on $\overline{q^{\mathbb{N}}}$ yet. In the future studies, we would intend to derive the multiplication rule of two generalized polynomials or to provide an approximation of other nonlinear q-partial dynamic equations by using the proposing method.

APPENDIX: BASIC IDEAS OF THE VARIATIONAL ITERATION METHOD

To clarify the ideas of the variational iteration method, we consider the following nonlinear equation:

$$Lu(t) + Nu(t) = g(t),$$

where L is a linear operator, N is a nonlinear operator and g is an inhomogeneous term. According to the variational iteration method, we can construct a correction functional as follows:

$$u_{n+1}(t) = u_n(t) + \int_0^t \lambda\left\{Lu_n(s) + N\tilde{u}_n(s) - g(s)\right\} ds,$$

where λ is a general Lagrange multiplier, u_0 is an initial approximation which must be chosen suitably and $\tilde{u}_n$ is considered as a restricted variation, that is, $\delta\tilde{u}_n = 0$. To find the optimal value of λ, we make the above correction functional stationary with respect to u_n, noticing that $\delta u_n(0) = 0$, and have

$$\delta u_{n+1}(t) = \delta u_n(t) + \delta \int_0^t \lambda L u_n(s)\, ds = 0.$$

Having obtained the optimal Lagrange multiplier, the successive approximations u_n, $n \geq 0$, of the solution u are determined upon the initial function u_0. Therefore, the exact solution is obtained at the limit of the resulting successive approximations.

REFERENCES

1. Bohner M, Peterson A: Dynamic Equations on Time Scales: An Introduction with Applications. Birkhäuser, Boston; 2001.
2. Hoffacker J: Basic partial dynamic equations on time scales. J. Differ. Equ. Appl. 2002, 8: 307-319. (in honor of Professor Lynn Erbe) 10.1080/1026190290017379
3. Ahlbrandt CD, Morian C: Partial differential equations on time scales. J. Comput. Appl. Math. 2002, 141: 35-55. 10.1016/S0377-0427(01)00434-4
4. Bohner M, Guseinov GS: Partial differentiation on time scales. Dyn. Syst. Appl. 2004, 13: 351-379.
5. Bohner M, Guseinov GS: Multiple integration on time scales. Dyn. Syst. Appl. 2005, 14: 579-606.
6. Jackson B: Partial dynamic equations on time scales. J. Comput. Appl. Math. 2006, 186: 391-415. 10.1016/j.cam.2005.02.011
7. Haile BD, Hall LM: Polynomial and series solutions of dynamic equations on time scales. Dyn. Syst. Appl. 2003, 12: 149-157.
8. Mozyrska D, Pawtuszewicz E: Hermite's equations on time scales. Appl. Math. Lett. 2009, 22: 1217-1219. 10.1016/j.aml.2009.01.033
9. Liu H-K: The formula for the multiplicity of two generalized polynomials on the time scale. Appl. Math. Lett. 2012, 25: 1420-1425. 10.1016/j.aml.2011.12.015
10. He J-H: Variational iteration method - some recent results and new interpretations. J. Comput. Appl. Math. 2007, 207: 3-17. 10.1016/j.cam.2006.07.009
11. Momani S, Abusasd S, Odibat Z: Variational iteration method for solving nonlinear boundary value problems. Appl. Math. Comput. 2006, 183: 1351-1358. 10.1016/j.amc.2006.05.138
12. Ganji DD, Afrouzi GA, Talarposhti RA: Application of variational iteration method and homotopy-perturbation method for nonlinear heat diffusion and heat transfer equations. Phys. Lett. A 2007, 368: 450-457. 10.1016/j.physleta.2006.12.086
13. Ganji DD, Afrouzi GA, Talarposhtib RA: Application of He's variational iteration method for solving the reaction-diffusion equation with ecological parameters. Comput. Math. Appl. 2007, 54: 1010-1017.

14. Hilger S: Analysis on measure chains - a unified approach to continuous and discrete calculus. Results Math. 1990, 18: 18-56. 10.1007/BF03323153
15. Koornwinder, TH: q-special functions: a tutorial. arXiv:math/9403216v1

INDEX

O

P

Q

R